1986

University of St. Francis
GEN 535.842 D961c

T3-BUM-132

3 0301 00075669 8

CHEMICAL, BIOLOGICAL and INDUSTRIAL APPLICATIONS of INFRARED SPECTROSCOPY

CHEMICAL, BIOLOGICAL and INDUSTRIAL APPLICATIONS of INFRARED SPECTROSCOPY

Edited by

James R. Durig
Department of Chemistry
University of South Carolina
Columbia, South Carolina 29208

A Wiley–Interscience Publication

JOHN WILEY & SONS
Chichester · New York · Brisbane · Toronto · Singapore

LIBRARY
College of St. Francis
JOLIET, ILLINOIS

Copyright © 1985, by John Wiley & Sons Ltd.

All rights reserved.

No part of this book may be reproduced by any means, or
transmitted, or translated into a machine language
without the written permission of the publisher.

British Library Cataloguing in Publication Data:

Chemical, biological and industrial applications of
   infrared spectroscopy.
   1. Infrared spectroscopy
   I. Durig, James R.
   535.8'42   QC457

   ISBN 0 471 90834 7

Printed in Great Britain

535.842
D961c

To Professor Richard Collins Lord

5-30-86   J. Webster   # 13.95

120,064

CONTENTS

ANALYTICAL APPLICATIONS OF VIBRATIONAL SPECTROSCOPY --
A HISTORICAL REVIEW

R. Norman Jones

THE IMPORTANCE OF GROUP FREQUENCIES IN PRACTICAL ANALYSIS

Clara D. Craver

INFRARED GROUP FREQUENCIES FOR STRUCTURE DETERMINATION IN ORGANO-
SILICON COMPOUNDS

A. Lee Smith

THE ROUTINE USE OF FT-IR AND RAMAN SPECTROSCOPY FOR SOLVING NON-ROUTINE PROBLEMS IN AN INDUSTRIAL LABORATORY

M. Mehicic, M. A. Hazle, J. R. Mooney, and J. G. Grasselli

THE QUANTITATIVE ANALYSIS OF COMPLEX, MULTICOMPONENT MIXTURES BY FT-IR; THE ANALYSIS OF MINERALS AND OF INTERACTING ORGANIC BLENDS

James M. Brown and James J. Elliott

VIBRATIONAL SPECTROSCOPIC STUDIES OF THIN FILMS

John F. Rabolt

VIBRATIONAL SPECTROSCOPY FOR THE STUDY OF ORDER-DISORDER PHENOMENA IN ORGANIC MATERIALS

Bernard J. Bulkin

STRUCTURAL AND DYNAMICAL PROPERTIES OF MODEL AND INTACT MEMBRANE
ASSEMBLIES BY VIBRATIONAL SPECTROSCOPY

Ira W. Levin

BIOLOGICAL FT-IR: INDUSTRIAL AND ACADEMIC APPLICATIONS

R. J. Jakobsen, F. M. Wasacz, and K. B. Smith

CHEMISTRY AND STRUCTURE OF FUELS: REACTION MONITORING WITH DIFFUSE
REFLECTANCE INFRARED SPECTROSCOPY

E. L. Fuller, Jr., N. R. Smyrl, and R. L. Howell

UNCONVENTIONAL APPLICATIONS OF FT-IR SPECTROMETRY TO THE ANALYSIS OF SURFACES

Peter R. Griffiths, Kenneth W. Van Every, Norman A. Wright, and Isaam M. Hamadeh

APPLICATIONS OF VIBRATIONAL SPECTROSCOPY TO THE STUDY OF PESTICIDES

Kathryn S. Kalasinsky

TIME RESOLVED STUDIES VIA FT-IR OF POLYMERS DURING STRETCHING AND RELAXATION

J. A. Graham, R. M. Hammaker, and W. G. Fateley

CHEMICAL UTILITY OF LOW FREQUENCY SPECTRAL DATA

J. R. Durig and J. F. Sullivan

PREFACE

It has been my pleasure, as the organizer of a symposium in commemoration of the Thirtieth Anniversary of the Founding of the Coblentz Society at the Eleventh Annual FACSS Meeting, to edit this book resulting from the invited presentations. The emphasis of the symposium was on the use of infrared spectral data to solve chemical, biological and industrial problems. Unfortunately, R. N. Jones and K. S. Kalasinsky were not able to attend the symposium but I nevertheless asked them to contribute their articles to the book. It is believed that this book provides examples of a wide variety of applications of infrared spectroscopy and, where appropriate, the complementary use of Raman data is described.

The first article is a historical review of the analytical applications of vibrational spectroscopy. The next two articles demonstrate the utility of group frequencies which have been so important in problem solving in industrial settings for the past four decades. The next four articles provide specific examples of the use of vibrational spectroscopy for industrial problems. There are then three articles which give examples of the utility of vibrational spectroscopy to the solution of biological problems. The final five articles demonstrate chemical applications of vibrational spectroscopy. These include the studies of coal, catalytic surfaces, pesticides, and polymers, as well as the use of far infrared spectroscopy for the determination of potential surfaces during conformational changes. Therefore this book provides the reader with a large number of different examples of the extensive utility of vibrational spectroscopy for the solution of chemical, industrial, and biological problems.

The Editor would like to thank his secretary, Mrs. Janice Long, for superbly typing all of the articles in camera-ready copy, Dr. J. F. Sullivan for her editorial assistance, Mr. Russell Jeffcoat for preparing the glossy prints of the figures, and his wife, Marlene, for the preparation of both the author and subject indices. The Editor would also like to thank Dr. W. C. Harris and the Coblentz Society for the invitation to organize the symposium in commemoration of the Thirtieth Anniversay of the Founding of the Society.

It is a pleasure to dedicate this book to Professor Lord, a true gentleman, an outstanding teacher, and a fine scientist, in this year of his 75th birthday.

James R. Durig
Columbia, South Carolina

LIST OF CONTRIBUTORS

J. M. BROWN, Exxon Research and Engineering Company, Analytical Division, Clinton Township, Route 22 East, Annandale, NJ 08801

B. J. BULKIN, Department of Chemistry, Polytechnic Institute of New York, Brooklyn, NY 11201

C. D. CRAVER, Chemir Laboratories, 761 West Kirkham, Glendale, MO 63122

J. R. DURIG, Department of Chemistry, University of South Carolina, Columbia, SC 29208

J. J. ELLIOTT, Exxon Research and Engineering Company, Analytical Division, Clinton Township, Route 22 East, Annandale, NJ 08801

W. G. FATELEY, Chemistry Department, Willard Hall, Kansas State University, Manhattan, KS 66506

E. L. FULLER, JR., Plant Laboratory, Department of Energy Y-12 Plant, Martin Marietta Energy Systems, Inc., P. O. Box Y, Oak Ridge, TN 37830

J. A. GRAHAM, Hercules Inc., Research Center, Wilmington, DE 19894

J. G. GRASSELLI, The Standard Oil Company (Ohio), Research & Development Center, Cleveland, OH 44128

P. R. GRIFFITHS, Department of Chemistry, University of California, Riverside, CA 92521

I. M. HAMADEH, Monsanto Corporation, 800 N. Lindbergh Blvd., St. Louis, MO 63167

R. M. HAMMAKER, Chemistry Department, Willard Hall, Kansas State University, Manhattan, KS 66506

M. A. HAZLE, The Standard Oil Company (Ohio), Research & Development Center, Cleveland, OH 44128

R. L. HOWELL, Plant Laboratory, Department of Energy Y-12 Plant, Martin Marietta Energy Systems, Inc., P. O. Box Y, Oak Ridge, TN 37830

R. J. JAKOBSEN, National Center for Biomedical Infrared Spectroscopy, Battelle's Columbus Laboratories, Columbus, OH 43201

R. N. JONES, Division of Chemistry, National Research Council of Canada, Ottawa, Ontario K1A OR6, Canada

K. S. KALASINSKY, Mississippi State Chemical Laboratory, Mississippi State University, Mississippi State, MS 39762

I. W. LEVIN, Laboratory of Chemical Physics, National Institute of Arthritis, Diabetes, and Digestive and Kidney Diseases, National Institutes of Health, Bethesda, MD 20205

M. MEHICIC, The Standard Oil Company (Ohio), Research & Development Center, Cleveland, OH 44128

J. R. MOONEY, The Standard Oil Company (Ohio), Research & Development Center, Cleveland, OH 44128

J. F. RABOLT, IBM Research Laboratory, San Jose, CA 95193

A. L. SMITH, Dow Corning Corporation, Midland, MI 48640

K. B. SMITH, National Center for Biomedical Infrared Spectroscopy, Battelle's Columbus Laboratories, Columbus, OH 43201

N. R. SMYRL, Plant Laboratory, Department of Energy Y-12 Plant, Martin Marietta Energy Systems, Inc., P. O. Box Y, Oak Ridge, TN 37830

J. F. SULLIVAN, Department of Chemistry, University of South Carolina, Columbia, SC 29208

K. W. VAN EVERY, Department of Chemistry, University of California, Riverside, CA 92521

F. M. WASACZ, National Center for Biomedical Infrared Spectroscopy, Battelle's Columbus Laboratories, Columbus, OH 43201

N. A. WRIGHT, Department of Chemistry, University of California, Riverside, CA 92521

HISTORY OF THE COBLENTZ SOCIETY

by A. Lee Smith[*]

During the late 1940's and early 1950's the then new field of
chemical infrared spectroscopy was the subject of rapidly growing
interest. Double-beam spectrometers had become commercially available; new techniques, accessories and instrumentation combined with
a growing body of applications kept the field in rapid ferment.
The summer symposium at Ohio State University, Columbus, and the
spring Pittsburgh Conference brought together groups of spectroscopists to exchange information and solutions to problems both by
formal papers and informal discussions.

It was apparent that while infrared was not a primary field of
research, it provided a common tool to attack diverse problems.
Further, a number of problems, such as reference spectra availability and retrieval, were of common concern. From informal
discussions held at Columbus in 1953, a Columbus Committee was
formed to consider the situation. This committee consisted of Drs.
Howard Cary, Bryce Crawford, Jr., R. A. Oetjen, Eugene Rosenbaum,
Van Zandt Williams and Norman Wright. The committee reported that
there was need of an association of infrared workers to provide the
field with a unified voice where necessary, and to keep the field
informed on matters of common interest. Recommendations as to the
organization and functioning of the society were also made, and
suggestions for activities were listed. It is interesting to
review this list, which was (a) publish reference spectra, (b) keep
abreast of related organizations which make decisions that affect
infrared, (c) produce an authoritative infrared book, (d) provide
consultant and speaker listings, (e) serve as a communication
channel between the field and instrument manufacturers, (f) publish
specific analyses in compact form, and (g) organize symposia on
infrared to concentrate IR papers at major scientific meetings.

By June of 1954, plans for the new society had been fully developed, including the formation of an executive committee and writing
of by-laws. The name chosen was "The Coblentz Society", after Dr.
W. W. Coblentz, whose post-doctoral studies at Cornell University
in 1904 and subsequent work at the Bureau of Standards had done so
much to demonstrate the potential application of infrared spectroscopy to the field of chemistry. Because Dr. Coblentz was a
physicist his work was not much read by chemists, and it was not
until the 1930's that rapid further development took place.

The objective of the new society was: "To foster the understanding
and application of infrared spectra". A corollary of this objective was to promote international interaction between academic and
industrial spectroscopy.

The prime movers in the organization of the Coblentz Society were
Drs. Norman Wright of Dow Chemical and Van Zandt Williams of
Perkin-Elmer. Dr. Wright, a physicist, joined Dow in 1937 with the

assignment of designing and building an infrared spectrometer. Four months later, he was recording spectra of chemical interest and in a few more months had demonstrated that infrared represented a powerful new tool for characterization of pure materials and for analysis of mixtures. His subsequent talks to ACS groups and his publications provided a real impetus to the growth of the field. Dr. Wright served as first President of the Coblentz Society.

Dr. Van Zandt Williams was also a physicist, and was initially a member of the infrared group at American Cyanamid, where development work was going on along the same lines as that done at Dow. He later moved to Perkin-Elmer, where he eventually became Executive Vice-President. He retained throughout his career a strong interest in infrared spectroscopy, and served as the first registrar of the Coblentz Society.

There was no lack of activities for the new society. Arrangements were made with _Analytical Chemistry_ to publish quantitative infrared methods in a condensed format. This program started in October, 1957, and continued for several years under Robert C. Wilkerson. It was subsequently managed by Don E. Nicholson and from 1960 by A. Lee Smith. When submissions became irregular, publication was transferred to _Applied Spectroscopy_. An index to the methods was published in _Applied Spectroscopy_, 18, 38 (1964).

A speakers bureau was organized under Mr. Harry Bowman. A list of special collections of spectral data was compiled. In 1956 a program of collecting, compiling, and selling spectra contributed from members' files was proposed. By 1960 almost 2000 spectra had been collected by Clara D. Smith Craver, and the first set of 1000 spectra had been published. A successful symposium on group frequencies had been organized at the 1959 Pittsburgh Conference by W. J. Potts. The Coblentz Symposium thereafter became an annual tradition of this meeting. The editors of several scientific journals were approached by Foil A. Miller regarding publication of infrared spectra.

In 1961-1962 a growing awareness that examiners in the U. S. Patent Office did not consider an infrared spectrum to be characteristic of a material for patent purposes culminated in Coblentz Society participation in a symposium at the 1962 National ACS Meeting in Washington, D. C. Ellis R. Lippincott, on behalf of the Society, presented a definitive paper on "The Limitations and Advantages of Infrared Spectroscopy in Patent Problems" which was published in the _Journal of the Patent Office Society_, May, 1963, Vol. XLV, pp. 380-415.

On March 1, 1961, a symposium in honor of Dr. Coblentz was held at the Pittsburgh Conference which featured talks by Earl Plyler, H. W. Thompson, Bryce Crawford, and R. C. Lord. The passing of Dr. Coblentz on September 15, 1962, in Washington, D. C., was noted in the February 18, 1963 Newsletter. In 1963 the Coblentz Award, to be given to an outstanding young spectroscopist, was established, and John Overend chosen as the first awardee. Also in 1963, Dr.

Coblentz's original monograph, "Investigations of Infrared Spectra, Parts 1-7" was reprinted under the direction of Dave Kendall.

The National Standard Reference Data Program of the National Bureau of Standards, instituted in 1963, defined a need for specifications for evaluation of infrared reference data. The Office of Standard Reference Data contracted with the Coblentz Society in 1965 to write such specifications, which were developed and published in Analytical Chemistry, Aug., 1966, p. 27A. With only modest modifications these have become adopted as international standards by IUPAC. For complete details see Analytical Chemistry, 47, 945A (1975).

Early educational activities of the Society included producing 8 mm films on techniques and instrument operation, and holding technique clinics at the Pittsburgh Conference. More recent courses, under the direction of Jeanette Grasselli, have been expanded to include full day workshops on computer assisted spectroscopy. In a cooperative venture between the Society and SAVANT, Howard J. Sloane began production of audio-visual courses on IR techniques in 1982.

There are now three major annual awards sponsored or co-sponsored by the Coblentz Society. These are the Coblentz Award, for an outstanding young spectroscopist under the age of thirty-six, the Ellis R. Lippincott medal, sponsored jointly by the Society for Applied Spectroscopy for presentation to an individual who is judged to have made a significant contribution to vibrational spectroscopy, and the Williams-Wright Award which honors an industrial spectroscopist who has made significant contributions in the application of infrared spectroscopy while working in industry.

Coblentz Award Recipients	Year
John Overend	1964
William Fateley/Robert Snyder	1965
Edwin Becker	1966
Peter Krueger	1967
Jon Hougen	1968
James Durig	1969
Guiseppi Zerbi	1970
Clive Perry	1971
George Leroi	1972
C. Bradley Moore	1973
C. K. N. Patel	1974
Bernard J. Bulkin	1975
Geoffrey Ozin/George Thomas, Jr.	1976
Peter Griffiths	1977
Lester Andrews	1978
Lionel Carreira	1979
Richard Van Duyne	1980
Laurence Nafie	1981
Christopher Patterson	1982
David Cameron	1983
Stephen Leone	1984
John Rabolt/Graham Fleming	1985

The Lippincott Award Recipients	Host Society	Year
Richard Lord	SAS	1976
Lionel Bellamy	Coblentz	1977
Bryce Crawford, Jr.	OSA	1978
E. Bright Wilson	SAS	1979
George Pimentel	Coblentz	1980
Ian Mills	OSA	1981
Michel Delhaye	SAS	1982
John Overend	Coblentz	1983
Jon Hougen	OSA	1984
Ira Levin	SAS	1985

Williams-Wright Award Recipients	Year
Norman Wright	1978
Norman Colthup	1979
Jeanette Grasselli	1980
Paul Wilks	1981
Robert Hannah	1982
Harry Willis	1983
Robert Jakobsen	1984
Clara Craver/Richard Nyquist	1985

The affairs of the Society are managed by an eight-member board
which elects its own presidents. The wide base of leadership is
shown by the following list of presidents of the group:

Coblentz Society Presidents	Year
Norman Wright	1955-1958
Foil Miller	1959
R. A. Friedel	1960-1961
R. N. Jones	1962-1963
A. Lee Smith	1964-1965
Nelson Fuson	1966-1967
Freeman Bentley	1968
Charles Angell	1969-1970
James R. Scherer	1971
Robert P. Bauman	1972-1973
James R. Durig	1974-1975
Robert J. Jakobsen	1976
Peter R. Griffiths	1977
Ira Levin	1978
Bernard J. Bulkin	1979-1980
William C. Harris	1981-1982
Albert B. Harvey	1983-1984
Bruce D. Chase	1985

*Taken from The Coblentz Society Desk Book of Infrared Spectra by
permission.

ANALYTICAL APPLICATIONS OF VIBRATIONAL SPECTROSCOPY -- A HISTORICAL REVIEW

R. Norman Jones[†]

Division of Chemistry
National Research Council of Canada
Ottawa, Ontario
Canada K1A 0R6

60 B.C. - 1800 A.D. HOW IT ALL BEGAN

> "Perhaps the sun, shining above with rosy lamp, is sur-
> rounded by much fire and invisible heat. Thus the fire
> may be accompanied by radiance which increases the power
> of the rays".[††]

So wrote Lucretius, the first infrared spectroscopist about 60 B.C.
[1]. The infrared region of the spectrum remained but a hypothesis
of this Roman scholar until the seventeenth century. In 1686
Mariotte [2], using a concave metal mirror observed that the heat
generated at the focussed image of a lamp disappeared when a plate
of window glass was inserted into the beam, and this experiment was
repeated with refinements by Scheele [3] in 1781 and later by
Pictet [4]. Thus by the end of the eighteenth century some
association between light and radiant heat had come to be
recognized.

The next step, and the real beginning of infrared spectroscopy, was
taken on a sunny spring day at an English country house at Slough
... now a few minutes ride from Heathrow Airport. Frederick
William Herschel, a professional musician, emigrated from war torn
Hannover to England in 1757. He prospered in England where he
acquired a fine home and became an amateur astronomer and professional telescope maker. With the help of his son John he constructed a large reflecting telescope and set it up in his garden.
Unlike refracting telescopes, reflectors transmit a good deal of
near infrared radiation which can cause discomfort for the
observer. This motivated Herschel to study the heat associated
with the various color ranges of the solar spectrum, presumably
with the intention of making it easier for the observer by

---

†Permanent Address: Ann Manor Apt. 601, 71 Somerset Street West,
                     Ottawa, Ontario, Canada K2P 2G2.

††A free translation of "Forsitan et rosae Sol alte lampade lucens,
                         Possideat multum caecis fervoribus ignem
                         Circum se, qui sit fulgore notatus
                         Aestifer ut tantum radiorum exaugeat
                         ictum".

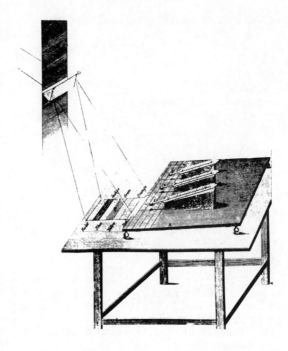

FIG. 1. The classical experiment of Sir William Herschel.

inserting a color filter into the telescopic beam.   Herschel's
classic experiment, carried out in March 1800, is illustrated in
Fig. 1.   He projected onto a table a solar spectrum generated by a
glass prism attached to a slot in the blind of his living room
window.   His heat detectors were mercury-in-glass thermometers with
blackened bulbs.   One was placed in the radiation beam and the
second was set slightly off to the side of the spectrum as a
control.   When equilibrium had been attained with the bulb in the
violet region of the spectrum that thermometer read 2° higher than
the control.   The difference increased progressively to 7° as the
probe thermometer was moved into the red.   It is indicative of an
astute experimentalist that Herschel continued to move the probe
thermometer into the non-illuminated region beyond the red end of
the visible spectrum and he observed a further rise of 2°.   He
correctly interpreted this as proof of the existence of heat
radiation less refrangible than visible light.   Many confirmatory
experiments followed including measurements through a variety of
colored glass filters with predictable effects on the temperatures
measured in the various regions of the spectrum.   These observations were reported to the Royal Society of London in a series of
papers [5].   They   substantiated   the   original   hypothesis   by
demonstrating the reflection of the invisible heat radiation by
plane mirrors and its focusing by solid metal concave mirrors.   The
first paper included the earliest record of an infrared emission
spectrum, that of the spectral distribution of solar radiation.

Almost all of Herschel's infrared measurements dealt with the transmission of glasses and other solids but he did make one set of absorption measurements on well-water, alcohol, sea-water, gin and brandy. His measured absorptions were 558/612/682/734/796, respectively, in his units of "rays of heat". At his path length of 3 inches in a cell with glass windows and no dispersion would a modern spectroscopist do much better?

In his third paper Herschel speculates about the nature of the heat rays. He could not accept the identification of heat with light and in his third paper he argued vehemently against this on physiological grounds

> "...it does not appear that nature is in the habit of using the same mechanism with any two of our senses; witness the vibrations of air that make sound; the effluvia that occasion smells; the particles that produce taste; the resistance or repulsive powers that affect the touch: all these are evidently suited to their respective organs of sense. Are we then here, on the contrary, to suppose that the same mechanism should be the cause of such different sensations, as the delicate perceptions of vision, and the very grossest of all affections, which are common to the coarsest parts of our bodies, when exposed to heat?"

Over the next thirty years there was much controversy about the relationship between heat and light and it was not until 1832 that a unitary theory was firmly established by Ampère [6].

After Herschel, infrared spectroscopy made little further progress until Thomas Seebeck discovered the thermoelectric effect in 1822 [7]. He constructed thermocouples of various metals with antimony or bismuth to establish a thermoelectric series [8]. Seebeck also devised a thermopile of several antimony/bismuth junctions which he used to replace the mercury-in-glass thermometers in establishing a new temperature scale [9]. The superiority of thermopiles for measuring radiant heat was recognized by Leopoldo Nobili, especially when used with Nobili's sensitive astatic galvanometer in which the damping effect of the earth's magnetic field was suppressed [10].

Nobili was a student of Ampère. After spending some time as an artillary officer he became Professor of Physics at the Archducal Museum at Florence where he constructed his first six element thermopile in 1829 and demonstrated its heat sensitivity by cooling it in an evacuated bell jar [11]. He spent much effort trying to increase the sensitivity by adding more elements but this ultimately failed. He recognized that with increasing size the thermal capacity of his thermopiles was increasing more rapidly than their thermo-electric activity. He reversed the process, reduced the overall size of the detector and achieved a more sensitive response with only four elements. Nobili collaborated in this with a younger colleague Macedonio Melloni. At this time Central Europe, and particularly the separate states that were soon to become Italy, were in revolutionary ferment. Melloni's involvement in these political events caused him to vacate his chair as Professor of Physics at the University of Parma. He fled to France, first to Paris and later to Montpellier, returning eventually to Naples as Professor of Physics and Director of the Meteorological Observatory on Mount Vesuvius. Melloni rather than Nobili was the main instigator of their spectroscopic studies. Their collaboration was most active between 1829 and 1839 including the period of Melloni's exile in France. It has been reviewed by W. W. Coblentz [12].

In his initial experiments in 1834 Melloni used a 27 element Nobili thermopile to study the infrared transmission of a wide range of substances [13]. At this time only glass prisms were available. Melloni's early measurements were made non-dispersively using four radiation sources, a wick oil lamp (without glass chimney), a platinum spiral, a thin copper plate heated to redness in an alcohol flame, a "Leslie cube" of hollow blackened copper through which boiling water was circulated. With these sources he discovered the high infrared transparency of rock salt (NaCl). Soon afterwards he made his first rock salt prism which he mounted on a graduated circle. This first "Melloni apparatus" was without benefit of lenses, mirrors or slits [14] but later he inserted a rock salt lens to focus the radiation onto a slit with provision to move the thermopile across the face of the fixed prism.

This "Melloni apparatus" remained the basic instrument for the study of radiant heat for some four decades. Thus in 1879 we find W. W. Jacques using a similar instrument with a rock salt prism and lenses in his studies of various radiation sources [15]. The introduction of front surface concave mirrors came slowly. Already in 1835 Liebig at Giessen was producing brilliant silver deposits on glass by the reduction of an ammoniacal solution of silver nitrate with formaldehyde but these were back surface mirrors of little use in the infrared. A front surface concave steel mirror was used by Herschel. Front surface mirrors of speculum metal were incorporated into heliostats in 1872 but the first use of a front surface mirror to replace the non-achromatic rock salt lenses in laboratory infrared spectrometers is attributed to Pringsheim [16] in 1883. Though infrared spectroscopy made slow progress between 1840 and 1880 it was a time of much activity in the application of visual spectroscopy to chemical analysis using the eye as a detector. In the laboratory, flame emission spectroscopy led to the discovery of caesium in 1860 [17], rubidium and thallium in 1861 [18,19], indium in 1863 [20], and gallium in 1875 [21]. In 1868 helium was discovered in the sun [22,23].

1880-1900 LAYING THE FOUNDATIONS

The next significant advance in infrared spectroscopy was the development of a more efficient detector. The Nobili thermopiles were cumbersome; they had a large target area and were difficult to make. Antimony and bismuth were available only as rectangular bars and the materials were brittle and hard to work with. When other detectors became available in the 1880's thermopiles became obsolete for a while. Their use was revived in 1910 when pliable bismuth wire became a market commodity and a pliable wire of antimony/tin alloy was developed. As a miniaturized single element thermocouple the thermo-electric detector continued to be used into modern times. Its later development was reviewed by V. Z. Williams in 1948 [24]. Changing demands on radiation detectors for FT-IR spectrometers, principally for faster response time, have now tended to their replacement by other solid state devices.

In 1851 A. F. Svanberg noted that a Wheatstone bridge in which one of the arms was a blackened spiral of copper wire went out of balance if the wire spiral were warmed by the hand [25]. He recognized that such a change in electrical resistance might replace the thermo-electric effect as a heat detector in the Melloni apparatus. However, Svanberg's coils were bulky and in practice had poor sensitivity. In 1881 the thermal changes in the resistance of a single thin platinum wire forming one arm of a Wheatstone bridge became the heat detecting element of the bolometer invented by S. P. Langley [26]. A similar platinum wire maintained at a constant temperature formed the compensating arm of the bridge. With this system a temperature difference of 10^{-5} °C could be detected after a 1 sec. exposure with 1% error.

S. P. Langley was born and educated in Boston, Massachusetts. He came from a New England merchant family. He was well read in history and literature as well as science and engineering though his formal education did not extend beyond high school. As befits his New England puritan heritage he was a man of strong religious convictions. To other than close friends he was a dour man and unapproachable. In 1865 he was appointed Assistant to the Faculty at Harvard University. Two years later he became Director of the Allegheny Observatory attached to the University of Pittsburgh. In 1887 he became Secretary of the Smithsonian Institution and he moved his laboratory from Pittsburgh to Washington in 1889. Apart from his spectroscopic work Langley was most widely acclaimed as an aeronautics engineer. In 1896 he built and successfully flew motor powered model airplanes across the Potomac river and came near to forestalling Wilbur and Orville Wright in first achieving manned flight in a heavier than air machine.

At the Allegheny Observatory Langley incorporated his bolometer into spectrometers using either a rock salt prism or a concave grating. An important feature of his bolometer was its small target area. The Nobili thermopile had a cross section of 2.5 mm. This required that the spectrometer slits be wide in order to provide an image that would fill the detector. The width of the platinum wire of the bolometer was less than 1 mm. This permitted the design of a spectrometer with much higher resolution since narrower slits could be used. His innovative use of a grating had important consequences which we shall discuss below. The art of ruling gratings for spectroscopy in the visible had been developed about 1870 in England by L. M. Rutherfurd [27]. His gratings were small (1" × 1") and were ruled on flat speculum metal [28]. In

1882 at Johns Hopkins University the machinist, Mr. Schneider built a ruling engine to the novel design of H. A. Rowland [29]. The gratings made on this machine were far superior to any constructed previously. They were much larger (up to 6¼" × 4¼") and being ruled on a concave speculum surface were self-focusing, eliminating the need for mirrors or lenses in the optical system. The early history of gratings was reviewed by G. R. Harrison in 1949 [27].

Much of Langley's work at Allegheny Observatory concerned measurements of infrared emission from atmospheric and astronomical sources. He studied atmospheric absorption and discovered the radiation window between 8 μm and 10 μm. These measurements were made in California on the summit of Mt. Whitney. He also measured the lunar spectrum. Some of his atmospheric spectra recorded in Pittsburgh show absorption that can now be recognized as due to sulfur dioxide ... evidence that environmental pollution is no new thing [30].

After his move to the Smithsonian Institution in 1889 Langley built a new infrared spectrometer incorporating a very large rock salt prism 19 cm high. He also devised a photographic recording system to track the motion of the light beam from the galvanometer mirror. The resulting Langley "bolographs" were the first automatically recorded spectra [31].

These astronomical and atmospheric radiation studies were novel and spectacular but Langley's major contribution was more fundamental. Prior to his time, spectroscopists could only record their infrared measurements in terms of some arbitrary wavelength scale, usually the angular setting of the prism; they had no means to convert the wavelength to a metric scale. The missing piece of information was the refractive index of the prism material. Once gratings became available, for which the number of lines per cm were known, theoretically it became possible to compute the wavelength directly from the known grating spacing and the angular deflection of the beam. Unfortunately the Rowland gratings were energetically inefficient and not effective replacements for prisms in the infrared. We may digress here to note that efficient gratings for the infrared only became available after 1910 when R. W. Wood and A. Trowbridge constructed "echelette" gratings in which the shape of the groove was tailored in such a way that it channeled most of the energy into one order [32,33]. Indeed infrared grating spectrometers only came into use in the hands of W. W. Sleator in 1918 [34]. They did not compete effectively with prism instruments for general analytical use until the 1950's, mainly because of difficulties in sorting out the signals from overlapping orders [35].

To measure the refractive index of rock salt Langley generated the yellow sodium D line using a powerful electric arc which also generated a strong continuous background of infrared radiation [36,37]. This was dispersed by his grating spectrometer to produce an infrared continuum superimposed on which was a series of yellow sodium D lines from nine successive orders of the grating. These overlaid the infrared spectrum at wavelengths corresponding to

0.5890n (n = 1 ... 9) μm. Langley directed this mixed radiation onto the entrance slit of his rock salt prism spectrometer. With the entrance slit lined up with the mth D line image the underlying infrared radiation was dispersed by the prism and the bolometer would record this as a transmission peak at a dispersion angle corresponding to 0.5890m μm. In this way he could determine the refractive index of the prism material at 0.5890 μm in the visible and at eight discrete wavelengths in the infrared. The complete rock salt dispersion curve could then be evaluated by interpolation. One cannot overstress the importance of these dispersion measurements in establishing the absolute metric wavenumber scale. Apart from a few measurements on fluorite (CaF_2) [38] Langley did not pursue these studies further and he became increasingly involved with his aerodynamic work.

The main centers of experimental infrared spectroscopy next shifted to Germany where the two independent research groups of Heinrich Rubens and Friedrich Paschen made advances in experimental technique that laid the ground work for the measurement of infrared spectra for the next forty years.

Following the custom of the times, Rubens studied at various universities and technical schools in Germany. First at Darmstadt and later at Berlin and at Strasbourg where he came under the influence of August Kundt. Rubens accompanied Kundt when the latter moved to Berlin and in 1889 he earned his doctorate in Kundt's laboratory. The rest of his professional life was spent in Berlin, first at the Technische Hochschule at Charlottenburg and from 1906 on at the University where he succeeded P. Drude as Professor of Physics.

Friedrich Paschen also studied under Kundt at Strasbourg where he acquired his doctorate in 1888, one year prior to Rubens. Between 1888 and 1891 Paschen held minor academic positions, first with W. Hittorf at Münster and later with H. Kayser at Hannover, where he secured a tenured post in 1895. In 1901 he became Professor at the University of Tübingen. He remained there until 1924 when he succeeded W. H. Nernst as President of the Physikalische-Technische Reichsanstalt in Berlin.

Independently between 1894 and 1908 both the Paschen and Rubens schools made extensive measurements of the dispersion of the common optical materials, picking up from where Langley left off. Paschen used the same method as Langley with a prism spectrometer in tandem with a grating [39]. Initially Rubens used a prism/interferometer combination but later he also adopted Langley's system [40]. Paschen measured fluorite (CaF_2) to 1060 cm^{-1} [39]. In the same year Rubens published on both rock salt and fluorite to 1117 cm^{-1} and on sylvine (KCl) to 1383 cm^{-1}. A series of papers followed from both laboratories extending the wavenumber ranges, refining the values and comparing interpolation formulae. By 1908 the dispersion curves of fluorite, rock salt and sylvine were known to 1030, 628 and 565 cm^{-1}, respectively. These dispersion values and their temperature coefficients were coordinated in a review by C. Schaefer and F. Matossi in 1930 [41]. At this time the prism materials all came from natural mineral sources. Potassium bromide only became available in 1930 when a method of growing alkali halide crystals of optical quality was developed by J. Strong [42]. This extended the infrared range to 400 cm^{-1}. Caesium iodide prisms were first introduced by E. K. Plyler and N. Acquista in 1953 with a further extension to 200 cm^{-1} [43].

By 1930 the calibration of the wavenumber scale by dispersion measurements using prisms came to be replaced by direct measurements on grating spectrometers or interferometers. This phase is principally associated with the names of C. J. Humphries, W. F. Meggers, H. H. Nielsen, E. K. Plyler, W. C. Price, D. H. Rank, K. N. Rao and G. R. Wilkinson over the period 1930-1960. The state of the art of grating infrared spectroscopy in 1938 was reviewed by H. M. Randall [44]. In 1961 these gas phase wavenumber calibration standards were collated and assessed by the Commission on Molecular Structure and Spectroscopy of the International Union of Pure and Applied Chemistry (IUPAC) [45].

In the laboratories of Paschen and Rubens the bolometer remained the favorite radiation detector, taking advantage of improvements in its sensitivity and speed of response made by several investigators [46]. At this time two new detectors also appeared. One was the radiometer described by C. V. Boys in 1887 [47]. Essentially this was an adaptation of the thermopile in which a single antimony/ bismuth junction was directly incorporated into the needle element of the galvanometer. The author stated that

> "this ... instrument is capable of showing the heat which would be cast by a candle flame on a halfpenny at a distance of 200 yards".

Another device was the repulsion radiometer, first described by W. Crookes [48,49]. It was improved by E. Pringsheim [50] and developed into its most sensitive form by E. F. Nichols in Rubens' laboratory in 1894 [51]. This detector came to be widely used during the next decade. E. F. Nichols was a graduate of Cornell University. He spent two years (1894-1896) at the University of Berlin with Rubens and Warburg. He brought these new techniques back to the United States where he held professorships at Dartmouth College, and at Yale and Columbia Universities. The propulsion radiometer consisted of two identical thin rectangular mica vanes suspended laterally on each side of a fine quartz fiber to which a small mirror was also attached. The torsional motion of the fiber could be recorded by a light beam reflected from the mirror. The whole system was mounted in an evacuated tube maintained at an accurately controlled air pressure of approximately 5×10^{-2} mm. The whole system was lightly constructed and weighed only 5 - 10 mg. In operation the radiation admitted through a transmitting window was directed onto one of the vanes which was blackened. Its heating effect set up a convection current in the rarefied gas normal to the vane; this generated a torque which was balanced by the rotatory displacement of the quartz fiber. This essentially mechanical device had the advantage that it was uninfluenced by the magnetic or thermo-electrical disturbances that occur with thermocouple-galvanometer or bolometer-galvanometer systems. It was also free from drift, had a fast time response, and the angular displacement of the fiber was linear with respect to the amount of incident radiation. Its main disadvantages were its lack of portability, its sensitivity to mechanical vibration, to electric charge on the vanes and to variation in the internal gas pressure. Coblentz compared these various detectors in 1908, following with updating revisions in 1912, 1914, 1918 and 1921 [52].

The Rubens school made extensive measurements of infrared reflection spectra and this led to a new technique for extending infrared measurements and opened up the far infrared beyond the cut off limit of KCl [53]. Their reflection studies on the alkali halides showed that below their cut off wavelengths the reflections strongly intensified and became metallic in character as predicted by Clerk Maxwell's electromagnetic theory. By multiple reflections from a set of flat crystalline plates a beam of pseudo-monochromatic radiation could be generated; usually four reflections were used to form a selective reflection filter for such "reststrahlen". By this means Rubens and co-workers, using a set of different optical materials extended the infrared range to 65.9 cm^{-1}. The materials used and the ranges of the selective beams are shown in Table 1. A diffraction grating or interferometer was used to measure the wavelengths passed by the filters. The detector was a bolometer attached to a "Panzer Galvanometer" shielded against electromagnetic disturbances by a massive steel casting. The sources were incandescent Nernst glowers or Welsbach mantles, the emission of which came from cerium oxide.

The experimental work of Rubens and Paschen had a great influence on theoretical spectroscopy between 1890 and 1910. N. Bjerrum, F. A. Lindemann, M. Planck, C. Runge and W. Wien formulated theories based extensively on experimental measurements eminating from their laboratories. In this article we mainly stress those aspects of nineteenth century spectroscopy that laid the ground work for the analytical applications that were to follow later. We should note however that most of the activity between 1890 and 1930 concerned basic molecular structure and radiation physics. This followed two principal lines of development.

One of these was the analysis of the shape of the band envelopes of continuous radiation sources, especially the black body radiation from the sources devised by O. Lummer and E. Pringsheim [54]. This posed a challenge for the theoreticians. There were particular problems with the far infrared wing where the intensities from Rubens' reststrahlen measurements were much greater than those predicted from Wien's equations based on classical physics. This was the major problem stimulating Planck's introduction of his quantum theory in 1901 [55]. The fascinating story of the personal relations between Rubens and Planck leading up to the formulation of the quantum theory has been retold by H. Kangro [56].

TABLE 1. Reflecting Filter Materials for Rubens' Reststrahlen[*]

Substance	Band Maxima (cm^{-1})
SiO$_2$ Quartz	482
CaF$_2$/CaCO$_3$ Fluorite + Calcite	366
CaCO$_3$ Marble	340
MgCO$_3$ Magnesite	333
ZnS Zinc Blend	324
CaF$_2$ Fluorite	305
CaCO$_3$ Aragonite	286, 244
SrCO$_3$ Strontianite	231
NH$_4$Cl Sal ammoniac	216, 194, 185[a]
BaCO$_3$ Witherite	215
NaCl Rock salt	210, 192, 185[a]
NaBr	200-182
NH$_4$Br	181, 168, 160[a]
KCl Sylvine	161[a], 158, 142
RbCl	136
AgCl	135, 123, 111[a]
KBr	134, 121, 115[a]
TlCl	109
HgCl$_2$ Calomel	109[a], 101, 85
AgCN	107
KI	106
HgCl$_2$ Sublimate	105
AgBr	86
TlBr	85
TlI	66

[*]Transcribed from pages 60-63 of Ref. [41].
[a]Main peak.

The second area of intensive theoretical study at this time was the explanation of the fine structure of gas phase infrared spectra. This became increasingly puzzling as the resolution of the spectrometers improved and the spectra increased in complexity. It is not easy for us to place ourselves in the situation of these early theoreticians. For us today the association of these complex band systems with changes in the vibrational and rotational energy levels of the molecule has become intuitive by a gestalt process. The initial step was taken by Deslandres who, in 1886 formulated an empirical set of parabolic functions to fit the emission spectrum generated by a nitrogen discharge tube in the visible [57]. Once the metric wavelength scale was extended into the infrared similar functions were found to fit the band spacings and this opened up a wide field of study on such molecules as H_2O, CO_2, HCl and NH_3. There was considerable confusion. This was partly rationalized at the famous First Solvay Conference in 1911 which was attended by most of the eminent theoretical physicists of the day. A major contributor was N. Bjerrum and much of the discussion about infrared band series centered on his speculative paper in which he introduced the concept of three forms of molecular energy, translational, vibrational and rotational [58]. Strong support for his hypothesis came soon afterwards from high resolution measurements on HCl by Eva von Bahr, using Rubens' spectrometer [59,60]. The development of the understanding of the origins of infrared band spectra has recently been reviewed in depth by C. Fujisaki [61].

1870 - 1900 PERIPHERAL CONTACTS WITH CHEMISTRY

In the foregoing we have traced the development of infrared spectroscopy to the end of the nineteenth century with a few excursions beyond. Both on the experimental and the theoretical side these nineteenth century pioneers were physicists; they had no perception of the future role that infrared spectroscopy was to play in chemical analysis. It was only with the development of electronic aids to spectral measurements that this would come about. Later we shall set this into perspective but here we should take note of a few chemically oriented spectroscopists of this earlier period.

Credit for the first suggestion of the potential use of infrared spectroscopy in chemical analysis goes to William Abney, a British army engineer who is more commonly acclaimed as one of the founders of modern photography. In 1869 Abney was invalided home from service with the British army in India and appointed Chemical Assistant at the Chatham School of Military Engineering. There he began chemical studies on the sensitization of photographic plates. During this work he developed a method to extend the range of photographic sensitivity into the infrared. This was based on treatment of the currently used silver bromide/collodion emulsions with nitric acid [62]. By this means he was able to extend the photographic sensitivity to the lower limit of his glass optics at about 1.2 μm (8333 cm^{-1}). Using a prism spectrometer designed by Adam Hilger, in collaboration with E. R. Festing he studied the absorption of 46 organic liquids in cells of up to 24" length. These spectra were tabulated on an arbitrary scale of wavelength

[63]. Most of the bands they observed we can now identify with second and higher overtones of C-H stretch vibrations. Abney and Festing [64] noted that whereas $CHCl_3$ showed a single sharp band, no bands were detected in CCl_4 or CS_2 and they therefore surmised that the presence of hydrogen was necessary for their appearance. They also noted structure specific to the ethyl group and to the benzene ring. The discussion section of their paper ends with a prophetic paragraph:

> "We should have liked to have said more regarding the detection of the different radicals, but it might seem presumptuous on our part to lay down any general law on the results of the comparatively few compounds which we have examined. In our minds there lingers no doubt as to the easy detection of any radical which we have examined, but it will require more energy and ability than we possess to thoroughly classify all the different modifications which may arise. We may say, however, it seems highly probable by this delicate mode of analysis that the hypothetical position of any hydrogen which is replaced may be identified, a point which is of prime importance in organic chemistry. ... We quit this part of our subject in hope that chemists will be able to help us to decipher more than has as yet been done".

Later in the century some measurements of mid-infrared spectra of organic compounds began to be reported. In 1889 Ångström recorded the infrared spectra of methane, ethylene, diethyl ether, benzene and carbon disulfide in both the vapor and liquid phases [65]. In 1892 Julius recorded the spectra of 20 organic liquids out to 10 µm (1000 cm^{-1}) [66]. Both investigators used rock salt prisms and lenses, and bolometer detectors. Julius used a rock salt cavity cell of about 0.2 mm path length. In all the compounds containing methyl groups he observed a band at 3.45 µm (2898 cm^{-1}) and he drew the general conclusion that the absorption is due to internal motions in the molecule and that the internal structure of the molecule determines the spectrum. In 1896 M. Ransohoff [67] measured the spectra of six alcohols recording bands at 1.71, 3.0 and 3.43 µm (5848, 3333, 2915 cm^{-1}). He noted that the 3.0 µm band agreed with one observed by E. Aschkinass [68] and by Paschen [69] in water and attributed it to the hydroxyl group. In 1899 and 1900 L. Puccianti [70,71] measured several benzene derivatives using a quartz prism; he identified a band at 1.71 µm (5848 cm^{-1}) with the C-H bond and bands at 2.18 and 2.49 µm (4587, 4016 cm^{-1}) with the benzene ring.

Thus by the turn of the century the predictions of Abney and Festing had begun to gain support from measurements extending further into the infrared and an empirical basis for the concept of group frequencies had been initiated. The ground work was being laid for the work of W. W. Coblentz which was soon to follow and which brings us to the beginning of the era of analytical infrared spectroscopy as we have come to realize it.

1900 - 1905 THE COBLENTZ AFFAIR

William Weber Coblentz was born at North Lima, Ohio in 1873. His parents were farmers of Swiss-German stock and Coblentz grew up in a rural environment. As a boy he showed ingenuity in building models of farm machinery, thus early demonstrating the inventiveness and manual dexterity that foretold his later skills as an experimental physicist. He attended the local Lutheran Seminary at Poland, Ohio and entered the Case School of Applied Science at Cleveland in 1896. He graduated in electrical engineering; his lifelong interest in radiation phenomena was already in evidence at Case where he sent what were probably the first radio signals to be transmitted in North America from one college dormitory across a court yard to another. On graduation in 1900 he went to Cornell University to study physics under Professor E. L. Nichols who suggested that as a doctoral project he should build an infrared spectrometer. It should be noted that E. L. Nichols was himself a Cornell graduate where he had studied under E. F. Nichols -- he who had constructed the sensitive radiometer in Rubens' laboratory. The two Nichols were not related. At this time E. F. Nichols was Professor of Physics at Dartmouth College but although Coblentz as part of his graduate project built a radiometer to E. F. Nichols' design there does not appear to have been much communication between them. The prospect of working on infrared spectroscopy was not too well received by the young Coblentz who at first considered it a dull and almost completed field [72].

At Cornell, Coblentz built two infrared spectrometers. In the first and smaller instrument the prism and telescope were fixed and the spectrum was scanned by rotating the collimator-source unit about the prism axis. This was constructed during his graduate study period and was used in the early work on iodine which formed the basis of his doctorate thesis [73]. Its range extended to 15 μm (667 cm^{-1}). The radiation source was a Nernst heater and the detector was a Nichols radiometer as noted above. After receiving his doctor's degree in 1903 Coblentz continued to work under E. L. Nichols' direction as a Carnegie Institution Research Associate and Honorary Fellow of the University. It was during this two year postdoctoral period that he made his classical measurements on 112 organic compounds [74]. Unfortunately, Coblentz has little to say in his autobiography [75] as to what motivated him to undertake this work. He does discuss the alternatives of either making a detailed analysis of these spectra of a few selected substances or making a survey of as many substances as he could obtain, and he records his decision to follow the latter course [76]. In this he may have been influenced by his friendship with Professor C. F. Mabery at Case who provided him with a valuable set of pure hydrocarbons from Mabery's work on petroleum distillation.

For this work Coblentz built his second and larger spectrometer which is shown in Fig. 2. This instrument had a 10 cm aperture and gave twice the dispersion with the same rock salt prism. Energy limitations however prevented its use below 7 μm (1428 cm^{-1}). Many of his spectra were first measured to 15 μm (667 cm^{-1}) on the

FIG. 2. Coblentz's larger spectrometer. The Nernst source is on the right and the radiometer detector on the left. The Geissler pump to evacuate the detector is in the left rear corner. Some absorption cells are on the left front table.

smaller instrument and then remeasured over the restricted range on the higher resolution instrument.

Coblentz's measuring techniques have been discussed previously, both by the present author [77] and by E. K. Plyler [72], so we shall only summarize them briefly here. As a typical example, in Fig. 3 his spectrum of pyridine is compared with that from the Coblentz Society's Collection (No. 4840). Having chosen a setting for his spectrometer circle, Coblentz measured the galvanometer deflection using a cell in, cell out technique. An iron shutter between the source and the sample was raised and lowered to expose the radiation beam. During the cell out measurement a rock salt plate was inserted to compensate for the cell windows. A single measurement took 1.5 minutes and the complete spectrum was scanned in about 4 hours. The operator sat in an outer room and manipulated the shutter and cells by cables, plotting the spectrum by hand at the same time.

That Coblentz had a limited knowledge of chemistry shows up occasionally in his atlas, as for instance in some confusion between "hydrocarbons" and "carbohydrates". Nevertheless he acquired good experimental skills in preparative chemistry and appreciated the

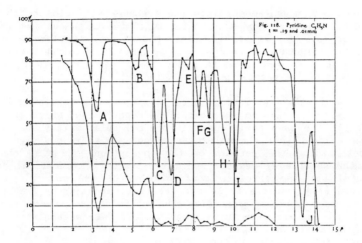

FIG. 3A. Coblentz's spectrum of pyridine.

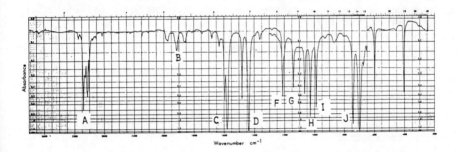

FIG. 3B. Pyridine spectrum from The Coblentz Society's Atlas collection (No. 4840).

importance of purifying his compounds before making spectral measurements. In preparing ethane from ethyl iodide he used the infrared spectrum in what is probably its first application in preparative organic chemistry. The crude product showed bands at 8.25 μm (1212 cm^{-1}) and 10.5 μm (952 cm^{-1}) which he attributed respectively to the starting material and to ethylene; he continued the purification until these bands disappeared [78].

We have noted previously that the earlier work on the infrared spectra of organic compounds was limited largely to the 1 - 3 μm region. It was natural that these investigators were impressed by the harmonic frequency relations they observed since many of the bands were overtones of the C-H stretch vibrations. As Coblentz extended his measurements into the fundamental region he observed the C-H scissoring modes at 6.83 μm (1464 cm^{-1}) and the rocking modes at 13.9 μm (719 cm^{-1}). This placed prominent bands in many aliphatic compounds in an approximate harmonic wavelength series of 1:2:4:8:16. We now know that the lower part of this series is fortuitous but it greatly puzzled Coblentz [79].

Coblentz' initial interest in organic spectra was motivated in part by a desire to examine the so-called "Kundt's Law" which attempted to relate the band positions with the molecular weight but it soon became evident that the molecular weight per se is not significant in determining the pattern of the absorption. He also observed that the spectra of dextro- and laevo-pinene are identical and he deduced that the absolute configuration of the molecule is not involved [79]. Initially he was skeptical about group frequencies but he soon came to recognize the correlation between specific bands and certain structural units as his data accumulated. Thus he noted that the replacement of -H by -NH$_2$ or -OH introduced characteristic changes in the spectrum and that

> "the vibration of the benzene ring is not destroyed by incorporation into a more complex molecule".

In his monograph he published a list of characteristic group frequencies which is reproduced in Table 2. His most telling illustration of the effect of local structure on the spectrum is his comparison of the spectra of CH$_3$-S-C≡N, CH$_3$-S-N≡C and C$_6$H$_5$-S-N≡C. In both of the methyl compounds he notes bands at 3.4 and 7 μm (2942 and 1428 cm^{-1}), and the presence of a broad band near 4.8 μm (2083 cm^{-1}) in all the compounds containing the -C≡N or -N≡C groups. Of the phenyl compound he says

> "This is the most important compound investigated since it shows the superposition of the spectra of benzene and the mustard oils".

It is curious however that although the spectra of many carbonyl compounds are included in the Coblentz monograph, he did not recognize the specificity of the C=O stretch vibration. He compares the spectra of eugenol and acetyl eugenol [82] and indeed notes the appearance of a new band at 5.8 μm (1724 cm^{-1}) but does not specifically associate it with the introduction of the C=O bond. As I have commented previously [77]

> "Perhaps here more than anywhere, Coblentz' limited appreciation of the problems of organic chemistry is in evidence. Though his observations on the specificity of the mustard oil spectra are significant spectroscopically, those compounds are to be classed as relatively unimportant chemical curiosities, or were so to the chemists of that era. If he had placed more emphasis on infrared spectroscopy as a means of characterizing the oxygen containing functional groups, he might have created more excitement in chemical circles, perhaps even

TABLE 2.   Positions of Characteristic Band Maxima

	Coblentz' Values [79]		Modern Values [80,81]
	μm	cm^{-1}	cm^{-1}
CH$_3$- or -CH$_2$-	3.43	2915	2962 - 2853
	6.86	1458	1465 - 1450
	13.6 - 14.0	735 - 714	750 - 720
-NH$_2$	2.96	3378	3500 - 3300
	6.1 - 6.15	1640 - 1626	1650 - 1590
C$_6$H$_6$[a]	3.25	3077	3030
	6.75	1481	1500
	8.68	1152	-
	9.8	1020	1225 - 960[b]
	11.8	847	860 - 800[c]
	12.95	772	770 - 730[d]
-NO$_2$	7.47 ?	1339	1350 - 1250
	9.08	1101	-
-OH	2.95	3400	3650 - 3200
-SNC	4.78	2092	2100 - 2085

[a]Coblentz did not distinguish between benzene and phenyl groups with various degrees of substitution.
[b]These bands depend upon the degree and position of substitution.
[c]Tetra-substituted with two adjacent hydrogen atoms.
[d]Mono-substituted.

       enough to have persuaded him to keep working in the
       field".

Instead Coblentz left Cornell in 1905 to join the newly established
U. S. Bureau of Standards as a research assistant to S. W. Stratton
in the Optics Division.   Here he was encouraged to build a new
infrared spectrometer although the main interest of Stratton's
Division was interferometry.   The suggestion to continue working in
the infrared did not please Coblentz who writes [83]

"Frankly my heart sank; for I wanted to quit infrared
work and resume interferometry, which formed my thesis at
Case".

In fact, a compromise evolved. Coblentz continued to work on
infrared optics, developing a Radiometry Section at the N.B.S.,
concerned initially with the comparison and improvement of various
types of infrared detectors [52]. This became his life's work
until his retirement from the Bureau forty-six years later.

1905 - 1925 THEORY THRIVES BUT LEAN YEARS IN THE LAB

The Coblentz monograph provoked little interest in chemical
circles, and it would not be until the mid 1920's that general
interest in the group frequency analysis of the infrared spectra of
complex molecules revived. During this quiescent period the most
notable work was a paper by W. Weniger at the University of
Wisconsin in 1910 [84]. He made an extensive study of alcohols,
esters, aldehydes and ketones and recognized that the band near 5.9
μm (1695 cm-^{1}) was associated with the carbonyl group, and a band
near 3.0 μm (3333 cm-^{1}) with the hydroxyl group. For primary,
secondary and tertiary alcohols he also noted bands at 9.6, 9.1,
and 8.6 μm (1042, 1090, 1163 cm-^{1}), respectively, which he
suggested could be used to distinguish among these structures.

During this period however much progress was being made in
interpreting the fine structure of vapor phase spectra. This was
stimulated by Bjerrum's 1912 paper followed by Eva von Bahr's high
resolution measurements on HCl and H_2O which we have noted previously [58,59,60]. By 1914 it was well understood that the fine
structure in the infrared spectra of gases was associated with
vibrational and rotational motions. At the same time the
physicists were much concerned with the quantitative treatment of
the specific heats of gases and it was appreciated that there was
an intimate association between the specific heat and the infrared
spectrum. We should not digress to discuss this in detail, but we
might note, as typical of the problems worrying theoretical spectroscopists at this time, was the absence of a Q branch in the
fundamental vibration band of HCl. This was noted by W. Burmeister
in 1911 [85], but with dispersion too low to resolve the separate
rotation peaks. E. S. Imes' 1919 spectrum of HCl did resolve the
rotational fine structure and prominently showed the absence of the
central band; it also showed marked asymmetry of the intensities in
the P and R branches [86]. The problem of the "missing band" was
very worrisome at that time; it was explained by Kratzer in 1920 as
being due to the lifetime in the rotationless state being vanish-
ingly small [87]. The review article by Fujisaki treats the theoretical developments during this period in lucid detail with
reference to the contributions of N. Bohr, P. Ehrenfest, A.
Einstein, T. Heurlinger, F. Reich and K. Schwartzchild as well as
those cited above [88].

1920 - 1940 ACTIVITY IN ACADEME

The work reviewed above was almost exclusively concerned with simple diatomic and triatomic molecules in the gas phase. In 1914 Bjerrum quantified his earlier hypothesis about molecular forces in a way more applicable to polyatomic molecules by treating the atoms as point masses connnected by harmonic Hooke's Law forces [89]. Symmetry considerations were introduced by C. J. Brester in 1924 [90], and in 1930 E. Wigner first introduced group theory to classify the vibrations [91]. Soon afterwards D. M. Dennison [92], J. Rosenthal and G. M. Murphy [93], and E. B. Wilson [94], and others, developed normal coordinate vibrational analysis, culminating in the monographs of G. Herzberg [95], and E. B. Wilson, J. C. Decius and P. C. Cross [96], in 1945 and 1955; these remain the standard reference texts today.

In principle normal coordinate analysis made it possible to predict the vibrational frequencies of polyatomic molecules provided that the geometry, the atomic masses and the interatomic forces were known. In practice the force fields could be estimated only imperfectly and the computational difficulties were formidable in this pre-computer age. Meaningful normal coordinate calculations could be made only for small molecules or larger ones with very high symmetry. The twelve atom benzene molecule represented an upper limit in molecular complexity and through the 1930's and 1940's it was a prime subject of study by theoretical spectroscopists. Many partly collaborative and partly controversial papers came from the laboratories of B. L. Crawford [97], C. K. Ingold [98], A. Langseth [99], R. C. Lord [100], F. A. Miller [101], and E. B. Wilson [102]. Recently [103], Lord has reminisced about the scientific and personal relations among these spectroscopists in an interview article which opens an illuminating window on the state of infrared spectroscopy at that time.

The normal coordinate calculations applied principally to molecules in the gas phase and they failed to take account of the intermolecular forces that broaden bands and displace the peak frequencies, particularly in liquids. This premium on molecular symmetry and low vapor pressure led most laboratories to concentrate their efforts on a small variety of molecules of low molecular weight. The gap between the theoretical studies of the physical chemists and the pragmatic interests of organic chemists in complex natural products remained wide open during this period. There was little cross communication. The narrowing of this intellectual hiatus in the 1960's will be discussed later but we must first look at what was happening at this time concerning group frequencies and see how the seeds sown by Coblentz belatedly came to fruition.

In the years leading up to World War II studies of the infrared spectra of all types of organic compounds were being made in several laboratories throughout the world. Among the first to reopen this field was J. Lecomte at the Sorbonne in Paris. His lifelong involvement with chemical spectroscopy stemmed from his

early interest in planetary atmospheres. His first infrared
spectra appeared in 1924 and 1925 [104-106]. He summarized his
early work in 1928 [107], and wrote an encyclopedic review on
infrared group frequencies for V. Grignard's Traité de Chimie
Organique [108], in 1936.

In Germany R. Mecke established a research group in infrared spectroscopy initially at Bonn (1923-1930) and Heidelberg (1930-1937).
He moved to Freiburg im Breisgau in 1937 where he remained as
Director of the Institut für physikalische Chemie until 1963.
Subsequently the work of his group continued as a separate research
institute -- the Institut für Elektrowerkstoffe der Fraunhofer-
Gesellschaft -- which he established on the Freiburg campus. His
earlier papers from Bonn and Heidelberg dealt mainly with high
resolution vapor phase spectra. His work on more complex compounds
proliferated during the Freiburg period. He and his students
published in excess of 200 papers, including several review monographs and an atlas of 1800 infrared spectra measured in the
Freiburg laboratories [109]. His publications have been catalogued
by the University of Freiburg [110,111].

In Great Britain the 1930's saw the establishment of two prominent
schools of chemical spectroscopy one at Cambridge under G.B.B.M.
Sutherland and the other at Oxford under H. W. Thompson. The
Cambridge group tended to be more physically oriented; Sutherland
did not undertake the extensive surveys of many kinds of organic
compounds as did the French and German schools. Like Mecke, his
early infrared studies mainly dealt with vapor phase spectra of
small molecules. In 1933-1934 he published on C_2H_2, N_2O_4 and NO_2
[112,113,114]. His work on larger molecules tended to be selective
thus he and his students made notable contributions to the interpretation of hydrocarbon spectra [115] and hydrogen bonding [116].
Sutherland graduated from St. Andrew's University in 1928, followed
by a doctorate degree at Pembroke College Cambridge in 1933. He
remained at Cambridge until 1949. A spell as Professor of Physics
at the University of Michigan followed until 1954 when he returned
to England as Director of the National Physical Laboratory. In
1964 he moved back to Cambridge as Master of Emmanual College. In
addition to his research contributions, Sutherland was an inspiring
teacher and many of the next generation of infrared spectroscopists
owe much to the stimulation of working in his laboratory. Among
these were D. M. Agar (née Simpson), G. K. T. Conn, M. M. Davies,
A. R. Philpotts, D. A. Ramsay, N. Sheppard and H. A. Willis in
England, and S. Krimm at Ann Arbor. Sutherland's short monograph
"Infrared and Raman Spectra", published in 1935 was a valuable
introduction to the subject at a time when few general texts were
available [117].

H. W. Thompson's association with Oxford University paralleled
Sutherland's with Cambridge and together they typify the role of
these universities in British academic life. Thompson arrived at
Oxford as an undergraduate in 1925 and graduated three years later.
His early research was in chemical kinetics with C. L. Hinshelwood.
Later he spent two years with F. Haber at the University of Berlin

where he acquired a doctorate. He returned to Oxford in 1930 where he began his long association with St. John's College which continued until his retirement in 1975. Thompson's publications in infrared spectroscopy began shortly before World War II. In 1938-1939 papers appeared on the infrared spectrum of methylamine [118], and on hydrogen bonding [119]. Most of his formative work at Oxford developed during the war years and a steady stream of papers followed in the post-war period. His earlier interest in chemical kinetics is apparent in some of the later work from his laboratory where he and his students developed quantitative relationships between chemical reactivity, as measured by rate and equilibrium constants (Hammett and Taft parameters) and the positions and intensities of infrared bands [120,121]. Such correlations between infrared band positions and other physical properties had been noted earlier by R. H. Gillette [122], and M. St. C. Flett [123,124], and later became the focus of extensive research by A. R. Katritzky and associates [125]. Thompson was a man with wide involvements in both national and international scientific affairs and he and his Oxford laboratory became a center for coordinating the advances in infrared spectroscopy on a world wide basis.

In the United States interest in group frequency analysis developed strongly at the University of Michigan in the laboratory of H. M. Randall. This Michigan group was primarily physically oriented and in the pre-World War II period its contributions to analytical infrared spectroscopy were mainly instrumental. In 1931 Randall, in collaboration with J. Strong built a prism spectrometer with several novel features [126]. It included a photographic recording system based on a periodic amplifier designed by F. A. Firestone [127]. In some ways this harked back to the Langley bolograph. Randall also introduced the use of a heated base plate for the prism which largely eliminated the fogging of the prism by atmospheric water vapor. He also designed a grating spectrometer in which the troublesome problem of order separation was overcome by the use of a KBr fore-prism [128]. Randall's involvement with group frequency analysis came later when the instrumental expertise of his Michigan group was applied profitably to the World War II research on penicillin. It is reviewed in a monograph by Randall and collaborators published in 1949 [129].

Another major center of analytical infrared spectroscopy was established by R. C. Lord at M.I.T. Lord obtained his doctorate at Johns Hopkins under E. H. Andrews in 1936. Postdoctoral periods followed with N. Wright at the University of Michigan and with A. Langseth at the University of Copenhagen. In 1942 Lord first became associated with the Spectroscopy Laboratory at M.I.T. where he worked on military applications of infrared systems with C. Squire and G. R. Harrison. He moved permanently to M.I.T. in 1946 and remained there until his retirement. During the Johns Hopkins period much of Lord's work concerned both experimental and theoretical studies on benzene and its deuterium substituted species, which we have noted previously. Subsequently Lord built up a preeminent center for research and teaching at M.I.T. Both during the Johns Hopkins and the M.I.T. periods he was a revered teacher and among his students are included J. R. Durig, E. R. Lippincott, D. Mayo, R. S. McDonald, and F. A. Miller. In collaboration with G. R. Harrison and J. R. Loufbourow, in 1948 he published "Practical Spectroscopy" which long remained a standard text [130].

Coming as a wind blowing from an unexpected quarter, the discovery of the Raman effect in 1928 introduced a new element into vibrational spectroscopy. From the early 1930's on it would be unrealistic to portray the history of infrared and Raman spectroscopy separately. On the theoretical side the two immediately became totally interactive. From the experimental point of view there have been periods of close coordination interspersed with intervals in which one or other (more commonly infrared) dominated the analytical field.

The scattering of monochromatic radiation with change of frequency was predicted theoretically by the Austrian quantum physicist A. Smekal in 1923 [131]. The first experimental observation was reported in India by C. V. Raman and K. S. Krishnan in a paper dated February 16th, 1928 [132], and independently on July 13th of the same year by G. Landsberg and L. Mandelstam in the Soviet Union [133]. In their earliest experiments Raman and Krishnan used filtered sunlight as a radiation source and detected the Raman lines of some sixty liquids and gases. They observed the scattered light visually, using a set of compensating colored filters to enhance the optical sensitivity. A more definitive spectrum of carbon tetrachloride exhibiting both the Stokes and the anti-Stokes lines recorded photographically, using 4358.3 Å mercury excitation was published in the following year [134]. Landsberg and Mandelstam's first paper included a photograph of the 483 cm^{-1} Raman line of quartz recorded simultaneously with the 2538, 3126 and 3650 Å lines of the mercury arc; a table comparing the measured band shifts and the calculated frequencies was included. Exposures of 2 to 14 hours were required. These pioneer observations of Raman scattering have been illustrated and discussed by D. A. Long [135].

According to J. Brandmüller [136] Raman's interest in light scattering was first aroused in 1921 when, on a voyage to Europe, he was fascinated by the blue opalescence of the Mediterranean, and then seven years study of light scattering preceded his 1928 paper. He seemed aware of the importance of his discovery, but subsequently he showed little interest in extending its application to molecular structure and chemical analysis. The designation "Raman effect" came into general use almost immediately though the expression "combination light scattering" was used in the earlier publications from the U.S.S.R.

It is perhaps appropriate that it was a fellow countryman of Smekal who first recognized that the Raman spectrum could be a powerful tool in chemical analysis. This was K. W. F. Kohlrausch at the University of Graz. Already in 1928, at the instigation of his junior colleague A. Dadieu, he began to measure the Raman spectra of a wide range of organic compounds. Unlike infrared spectra, the Raman spectra could be obtained with the relatively simple apparatus available in a university teaching laboratory. Over the next few years a great flow of papers on the Raman spectra of all kinds of organic compounds issued from Kohlrausch's laboratory. These paralleled the publications on infrared spectra coming from Lecomte's laboratory. Kohlrausch summarized this work in a comprehensive review in 1931, followed by a supplement in 1938 and a revised edition in 1943 [137]. This was reissued in 1972 with introductory text by A. Dadieu, J. Brandmüller, H. W. Thompson and the present author [138]. The following passage from Dadieu's introductory text is worth quoting since it illuminates an interesting human aspect of life in the Graz laboratory. Kohlrausch was at first sceptical about the possibility of obtaining a Raman spectrum with the simple equipment at hand, but Dadieu's enthusiasm prevailed and as he recounts

> "I vividly remember Kohlrausch's characteristic look of ironic scepticism and his reply 'You want to achieve that with our teaching spectrometer?' ... After an exposure of 30 minutes I developed the plate and placed it in a fixing solution. Kohlrausch who had followed me to the dark room pushed up his glasses maliciously: he saw no Raman line! I 'divined' one and predicted that it would become visible on the dry plate. And that is exactly what happened. ... From then on the Raman lamp burned night and day. We slept in turns in the Institute, I myself distilled and cleaned until midnight the liquids we meant to analyse, made more exposures and measured the spectra. Kohlrausch saw to the marking of the lines, worked out the resulting Raman spectra and tried to interpret the structure. After two months we sent our first paper to the Austrian Academy of Science. Another five papers 'right across organic chemistry' followed to the end of 1929".

Access to the Kohlrausch publications was not widespread and an important contribution was made in 1939 by J. H. Hibben whose monograph "The Raman Effect and its Chemical Applications" appeared in the American Chemical Society Monograph series with a theoretical contribution by E. Teller. This is notable for the inclusion of a cross-indexed bibliography of 1757 references [139]. Thus at the end of the 1930's a situation developed in which those working on infrared group frequency analysis had often to resort to these encyclopedic Raman data collections for reference material since the Raman spectra were more extensive and better catalogued than the corresponding infrared data.

Studies of molecular structure based on spectroscopic techniques developed early in Japan but the workers followed a different approach from their contemporaries in Europe and North America. We might summarize this by saying that while the western scientists approached the problems of molecular vibrations by moving down the spectrum from the visible as new instrumental techniques made this

possible, their Japanese colleagues approached the mid-infrared
from studies based on static electromagnetic fields and up from the
radio-frequency end of the spectrum. This may seem a superficial
comment but it helps to clarify what can otherwise be a rather
confusing situation.

Physical chemistry in Japan at this period owed much to M. Katayama
at the Univerity of Tokyo [140]. Research on molecular structure
was pursued principally by his students S. Mizushima and Y. Morino
and by Mizushima's student, T. Shimanouchi. After graduation in
1923 Mizushima worked under the direction of Katayama on the
general topic of the effects of magnetic fields on chemical substances. He measured the radio-frequency absorption in the 1-50 m
wavelength range at various temperatures, principally for alcohols.
By comparing these with the dielectric constant measured in a
static field he confirmed Debye's theory relating the dielectric
constant to the square of the refractive index for visible light
[141]. Mizushima gained his doctorate in 1929. A period of study
in Europe followed, much of it spent with Debye at the University
of Leipzig. He returned to the University of Tokyo in 1931 and was
appointed Professor of Chemistry on Katayama's retirement in 1958.

Morino graduated in 1931. From 1943 until 1948 he was Professor at
the University of Nagoya when he was recalled to a separate chair
at the University of Tokyo. After Mizushima's return from Leipzig
he and Morino began a classical series of studies of the stereochemistry of the 1,2-dihalogenated ethanes (CH_2X-CH_2X, X = Cl,Br).
Using a combination of dipole moment measurements with Raman
spectroscopy they established the existence of an equilibrium
between gauche and trans conformers having restricted rotation
about the C-C bond. Their first papers with K. Higasi appeared in
1934 [142,143]. Later they extended these combined dipole moment
and Raman spectra measurements to a variety of organic compounds.
The concept of conformational isomerism had a profound influence on
structural organic chemistry and two years later it received strong
substantiation from the thermodynamic measurements of J. D. Kemp
and K. S. Pitzer [144].

Shimanouchi graduated in 1941 and obtained his doctorate in 1949.
He succeeded to Mizushima's chair when the latter retired in 1959.
Shimanouchi's early publications dealt principally with normal
coordinate calculations on a variety of complex molecules and he
also participated with Mizushima and Morino in the studies of
rotational isomerism. Later his interests widened to cover many
aspects of vibrational spectroscopy. In an article of this length
we cannot expand further on the ramifications and the interactions
among the research activities of these three prolific spectroscopists. Their individual contributions have been indexed in three
publications from the University of Tokyo [145-147] listing in
chronological order 193 papers authored or co-authored by Mizushima
between 1926 and 1959; 242 papers authored or co-authored by Morino
between 1932 and 1971, and 293 papers authored or co-authored by
Shimanouchi between 1942 and 1976. Taken collectively they impinge
on almost all aspects of vibrational spectroscopy, theoretical,

experimental and applied. Their students now occupy prominent university, governmental and industrial positions throughout Japan. The original school at the University of Tokyo continues under the direction of M. Tasumi, a student of Shimanouchi. He succeeded to the chair of physical chemistry when Shimanouchi retired in 1977.

1940 - 1960 INDUSTRY MOVES IN

During the period we have just considered each laboratory constructed its own spectrometers and the research centered on the study of infrared and Raman spectra in a spirit of basic research. By the end of the 1930's the potential of vibrational spectroscopy as an analytical tool in industrial and medical research came to be appreciated. This created a demand for off-the-shelf manufactured instruments, some standardization of the measuring techniques, and the training of technicians able to carry out repetitive measurements with skill and understanding. This emerging interest in applied infrared spectroscopy coincided with World War II, particularly during the latter years in the U.K. and the U.S.A. It developed under conditions of war time restrictions on publication; in the U.K. under the sponsorship of the Ministries of Supply and of Aircraft Production and in the U.S.A. of the National Defense Research Committee.

It would require a detailed analysis of the subsequently declassified reports to set these developments into their correct historical perspective. However, the situation was in no way comparable with the secrecy surrounding the contemporary developments in nuclear technology. Already by 1945 much of the information on new instrumental techniques was declassified in the U.S.A. In the U.K. this largely occurred under the auspices of the Faraday Society which held a General Discussion on the Application of Infra-red Spectra to Chemical Problems in January 1945 [148]. The tone of this meeting was set by the President of the Society in his opening address.

> "Here in this country much work in the field of infra-red spectroscopy has still to be withheld from publication. ... We have to thank Drs. Sutherland and Thompson for the trouble they have gone to and the great care they have exercised in separating the secret from the non-secret, permitting here and there a peep behind the scenes".

Several British spectroscopists made their first research contributions under these restrictive conditions in the laboratories of Sutherland and Thompson and their work first saw the light of day at this Faraday Discussion Meeting. Among these are D. M. Agar (née Simpson), G. K. T. Conn, A. R. Philpotts, R. E. Richards, N. Sheppard, P. Torkington, D. H. Whiffen, and H. A. Willis.

However it should not be inferred that industry's involvement in infrared spectroscopy arose solely from the exigencies of solving wartime problems. R. B. Barnes and V. Z. Williams at the Stamford

120,064

LIBRARY
College of St. Francis
JOLIET, ILLINOIS

Research Laboratories of the American Cyanamid Co. built an
industrial infrared spectrometer in 1936. It was described in two
publications in 1944 [149,150]; the second includes a bibliography
of 2701 references and an atlas of 363 spectra measured on the
instrument. The oil companies were also aware of the analytical
applications of infrared spectroscopy prior to 1940, particularly
the Shell Development Co. Research Laboratories at Emeryville,
California, under the direction of R. R. Brattain and R. S.
Rasmussen. In the U.K. publications from the Government
Laboratories by J. J. Fox and collaborators, from 1928 on were
evidence of the awareness of infrared spectroscopy beyond the
university sphere [151,152].

Writing in 1949, Lecomte estimated that the number of operational
infrared spectrometers in the United States had increased from 15
in 1938 to over 500 in 1947 of which 400 were being used in analytical laboratories [153]. I have no figures for Europe but German
companies were already using infrared analytical techniques in the
pre-war period, particularly the I. G. Farben group [153]. Because
of the political situation in France under German occupation,
Lecomte was not able to participate fully in these industrial
activities. Mecke's laboratory at Freiburg was largely destroyed
by allied bombing in 1944 and his institute was evacuated to the
small fishing village of Wallhausen on the Bodensee [154]. In the
immediate post-war period Freiburg came under French occupation.
This enabled Lecomte to reestablish scientific contact with Mecke.
Soon afterwards in collaboration with H. W. Thompson he organized
in 1947 the first European Molecular Spectroscopy Congress at
Constance and these meetings have continued to be held biennially
ever since. Full establishment of post-war collaboration with
vibrational spectroscopists in Japan came in 1962 with the holding
of the International Symposium on Molecular Structure and Spectroscopy at Tokyo under the chairmanship of Mizushima. It was
attended by 169 spectroscopists from overseas and 709 scientists
from Japan.

Among the wartime applications of infrared spectroscopy those concerned with petroleum fractionation and the analysis of hydrocarbon blends were important. There was need for accurate knowledge of the infrared spectra of all types of straight, branched
chain and cyclic hydrocarbons. This led to the inclusion of
infrared spectra in the broader based Project 44 of the American
Petroleum Institute sponsored by the U. S. National Bureau of
Standards in 1942 under the direction of F. D. Rossini [155]. Most
of the U. S. Petroleum companies participated in measuring these
spectra and the first set were released publicly soon after the end
of the war. The A.P.I. Project 44 was continued in peace time. It
was tranferred in 1950 to the Carnegie Institute of Technology at
Pittsburgh and again in 1961 to the Texas A&M University under the
direction of B. J. Zwolinski. Another wartime project making
extensive use of infrared analytical techniques was that concerned
with synthetic rubber manufacture in which there was need for the
rapid analysis of butadiene isomers and polymers.

From a more academic point of view the most challenging of these wartime induced problems was the use of infrared group frequency analysis in the determination of the structure of penicillin. This developed as a trans-Atlantic co-operative project between Sutherland and Thompson in the U.K. and Randall and four industrial laboratories in the U.S.A. (Merck, Pfizer, Sage and Shell). The classical methods of organic chemistry were unable to distinguish among the four structures shown in Table 3. Extensive group frequency studies on model compounds were made. These established that only the β-lactam structure I could satisfactorily explain the four bands observed at 1785, 1740, 1667 and 1538 cm^{-1}. The classical organic chemists had been sceptical about the β-lactam structure because such four-membered rings had not previously been observed in natural products. This work was pursued with great urgency in the hope that it would lead to a practical synthetic route to penicillin. However, this was not to be. A synthesis was only achieved in 1957 [156] and wartime production remained based on a fermentation process. An overview of the wartime research on the structure of penicillin was released by U. S. and U. K. government agencies in 1945 [157]. The details of the infrared work were published jointly by Thompson, Brattain, Randall and Rasmussen in 1949 [158].

In the late 1930's biochemical and medical research laboratories became aware of the analytical potentials of both infrared and Raman spectroscopy. In 1936 at the Harvard Medical School, J. T. Edsall and his collaborators began a series of studies of the Raman spectra of amino-acids and simple peptides which continued into the 1950's. The great advantage of the Raman spectrum in studying bio-systems was the possibility of making measurements over a range of pH. In this way the Edsall group was able to establish the existence of α-amino acids and peptides as zwitterions and determine the pH conditions under which charge transfer to the zwitterion form occurred [159]. A secondary spin-off from the penicillin research was the mass of information it provided about the group frequencies of the amide linkage and other nitrogenous base units. This laid the foundations for much of the later work on both the Raman and the infrared spectra of proteins and nuclei acids. An extensive investigation of the infrared spectra of amino acids, polypeptides and proteins was made by Darmon and Sutherland between 1945 and 1948 [160-162] and by E. J. Ambrose and A. Elliott in 1951 [163]. The subject was reviewed by Sutherland in 1952 [164]. The Raman measurements required large samples and were frequently inapplicable to natural bio-systems because of background fluorescence. Infrared measurements were restricted by the intense absorption of water. Only with the advent of laser Raman sources and Fourier transform infrared spectrometers did it become practicable to examine natural bio-systems. This field was reviewed in 1977 [165].

TABLE 3. Alternative Structures for Penicillin

I. <u>Beta-lactam structure</u>

Observed strong bands (cm^{-1})

1785 beta-lactam ring

1740 -COOH

1667 -O=C-N-

1538 Secondary amide

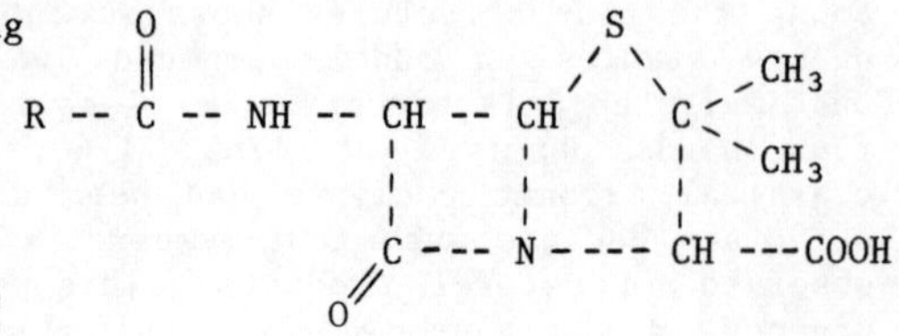

$$R--\overset{\overset{\textstyle O}{\|}}{C}--NH--CH--CH\quad\overset{S}{\diagup}\overset{CH_3}{C\diagup_{CH_3}}$$

$$\underset{O}{\diagdown}C---N----CH---COOH$$

II. <u>Tricyclic Structure</u>

Lacks the secondary amide band

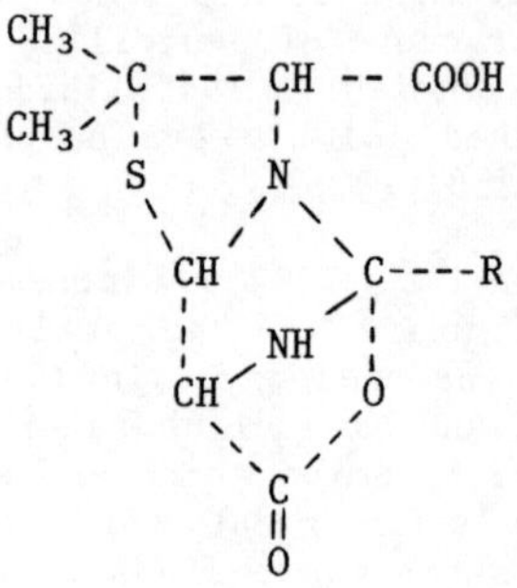

$$\overset{CH_3}{\underset{CH_3}{\diagdown}}C---CH--COOH$$

$$\underset{S}{}\quad\underset{N}{}$$

$$CH\qquad C----R$$

$$\underset{CH}{}\;NH\quad O$$

$$\underset{\underset{O}{\|}}{C}$$

III. <u>Zwitterion structure</u>

Bands at 1625, 2568 instead
of 1538 cm-1

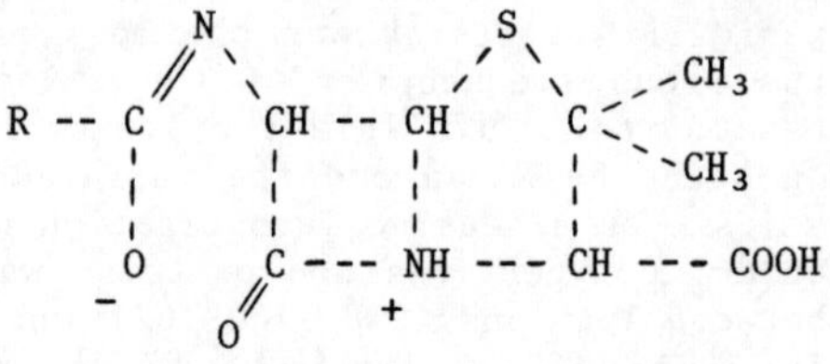

$$R--\overset{N}{C}\qquad CH--CH\qquad\overset{S}{C}\overset{CH_3}{\diagup}_{CH_3}$$

$$\underset{-}{O}\qquad\underset{\underset{O}{\|}}{C}---\underset{+}{NH}---CH---COOH$$

IV. <u>Az-lactone structure</u>

Band at 1811 instead
of 1785 cm-1

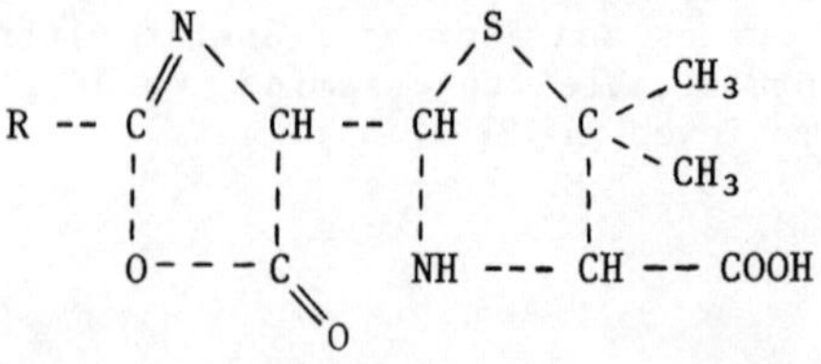

$$R--\overset{N}{C}\qquad CH--CH\qquad\overset{S}{C}\overset{CH_3}{\diagup}_{CH_3}$$

$$O^-\;--\;\underset{\diagdown O}{C}\qquad NH---CH--COOH$$

The application of infrared spectroscopy to clinically oriented medical research also began in the 1940's. Notable pioneer work was done by K. Dobriner at the Sloan-Kettering Institute for Cancer Research in New York City. Dobriner was a clinical pathologist who emigrated to the United States from Munich in the mid 1930's. In Germany he had specialized in the diagnosis of blood diseases associated with haemoglobin metabolism using visual spectroscopy. In the United States he worked initially at the Rockefeller Institute for Medical Research on carcinogenic hydrocarbon metabolism using ultraviolet spectroscopy. In 1939 he joined C. P. Rhoads at Memorial Hospital, New York, which later became the Sloan-Kettering Institute for Cancer Research. At Sloan-Kettering, Rhoads and Dobriner established a laboratory for clinical research on the metabolism of steroids. The steroids, being fat soluble hydrophobic substances, were eminently suitable for infrared analysis, principally as solutions in carbon disulfide. They possess exceptionally rich spectra in the range 1800-650 cm^{-1}.

Over the period 1946-1954 Dobriner and his associates at Sloan-Kettering in collaboration with the present author at the National Research Council of Canada mapped and codified some 460 group frequencies in steroid infrared spectra [166]. A review article was published in 1949 [167] and a two volume atlas of 760 steroid spectra in 1953 and 1958 [168,169]. The steroid framework built on four rings is rigid and for most compounds both the structure and the stereochemistry were known precisely. Small displacements of group frequencies, particularly those associated with hydroxyl and carbonyl groups, could be correlated with structural and conformational differences. Much of this information could be extended to other types of lipoid compounds. Some parallel can be drawn between the knowledge gained from this work about structure-spectra correlations in lipids with that acquired about proteins and nucleic acids from the survey measurements on nitrogen containing compounds in connection with the penicillin structure determination.

The instrumentation assembled at the Sloan-Kettering Institute is shown in Fig. 4. It represents a transition stage between the home built instruments of the 1930's and the commercially designed spectrometers which became available soon afterwards. It was based on the system developed in 1945 at the Research Laboratories of the American Cyanamid Co. by Barnes, McDonald, Williams and Kinnaird [170]. The negotiations leading to this collaboration between an industrial corporation and a private hospital to install infrared analytical facilities in a clinical laboratory were unique at this time and have been documented by Rhoads [171]. The spectrometer itself was one of the first batch of six manufactured by the Perkin-Elmer Co. at Glenbrook, Connecticut. A sodium chloride prism was used with a Globar source and a single element thermocouple detector. No recording system was included along with the Perkin-Elmer instrument but an automatic recorder was designed and installed by Barnes and Williams. The d.c. signal from the thermocouple, without benefit of pre-amplification, was fed to a wall mounted mirror galvanometer. The light beam from the mirror

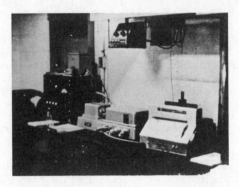

FIG. 4. The Perkin-Elmer Model 12A infrared spectrometer at
the Sloan-Kettering Institute for Cancer Research, Memorial
Hospital, New York in 1944. The power supply is on the left
of the spectrometer and the "light chaser" recorder is on the
right. The two mirror galvanometers are above the photograph.
The light beam from the mirror of the second galvanometer
shines vertically downwards and sweeps in an arc across the
back of the recorder.

reflected a focussed grid onto a second grid in front of a photo-
cell, whose output was fed to a second mirror galvanometer. This
optical amplification system was a development by Barnes and
Matossi [172] of a thermoelectric relay amplifier designed in 1925
by Moll and Burger [173]. The light beam from the second galvanometer was incident on a "light chaser" strip chart recorder first
described by D. J. Pompeo and C. J. Penther in 1942 [174]. These
and other ingenious detecting and recording systems devised at this
time have been reviewed by Williams [24].

The chart obtained was reproducible with good signal to noise ratio
but it was non-linear in wavenumber or wavelength and not baseline
corrected, being a sloping curve superimposed on the quasi black
body radiation curve of the Globar source. A manually executed
graphical method was devised to convert the chart to a linear
baseline, but this was necessary only for quality spectra for
publication. In the routine work of Dobriner's laboratory the
spectra were obtained as part of a system for characterizing
fractions from a liquid phase chromatograph, as illustrated in Fig.
5. On each chart a signature spectrum from a cell containing
acetone vapor was first recorded and the charts were aligned by
superimposing the acetone peaks at 1230, 1219, and 1207 cm^{-1}. Only
the most characteristic part of the steroid spectrum between 1180
and 875 cm^{-1} was recorded. The clinical samples were obtained from
24 hour urine collections, each of which provided about 200 chromatograph fractions, mostly as non-crystalline oils. The system
operated on a two shift basis and two technicians working together
could screen about 200 samples per day. Some 35 different steroid

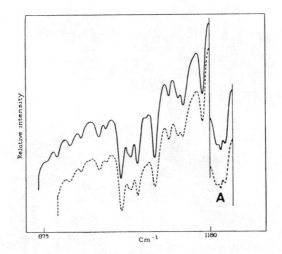

FIG. 5. Identification of a chromatographic fraction by infrared spectroscopy. ———— Steroid reference standard; ----- Urinary fraction. The wavenumber range (1180 - 875 cm^{-1}) was not recorded on the original chart. Charts were aligned by reference to the triplet band system A of acetone vapor at 1230, 1219 and 1207 cm^{-1} which was encoded routinely on each chart. The scanning direction was right to left and absorption is downward.

metabolites could be identified by this LC-IR system, though rarely would all be present in any one sample [167,175]. To my knowledge, this was the first published account of an LC-IR technique, though it is reasonable to surmise that similar systems were in operation in industrial laboratories at that time.

By the late 1940's the analytical applications of infrared spectroscopy had become widely recognized and a market for commercially manufactured spectrometers equipped with recording systems had been created. Much earlier in England Adam Hilger Ltd. had introduced their Model D-83 infrared spectrometer. I am uncertain when this first appeared in the Hilger catalog, but it was used by Lecomte in 1924 [104] and by Robertson and Fox in 1928 [151]. Its designer, F. Twyman recounts that as early as 1911 Adam Hilger Ltd. had consulted Paschen concerning the sales potential for a commercially manufactured infrared spectrometer. Not surprisingly the reply had been

> "I do not think a spectroscope for the infra-red will be a much purchased instrument. Everyone who makes these most difficult measurements will construct his own apparatus" [176].

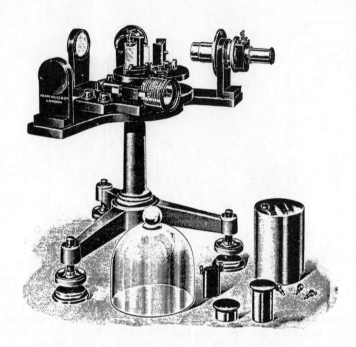

FIG. 6.   The Adam Hilger Model D83 spectrometer.

Adam Hilger offered a range of quartz, fluorite, rock salt and
sylvine prisms.   They recommended the use of a Nernst filament
source and a Paschen type mirror galvanometer.   Updated versions of
this instrument continued to be used into the 1940's.   It is illustrated in Fig. 6.

In the U.S.A. to cope with demands for increased infrared measuring
facilities during the war both the Perkin-Elmer Co. and the
National Technical Laboratories (later Beckman Instruments, Inc.)
batch produced infrared spectrometers under the sponsorship of the
U.S. government, and their first commercial models became available
soon after the end of the war (Figs. 7,8).   Also at this time
Baird Associates marketed an infrared spectrometer that became
widely used.   In the U.K., Adam Hilger produced new models and an
instrument manufactured by Sir Howard Grubb Parsons & Co. Ltd. came
into general use.   In the German Democratic Republic Carl Zeiss at
Jena produced their UR-10 spectrometer that was used in many of the
Socialist countries.   The earliest commercial instruments were
single beam with thermocouple or Golay detectors and chart
recorders.   They incorporated some kind of mechanical or electrical
cam system to linearize the wavenumber scale and the baseline.
Most had provision for purging with dry air or nitrogen but the
British Unicam instrument, which arrived on the scene somewhat
later, was evacuated.   The second generation instruments had
double-beam optics, using an optical wedge servo-system.   This

FIG. 7. The Perkin-Elmer Model 12C infrared spectrometer.

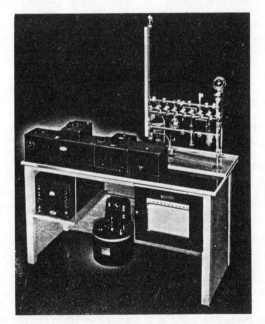

FIG. 8.   The National Technical Laboratories Model IR-2 infrared spectrometer.

removed the need for strict detector linearity since the detector served only as a null recorder to maintain a balance between the energy in the sample beam and the reference beam. The spectrum was measured by the motion of the comb attenuator. In the mid-1950's prisms began to be replaced by gratings using KBr fore-prisms or filters to provide order separation. Space does not allow more detailed discussion of these instruments but the state of the art in the late 1940's has been reviewed by Williams [24] and by Harrison, Lord and Loofbourow [130]. It was reviewed in the mid 1950's by Brügel [177] and in the early 1960's by Brügel [178] and by Martin [179].

By the 1950's a need arose to co-ordinate the diverse interests of infrared spectroscopists in industry, in government and in the universities. In the U.S.A. this led to the foundation of the Coblentz Society in 1955 with N. Wright of the Dow Chemical Co. as its first President. Similarly motivated societies evolved in the other industrially developed countries, among these were The Infrared and Raman Discussion Group in the U.K., The Groupement pour l'Advancement des Méthodes Spectrographiques (GAMS) in France, The Spectroscopy Society of Canada, and The Raman and Infrared Analytical Committee (RIAC) in Japan. On the international scene The Commission on Molecular Structure and Spectroscopy was established in 1957 under the aegis of the International Union of Pure and Applied Chemistry (IUPAC). It held its first meeting in Paris with H. W. Thompson as Chairman and has since met biennially in various countries. Indicative of the activities of these organizations are the formulation of specifications for the screening of infrared spectra intended for permanent retention in atlas collections, first set up by the Coblentz society in 1966 [180] and endorsed internationally by the IUPAC Commission in 1969 [181]. Notable also are the "Tables of Wavenumbers for the Calibration of Infrared Spectrometers" first published by the IUPAC Commission in 1961 and revised in 1977 by A.R.H. Cole [45]. Through such activities and by organizing symposia and awarding merit prizes these technical and professional organizations have done much to maintain uniformity and high standards in a field where diversification of interests could easily lead to conflict and duplication of effort.

The lack of trained and understanding personnel to operate these rapidly proliferating instruments and to interpret the spectra became acute in the late 1940's and in 1950 the first North American summer school in infrared spectroscopy was organized by Lord at M.I.T. Other one week or two week training courses combining hands on laboratory experience with practice in analysing spectra soon followed. Notable among these were the annual schools at the University of Minnesota, organized by B. L. Crawford and at Fisk University, Nashville Tennessee by N. Fuson. Participation in one or other of these training schools became a sine qua non for budding young infrared spectroscopists, especially for those seconded into the field in governmental or industrial laboratories who lacked the benefit or prior experience at a university. Later such courses have come to be replaced by instrument oriented courses arranged by the individual instrument manufacturers or by such organizations as the Coblentz Society. The original M.I.T. course however has continued to be held annually. With the retirement of Lord from M.I.T. its location moved to Bowdoin College, Brunswick, Maine where it continues as the Bowdoin International Infrared Course directed by D. Mayo.

Also in the 1950's several text books appeared directed specifically to the needs of analytical infrared spectroscopists. L. J. Bellamy's "The Infra-red Spectra of Complex Molecules" first published in 1954 [182] became a kind of bible carried in everyone's brief case. By many this was supplemented by the text published in 1956 by R. N. Jones and C. Sandorfy [183]. This appeared as a chapter in "Chemical Applications of Spectroscopy" edited by A. Weissberger. Being a heavier book it tended more to be kept on the shelf and consulted to supplement the more portable "Bellamy". Also much used were the frequently updated "Colthup Charts" issued with descriptive texts by N. B. Colthup. It was fortunate that Bellamy, Jones and Sandorfy collaborated before the publication of their books and agreed on some basic principles, such as the use of wavenumber (cm^{-1}) and to plot their curves with absorption peaks up rather than down.

1960 - 1985 THE AGE OF THE ACRONYM

As we approach the present, the task of the historian gets increasingly more difficult. Since 1960 there have been many new developments that have profoundly altered the practice of analytical infrared and Raman spectroscopy. As yet however we are too close to these to evaluate their impact on a long term basis with the objectivity which hindsight brings to earlier historic events. In these concluding sections we will drop the historic approach and restrict ourselves to some general observations. The major developments since 1960 have been: (a) digital data recording, first on punched paper tape and later on magnetic tape and disk; (b) data reduction using mini- and macro-computers; (c) interferometric techniques for measuring infrared spectra; (d) tunable lasers as sources for high resolution gas phase infrared spectroscopy; (e) a whole developing field of laser Raman spectroscopy with its extension to coherent anti-Stokes Raman scattering and related multi-photon processes.

We shall discuss these, but first we should take note of some broader considerations. A tendency has now developed for extreme specialization in instrumental technique which is affecting all analytical spectroscopy. This has encouraged the use of new jargons which make it increasingly difficult for analytical spectroscopists to communicate effectively across these self-erected barriers. The custom has also developed to identify these narrow fields by acronyms or lettered abbreviations, often to the extent of using them exclusively and dropping the descriptive name, even in the titles of publications. To some it may appear that a crowning success may be to devise some further sub-division of a spectroscopic technique which will enhance the reputation of the authors by adding yet another acronym to the collection. An assortment of some in current use is listed in Table 4.

In 1955 at a Faraday Society Discussion Meeting papers by J. N. Shoolery and H. S. Gutowsky signalled the advent of a new instrumental technique for chemical analysis -- nuclear magnetic resonance spectroscopy [184,185]. Some time elapsed before commercially manufactured NMR spectrometers became available in sufficient numbers to have an impact on the analytical laboratory, but a stream of papers on proton NMR applications to organic structure analysis began to appear in the late 1950's and these snowballed in the 1960's.

The coming of NMR had two important effects on infrared spectroscopy. In the first place it suppressed the incentives to seek for new infrared group frequencies. This had some merit since by 1960 most of the reliable infrared group frequencies were already known and there was danger that excessive zeal to refine these could lead to more "exceptions" with loss of confidence in the method as a whole. The second effect of the arrival of NMR was logistical. These instruments were very expensive and the funds to acquire them in many cases would come from the same budget appropriation that might otherwise have been used to update infrared equipment. It also had an effect on the personnel of the analytical laboratory. A new fraternity of NMR technicians and experts came into the laboratory. Hitherto in many laboratories the infrared spectroscopists had gained a unique status as oracles of molecular structure interpretation; now this had to be shared with their NMR colleagues. In the early days of NMR spectroscopy differences could arise about structure assignments based on infrared or NMR data such, for example, as different behaviour of structures exhibiting hydrogen bonding equilibria. This was due in large part to a failure to take account of the difference in the time scales of infrared and NMR events. Later with a better awareness that proton NMR is looking at events with rates of 10^6 - 10^{10} Hz while molecular vibrations occur on a time scale of 10^{12} - 10^{14} Hz these differences in themselves conveyed useful information.

Few Raman spectrometers were to be found in analytical laboratories prior to the introduction of laser sources. In 1958 the Applied Physics Corp. marketed the Cary Model 81 Raman spectrometer and the state of the art at the end of the pre-laser period, using this instrument, was reviewed by the present author and collaborators in 1965 [186]. The Cary Raman spectrometer used an ingenious image slicing optical device to collect the scattered radiation from the end of a long capillary tube which acted on a light pipe principle; the image slicer rearranged this circular beam into a narrow strip to be focussed on the 10 cm entrance slit of the spectrometer. The powerful Toronto arc mercury source needed a massive power supply. With this system the sample size was reduced to 10 - 100 mg but it did little to alleviate the problem of the fluorescent background. The Toronto arc was temperamental to run and difficult to ignite.

TABLE 4. Some Acronyms and Lettered Abbreviations Used in Analytical Spectroscopy

AAS	Atomic absorption spectroscopy
AES	Atomic emission spectroscopy
AES	Auger electron spectroscopy
AFS	Atomic fluorescence spectroscopy
ATR	Attenuated total reflection
CADI	Computer assisted dispersive infrared
CARS	Coherent anti-Stokes Raman Spectroscopy
CD	Circular dichroism
CIR	Cylindrical internal reflection
CSN-ICP	Conductive solids nebulizer inductively coupled plasma
DCP-AES	Direct current plasma atomic emission spectroscopy
EDS	Energy dispersive spectroscopy
EDXRF	Electron diffraction x-ray fluorescence
EELS	Electron energy loss spectroscopy
ESCA	Electron spectroscopy for chemical analysis
ESD	Electron stimulated desorption
ESIE	Electron stimulated ion emission
ESR	Electron spin resonance
FAB-MAS	Fast atom bombardment mass spectroscopy
FIR	Far infrared
FT-IR	Fourier transform infrared
FTIR	Frustrated total internal reflection
FTR	Frustrated total reflection
GC-IR	Gas chromatograph infrared
GC-MS	Gas chromatograph mass spectroscopy
GC-NMR	Gas chromatograph nuclear magnetic resonance
GF-AAS	Graphite furnace atomic absorption spectroscopy
GPMAS	Gas phase molecular absorption spectroscopy
HIXSE	Heavy ion induced x-ray satellite emission
ICAP	Inductively coupled argon plasma
ICPAES	Inductively coupled plasma atomic emission spectroscopy
ICP	Inductively coupled plasma

TABLE 4. (continued)

ICPES	Inductively coupled plasma emission spectroscopy
ICPMS	Inductively coupled plasma mass spectroscopy
ICRS	Ion cyclotron resonance spectroscopy
IRS	Internal reflection spectroscopy
ISS	Ion scattering spectroscopy
LAMMA	Laser microprobe analysis
LC-IR	Liquid chromatography infrared
LIMA	Laser ionization mass analysis
LIMS	Laser ionization mass spectroscopy
LRS	Laser Raman spectroscopy
MIP	Microwave induced plasma
MIR	Multiple internal reflection
MIR	Mid infrared
NIR	Near infrared
NIRA	Near infrared reflectance analysis
NMR	Nuclear magnetic resonance
OAS	Opto-acoustic spectroscopy
PAS	Photo-acoustic spectroscopy
PASCA	Positron annihilation spectroscopy for chemical analysis
PENIS	Proton enhanced nuclear induction spectroscopy
PES	Photo-electron spectroscopy
PNMR	Proton nuclear magnetic resonance
ROA	Raman optical activity
RRS	Resonance Raman spectroscopy
SAX	Selected areas x-ray photo-electron spectroscopy
SEM	Scanning electron microscopy
SIM	Scanning ion microscopy
SIMS	Secondary ion mass spectroscopy
WDS	Wavelength dispersive spectroscopy
XPS	X-ray photo-electron spectroscopy
XRD	X-ray diffraction
XRF	X-ray fluorescence

The introduction of the gas laser in the mid 1960's changed all that, and instrument manufacturers were prompt to produce Raman spectrometers with laser sources. The early models used weak but stable helium-neon lasers giving one excitation frequency in the red. These were soon superceded by more powerful krypton and argon ion lasers. These provided a range of stable frequencies throughout the visible and could be extended into the ultraviolet by frequency doubling devices. Both organic chemists and bio-chemists were quick to apply the new techniques, particularly in studies of proteins, nucleic acids and enzymes. Applications of Raman spectroscopy in industrial analysis were reviewed by Grasselli, Snavely and Bulkin in 1981 [187]. Laser Raman spectroscopy has not succeeded in competing on a large scale as a general analytical technique with either infrared or NMR spectroscopy. This might have been different if the introduction of the laser Raman technique had not been quickly followed by the arrival of the Fourier transform infrared which could also address some of the same problems which the older dispersion infrared failed to cope with, such as measurements on aqueous solutions.

When the laser Raman spectrometers first appeared much attention was given to measurements of organic and inorganic crystals, taking advantage of the additional depolarization data and information about symmetric vibrations. This was often essential in elucidating crystal structure. Thus the combination of infrared and Raman spectra brought about a closer liaison between vibrational spectroscopy and x-ray crystallography. Since the 1970's interest in Raman spectroscopy has tended to be diverted to resonance Raman spectroscopy (RRS), coherent anti-Stokes Raman spectroscopy (CARS) and other multi-photon processes. These have principally remained areas of academic research though they undoubtedly have applications as analytical tools for measurements on molecules in electronically excited states. CARS may also help to overcome the handicap of the background fluorescence in Raman measurements [188].

The introduction of the electronic computer had a salutory effect on infrared spectroscopy. Initially the influence was slow. The early computers were cumbersome and programming was an agonizing operation. The technique of writing programs in FORTRAN or other secondary languages has not changes fundamentally since the 1960's but the early compilers had little provision for error diagnosis. If the program aborted it was up to the programmer to trace it step by step through the algorithm to locate the source of the trouble.

One benefit of the electronic computer has been to narrow the intellectual gap between theoretical and analytical spectroscopists. Once the numerical computations of normal coordinate analysis could be transferred from a hand cranked desk calculator to an electronic computer it became possible to extend vibrational calculations from simple molecules of high symmetry to the wider range of substances encountered in industry and the bio-sciences. Already by 1960 Miyazawa, Mizushima and Shimanouchi at the University of Tokyo, collaborating with E. Blout at Harvard

University, began extensive calculations on polypeptides [189,190].
Later Shimanouchi and Tasumi made considerable progress in
automating the formulation of the F matrix which is an essential
and complex step in normal coordinate calculations. A major
advance was the publication in 1966 of the FORTRAN program of J. H.
Schachtschneider at Shell Development Co. [191]. In various
versions this is still widely used today. The calculations of R.
G. Snyder and J. H. Schachtschneider on hydrocarbons [192] in most
cases confirmed the group frequency assignments made earlier.
Though these developments widely extended the contacts between the
analytical and theoretical spectroscopists the routine expansion of
normal coordinate analysis to all types of organic compounds still
eludes us.

Infrared spectrometers recording digitally on punched paper tape
came into use in the 1960's [193]. Computer data processing
encouraged the application of statistical theory and numerical
analysis, and some understanding of information and communication
theory on the part of the spectroscopist became necessary. A.
Savitzky and M. J. E. Golay recognized this in 1964 [194] and drew
attention to algorithms for improving signal to noise ratio by
convolving the digitized experimental data with polynomial smoothing functions. This technique had long been in the text books of
numerical analysists but was unknown to spectroscopists. Later the
procedure of convolving the digitized data with filter functions of
increasing sophistication permitted the correction of the spectra
for systematic measuring errors or systematic optical distortion.
Actual enhancement of the resolution beyond that intrinsic in the
observed data could be achieved by such means though this could
prove to be a two edged weapon when applied injudiciously
[195-197].

In 1963 the first commercially available Fourier transform infrared
spectrometer for the mid infrared range was marketed by Block
Engineering Inc. (later Digilab Inc.) [198]. Its essential feature
was the replacement of the grating dispersion system by a Michelson
interferometer with provision for one of the parallel mirrors to
move uniformly along the optical axis. With this system the
complete spectral range could be recorded simultaneously with great
gain in energy throughput. This "Fellgett advantage" was pointed
out by P. B. Fellgett in Sutherland's laboratory in 1952 [199] and
by P. Jacquinot in 1954 [200]. The principal had been applied by
Rubens and Wood in 1911 [201] to enhance resolution over a small
range, but these early workers could not have realized its future
implications. The rapid progress in FT-IR technology can be
appreciated by comparing the review articles by J. Connes, E. V.
Loewenstein, P. R. Griffiths and J. E. Bertie in 1961, 1966, 1975
and 1985, respectively [202-205].

FT-IR spectroscopy was first engineered as a practical technique by
H. A. Gebbie at the National Physical Laboratory in the U.K. in
1956 [206] and marketed by Sir Howard Grubb Parsons Co. This early
instrument was limited to the far infrared. The primary output of
an FT-IR spectrometer is an interferogram, and three technical

problems had to be solved before it could be used effectively in
the mid infrared. One was the design of a precision monitoring
system to control the mirror motion. This had to achieve the
precision of a ruling engine, as the moving mirror can be considered to be generating a virtual grating in space during each
traverse. A second requirement was a fast detector capable of
maintaining a linear response without saturation at high signal
densities. The third requirement was efficient computer software
for rapid and accurate Fourier transformation to achieve the conventional spectrum. These were all solved in the mid 1960's. A
visible laser interferometer was attached in parallel with the
infrared beam to monitor the mirror motion. Fast solid state
detectors became available to replace thermocouples and a sophisticated factoring algorithm was developed by J. W. Cooley and J. W.
Tukey [207] which greatly reduced both the time and the computer
costs of the Fourier transformation.

The FT-IR spectrometer opened up new vistas in the analysis of
liquid and solid samples but the earlier instruments were limited
in resolution by the length of the mirror motion which was of the
order of a few cm. This could provide a resolution of about 0.5
cm^{-1}. It was adequate to reveal the true band profiles of condensed phase samples but was inadequate for high resolution
measurements on gases. Such gas phase spectra were important
theoretically and in applications to such problems as isotope
separation by selective photolysis. Later the mirror motions were
extended to 125 cm and high resolution FT-IR spectrometers capable
of a resolution of 2.6×10^{-3} cm^{-1} were developed commercially by
Bomen Inc. [208].

A somewhat inverse situation occurred in the 1970's with the introduction of tunable lasers. These have the potential to eliminate
the spectrometer completely by scanning the spectrum by varying the
source frequency. In 1976 using tunable diode lasers, Laser
Analytics Inc. developed a spectrometer with a resolution of 10^{-5}
cm^{-1} and capable of operating over the range 3 - 30 μm (3333 - 333
cm^{-1}) [209]. Unfortunately, in the present state-of-the-art each
diode can tune over only a small frequency span and a battery of
diodes would be needed to cover the conventional mid infrared.
Tunable laser sources were reviewed by A. S. Pine in 1982 [210].

WHERE DO WE GO FROM HERE?

Experimental infrared spectroscopy is still in a state of euphoric
excitement generated by the technical breakthrough which has
provided us with FT-IR spectrometers. This makes it an unhappy
time for the historian to have to bring his story to a close. It
will undoubtedly be some years yet before we will be able to view
the development of FT-IR spectroscopy with the objectivity of
hindsight. We are at present enamoured by the great gain in
optical efficiency which it provides and which we can at will use
vastly to increase the signal to noise ratio, the rate of data
acquisition, resolution, or sensitivity to small intensity differ-

ences, or other practical matters based on a trade off among these various possibilities.

Taking the long term view there are some inherent weaknesses in the FT-IR technique which we tend to overlook. With the changeover from dispersion optics to FT optics the desirable characteristics of the detector altered greatly. In dispersion systems the detector measures a weak signal and is limited in sensitivity by the random noise. In the FT system it receives the full radiation transmitted by the optical system which can cause distortion by saturation and associated non-linearity in the response. It is tacitly assumed that the instantaneous recording of the complete spectral range is a desirable attribute, but in some applications the analyst may only be interested in a narrow section of the completely recorded spectrum. In this case it is customary to restrict the measured signal by electronic or computer based filtering of the interferogram which does not reduce the load on the detector. In this circumstance the Fellgett advantage turns into a disadvantage. Furthermore most present day FT-IR spectrometers operate on a single beam principle with correction for the non-linearity of the baseline being made only before or after the measured run. This discards one of the main features of the ratio-recording double beam systems in which the base line is sampled several times per minute at the frequency of the beam switching mirror. As of the present there are also unresolved problems in the evaluation of the absolute photometric accuracy of FT spectrometers [211].

Historians are supposed to look backwards, not forwards, but not being a professional historian I can venture into science fiction and conclude with some speculations about another major revolution in infrared technique which could occur if the present narrow frequency ranges of tunable lasers were to be overcome by some technical breakthrough in applied physics. Then the whole infrared spectrometer/interferometer could be discarded, leaving us with only a monochromator source, a sample holder and a detector. Ideally one might conjure up a single solid state device to do this. Alternatively the same might be achieved by heterodyning the signal from one fixed frequency laser operating in the visible or ultraviolet with a signal from a second, possibly identical laser tunable over a range of 3000-4000 cm^{-1}. Continuous wave difference frequency tunable lasers operating in the infrared have been known since 1974 [210,212]; their tuning ranges are wide when pumped with a series of visible ion and dye lasers, but clearly these do not fall within our specifications for a single element widely tunable source. As of the present the main efforts in developing tunable laser infrared spectrometers are being directed to ultra high resolution measurements on gas phase spectra. There remains a wide R and D gap between these and a mass produced bench type tunable laser infrared spectrometer capable of performing throughout the mid infrared with a fast sweep and resolution of the order of 0.1 cm^{-1}; such as is hypothesized here.

Detractors can surely find arguments critical of this closing idea but I would like to leave you with two comments. Firstly, I would like to remind you that a decade ago few would have predicted that by now we would be rapidly replacing copper wire by optical fibers in land line communication. Secondly, please recall that Dr. Coblentz, whom we are honoring in this book, felt discouraged when Professor Nichols suggested he should build an infrared spectrometer in 1901. To quote myself from an earlier page

"the young Coblentz at first considered it a dull and almost completed field".

APOLOGIA

In a short article such as this it is not possible to acknowledge all the many spectroscopists who have made important contributions to our subject. I have had to be selective, and no doubt this selectivity has reflected my own bias and preoccupations with certain aspects of the field, to which I may have given undue weight. I have tried where I can to cite contemporary review articles which set the subject in better perspective as seen at the time. I apologize to the spectroscopic community as a whole for my sins of omission.

ACKNOWLEDGMENTS

I am grateful to Dr. H. H. Mantsch and the Division of Chemistry of the National Research Council of Canada for providing the facilities of his laboratory which have greatly eased my task in completing this article. I also wish to acknowledge numerous helpful discussions with colleagues both in the Division of Chemistry and in the Herzberg Institute of Astrophysics at N.R.C. Dr. Herzberg himself kindly read the manuscript and made several valuable suggestions. The willing help of the librarians at the Canada Institute for Scientific and Technical Information (CISTI) in locating and photocopying the early reference material is also warmly appreciated.

BIBLIOGRAPHY

1. Lucretius, De rerum natura. V. 610 (circa B.C. 60).
2. Mariotte, Traité de la Nature des Couleurs, (Pt. 2, Introduction), Paris (1686).
3. Scheele, Traité de l'Air et du Feu, p. 56, Paris (1781).
4. Pictet, Essai sur le Feu, p. 52 (incomplete).
5. F. W. Herschel, Phil. Trans. Roy. Soc. (London), 90, 284, 293, 437 (1800).
6. A. Ampère, Ann. Chim. Phys. (2), 58, 432 (1835).
7. T. J. Seebeck, Abhand. Preuss. Akad. Wiss., 265 (1822-1823).
8. T. J. Seebeck, Pogg. Ann. der Physik, 6, 1 (1823).
9. T. J. Seebeck, see Ref. [12].
10. L. Nobili, Bibl. Univ. Science et Arts Genève, 29, 119 (1825).
11. L. Nobili, Bibl. Univ. Science et Arts Genève, 44, 225 (1830).
12. W. W. Coblentz, Scientific Monthly, 68, 102 (1949).

13. M. Melloni, Annales de Chimie et Physique, <u>53</u>, 5 (1833); <u>55</u>, 337 (1835).
14. M. Melloni, Pogg. Ann. der Physik, <u>35</u>, 112 (1835).
15. W. W. Jacques, Proc. Amer. Acad. Arts and Sci., <u>14</u>, 142 (1879).
16. E. Pringsheim, Wied. Ann. der Physik, <u>18</u>, (3), 32 (1883).
17. G. R. Kirchhoff and R. Bunsen, Annales de Chimie et Physique, <u>62</u>, 452 (1861).
18. G. R. Kirchhoff and R. Bunsen, Phil. Mag., <u>22</u>, 329 (1861).
19. W. Crookes, Chem. News, <u>3</u>, 193, 303 (1861).
20. F. Reich and T. Richter, J. Pr. Chem., <u>89</u>, 441; <u>90</u>, 172, 175 (1863).
21. P. E. Lecoq de Boisbaudran, Phi. Mag., (5), <u>2</u>, 398 (1876).
22. P. J. C. Janssen. See J. W. Mellor, <u>A Comprehensive Treatise on Inorganic and Theoretical Chemistry</u>, Longmans, Green and Co., London (1947). Vol. VII, p. 890.
23. E. Frankland and J. N. Lockyer, Proc. Roy. Soc. (London), <u>17</u>, 91, 288 (1869).
24. V. Z. Williams, Rev. Sci. Instr., <u>19</u>, 135 (1948).
25. A. V. Svanberg, Pogg. Anal. der Physik, <u>84</u>, 411 (1851).
26. S. P. Langley, Proc. Amer. Acad., <u>16</u>, 342 (1881).
27. G. R. Harrison, J. Opt. Soc. Am., <u>39</u>, 413 (1949).
28. H. A. Rowland, Phil. Mag. <u>13</u>, Supplement No. 84 (1882).
29. H. A. Rowland, Phil. Mag., <u>13</u>, 469 (1882); <u>16</u>, 297 (1883).
30. Private communication from E. K. Plyler.
31. S. P. Langley, Annals. Smithsonian Obs., <u>1</u> (1900).
32. R. W. Wood, Phil. Mag., <u>20</u>, 770 (1910).
33. R. W. Wood and A. Trowbridge, Phil. Mag., <u>20</u>, 886 (1910).
34. W. W. Sleator, Astrophys. J., <u>48</u>, 125 (1918).
35. A. E. Martin in <u>Infra-red Spectroscopy and Molecular Structure</u>. Ed. M. Davies, Elsevier Pub. Co., Amsterdam, London, New York (1963), Chap. 2.
36. S. P. Langley, Phil. Mag., <u>17</u>, 194 (1884).
37. S. P. Langley, Phil. Mag., <u>22</u>, 149 (1886).
38. S. P. Langley, Annals. Smithsonian Inst. <u>1</u>, 221 (1902).
39. F. Paschen, Wied. Ann. der Physik, <u>53</u>, 301 (1894).
40. H. Rubens, Wied. Ann. der Physik, <u>53</u>, 267 (1894).
41. C. Schaefer and F. Matossi, <u>Das Ultrarote Spektrum</u>, Julius Springer, Berlin (1935). Reprinted by Edwards Bros., Ann Arbor, U.S.A. (1943).
42. J. Strong, Phys. Rev., <u>36</u>, 1663 (1930).
43. E. K. Plyler and N. Acquista, J. Opt. Soc. Am., <u>43</u>, 212 (1953).
44. H. M. Randall, Rev. Mod. Physics, <u>10</u>, 72 (1938).
45. <u>Tables of Wavenumbers for the Calibration of Infrared Spectrometers</u>, Butterworths, London (1961). Reprinted from Pure and Applied Chemistry, <u>1</u>, No. 4; 2nd Edition (revised), ed. A. R. H. Cole, Pergamon Press (1977).
46. See pp. 23-29 of Ref. [41].
47. C. V. Boys, Proc. Roy. Soc., (London), <u>42</u>, 189 (1887).
48. W. Crookes, Proc. Roy. Soc., (London), <u>22</u>, 32 (1874).
49. W. Crookes, Phil. Trans. Roy. Soc., (London), <u>166</u>, 325 (1876).
50. E. Pringsheim, Wied. Ann. der Physik, <u>18</u>, 1 (1883).
51. E. F. Nichols, Phys. Rev., <u>4</u>, 297 (1897).

52. W. W. Coblentz, Bull. Bur. Stds., 4, 391 (1908); 9, 7 (1913); 11, 131 (1914); 14, 507 (1918); J. Opt. Soc. Am., 5, 259 (1921).
53. H. Rubens and E. F. Nichols, Wied. Ann. der Physik, 60, 418 (1897).
54. O. Lummer and E. Pringsheim Wied. Ann. der Physik, 63, 395 (1897); Ann. der Physik, 3, 159 (1900).
55. M. Planck, Ann. der Physik, 4, 553 (1901).
56. H. Kangro, Vorgeschichte des Planckschen Strahlungsgetzes (Wiesbaden), (1970), see also under Rubens in Dictionary of Scientific Biography, Vol. XI, p. 582.
57. H. Deslandres, Compt. rend., 103, 375 (1886).
58. N. Bjerrum, Z. Elektrochemie, 17, 731 (1911).
59. Eva von Bahr, Verhand. Deutsche Phys. Ges., 15, 1154 (1913).
60. Eva von Bahr, Phil. Mag., 28, 71 (1914).
61. Chiyoko Fujisaki, Historia Scientiarum, Ed. The History of Science Soc. of Japan, Chuo University Tokyo, 192-03. No. 24, pp. 53-75 (1983); No. 25, pp. 57-86 (1983).
62. W. de W. Abney, Phil. Trans. Roy. Soc. (London), 171A, 753 (1880).
63. W. de W. Abney and E. R. Resting, Proc. Roy. Soc. (London), 31, 416 (1881).
64. W. de W. Abney and E. R. Festing, Phil. Trans. Ro. Soc. (London), 172A, 887 (1881).
65. K. Ångström, Ofv. Kongl. Vet. Akad. Förh. Stockholm, 46, 549 (1889), 47, 331 (1890).
66. W. H. Julius, Verhandl. Akad. Wetenschappen Amsterdam, 1, 1 (1892).
67. M. Ransohoff, Dissert. Berlin (1896): see p. 7 of Ref. [74].
68. E. Aschkinass, Wied. Ann. der Physik, 55, 401 (1895).
69. F. Paschen, Wied. Ann. der Physik, 53, 334 (1894).
70. L. Luccianti, Zeit. Physik, 1, 49 (1899).
71. L. Luccianti, Nuovo Cimento (4), 11, 141 (1900).
72. E. K. Plyler, Appl. Spec., 16, 73 (1962).
73. W. W. Coblentz, Phys. Rev. 16, 35, 72, 17, 51 (1903).
74. W. W. Coblentz, Investigations of Infra-red Spectra. Part I. Infrared Absorption Spectra. Part II. Infrared Emission Spectra. Carnegie Institution Publication No. 35 (1905). Republished by The Coblentz Society and the Perkin-Elmer Corp. (1962).
75. W. W. Coblentz, From the Life of a Researcher, Philosophical Library, Hallmark-Hubner Press Inc., New York (1951).
76. W. W. Coblentz, p. 105 of Ref. [74].
77. R. N. Jones, Appl. Optics, 2, 1090 (1963).
78. W. W. Coblentz, p. 46 of Ref. [74].
79. W. W. Coblentz, p. 116 of Ref. [74].
80. L. J. Bellamy, see Ref. [182].
81. L. J. Bellamy, Advances In Infrared Group Frequencies, Methuen & Co., Ltd., London (1968).
82. W. W. Coblentz, pp. 88, 89 of Ref. [74].
83. W. W. Coblentz, p. 133 of Ref. [75].
84. W. Weniger, Phys. Rev., 31, 388 (1910).
85. W. Burmeister, Zerhand. deut. Physik. Ges., 15, 589 (1913).
86. E. S. Imes, Astrophys. J., 50, 251 (1919).

87. A. Kratzer, Zeit. Phys., 3, 289 (1920).
88. Chiyoko Fujisaki, No. 25 of Ref. [61].
89. N. Bjerrum, Zerhand. deut. Physik. Ges., 16, 737 (1914); J.
     Chem. Soc. (ii), 110, 505 (1916).
90. C. J. Brester, Z. Physik, 24, 324 (1924).
91. E. Wigner, Göttinger Nachr. 133 (1930).
92. D. M. Dennison, Rev. Mod. Phys., 3, 280 (1931).
93. J. Rosentahl and G. M. Murphy, Rev. Mod. Phys., 8, 317 (1936).
94. E. B. Wilson, J. Chem. Phys., 2, 432 (1934); Phys. Rev., 45,
     427 (1934).
95. G. Herzberg, Infrared and Raman Spectra of Polyatomic Mole-
     cules, Van Nostrand Co., Inc., New York (1945).
96. E. B. Wilson, J. C. Decius and P. C. Cross, Molecular Vibra-
     tions, The Theory of Infrared and Raman Vibrational Spectra.
     McGraw-Hill Book Co., Inc., New York, Toronto, London (1955).
97. B. L. Crawford, Jr., and F. A. Miller, J. Chem. Phys., 17, 249
     (1949).
98. C. K. Ingold et al., J. Chem. Soc., pp. 912-987 (1936).
99. A. Langseth and R. C. Lord, Kgl. Danske Vid. Selskab. Mat-
     fys., Medd., 16, 6 (1938).
100. R. C. Lord and D. H. Andrews, J. Chem. Phys., 41, 149 (1937).
101. F. A. Miller and B. L. Crawford, Jr., J. Chem. Phys., 14, 282
     (1946).
102. E. B. Wilson, Phys. Rev., 45, 706 (1934); 46, 146 (1934).
103. R. C. Lord, European Spec. News, 56, 10 (1984).
104. J. Lecomte, Compt. rend., 178, 1530, 1698, 2073 (1924).
105. J. Lecomte, Compt. rend., 180, 825, 1481 (1925).
106. J. Lecomte, Trans. Faraday Soc., 25, 864 (1929).
107. J. Lecomte, Le Spectre Infrarouge, University of Paris Press
     (1928).
108. J. Lecomte, Spectres Dans L'Infrarouge. In Traité de Chimie
     Organique, Ed. V. Grignard, Masson, Paris (1936). Vol. II, pp.
     143-293.
109. R. Mecke and F. Langenbucher, Infrared Spectra of Selected
     Chemical Compounds. Heyden and Sons, Ltd., London.
110. R. Mecke, Wissenschaftliche Veröffentlichungen, 1937-1960.
     Albert Ludwigs University, Frieburg im Breisgau, F. R. D.
     (1960).
111. R. Mecke, Wissenschaftliche Veröffentlichungen, 1960-1965,
     Albert Ludwigs University, Freiburg im Breisgau F. R. D.
     (1965).
112. G. B. B. M. Sutherland, Phys. Rev., 43, 883 (1933).
113. G. B. B. M. Sutherland, Proc. Roy. Soc. (London), A141, 342
     (1933).
114. G. B. B. M. Sutherland, Proc. Roy. Soc. (London), A145, 278
     (1934).
115. N. Sheppard and D. M. Simpson, Quart. Rev. Chem. Soc., 7, 19
     (1953).
116. G. B. B. M. Sutherland, Trans. Faraday Soc., 36, 889 (1940).
117. G. B. B. M. Sutherland, Infrared and Raman Spectra, Methuen
     and Co., London (1935).
118. H. W. Thompson, J. Chem. Phys., 7, 448 (1939).
119. H. W. Thompson, J. A. C. S. , 61, 1396 (1939).

120. H. W. Thompson and G. Steel, Trans. Faraday Soc., 52, 1451 (1956).
121. P. J. Krueger and H. W. Thompson, Proc. Roy. Soc. (London), A250, 22 (1959).
122. R. H. Gillette, J. A. C. S., 58, 1143 (1936).
123. M. St. C. Flett, Trans. Faraday Soc., 44, 767 (1948).
124. M. St. C. Flett, J. Chem. Soc., 962 (1951).
125. A. R. Katritzky and R. D. Topsom, Angew. Chem. Internat. Edit., 9, 87 (1970) (review).
126. H. M. Randall and J. Strong, Rev. Sci. Instr., 2, 585 (1931).
127. F. A. Firestone, Rev. Sci. Instr., 3, 162 (1932).
128. H. M. Randall, J. Appl. Phys., 10, 768 (1939).
129. H. M. Randall, R. G. Fowler, N. Fuson and J. R. Dangl, Infrared Determination of Organic Structures. Van Nostrand Co., Inc., Toronto, New York, London (1949).
130. G. R. Harrison, R. C. Lord, and J. R. Loofbourow, Practical Spectroscopy, Prentice-Hall Inc., (1948).
131. A. Smekal, Naturwiss., 11, 873 (1923).
132. C. V. Raman and K. S. Krishnan, Nature, 501 (1928).
133. G. Landsberg and L. Mandelstam, Naturwiss., 16, 557, 722 (1928).
134. C. V. Raman and K. S. Krishnan, Proc. Roy. Soc. (London), 122, 23 (1929).
135. D. A. Long, Raman Spectroscopy, McGraw-Hill Internat. Book Co. (1977).
136. J. Brandmüller. See Introduction to Ref. [138] (p. xii).
137. K. W. F. Kohlrausch, Ramanspektren, Akademische Verlag. Becker und Erler Kom. -Ges., Leipzig (1943).
138. K. W. F. Kohlrausch, Ramanspektren, Heyden and Sons Ltd., London, New York, Rheine. (1972) (Contains introductory text by A. Dadieu, J. Brandmüller, H. W. Thompson and R. N. Jones).
139. J. H. Hibben, The Raman Effect and Its Chemical Applications, Reinhold Pub. Corp., New York (1939).
140. S. Mizushima, Ann. Rev. Phys. Chem., 23, 1 (1972).
141. S. Mizushima, Bull. Chem. Soc. Japan, 1, 47, 83, 115, 143, 163 (1926).
142. S. Mizushima, Y. Morino and K. Higasi, Physik. Zeit., 35, 905 (1934).
143. S. Mizushima and Y. Morino, J. Chem. Soc. Japan, 55, 131 (1934).
144. J. D. Kemp and K. S. Pitzer, J. Chem. Phys., 4, 749 (1936).
145. The Scientific Papers of Professor S. Mizushima--A Collection. The University of Tokyo (1959) (limited edition).
146. The Scientific Papers of Professor Yonezo Morino. The University of Tokyo (1971) (limited edition).
147. Vibrational Spectroscopy and Its Chemical Applications--Collection of the Scientific Papers of Takehiko Shimanouchi, The University of Tokyo (1977) (limited edition).
148. The Application of Infra-red Spectra to Chemical Problems --A General Discussion. Trans. Faraday Soc., 41, 171-297 (1945).
149. R. B. Barnes, U. Liddel and V. Z. Williams, Ind. Eng. Chem. Anal. Edit., 15, 659 (1943).

150. R. B. Barnes, R. C. Gore, U. Liddel and V. Z. Williams, Infra-
     red Spectroscopy -- Industrial Applications and Bibliography.
     Reinhold. Pub. Corp., New York (1944). (Contains text of Ref.
     [149] plus atlas and bibliography).
151. R. Robertson and J. J. Fox, Proc. Roy. Soc. (London), 120,
     128, 149, 161 (1928).
152. J. J. Fox and A. E. Martin, Proc. Roy. Soc. (London), A162,
     419 (1937); A175, 208 (1940); J. Chem. Soc., 318 (1939).
153. J. Lecomte, Le Rayonnement Infrarouge. Gauthier-Villars, Paris
     (1949). Vol. II, p. 395.
154. G. Scheibe, Zeit. Elektrochem., 64, 549 (1960).
155. Selected Infrared Spectral Data. American Petroleum Institute
     Research Project 44, Ed. B. J. Zwolinski. Texas A&M Univer-
     sity, Texas, U.S.A. The early volumes provide detailed
     descriptions of the spectrometers used by the various compan-
     ies generating the first set of spectra in 1945-1946.
156. J. C. Sheehan and J. P. Ferris, J. A. C. S., 79, 1262 (1957).
157. Joint Report of the Committee on Medical Research and the
     Office of Scientific Research and Development (U.S.A.) and
     the Medical Research Council (U.K.). Science, 102, 627 (1965).
158. H. W. Thompson, R. R. Brattain, H. M. Randall and R. S.
     Rasmussen in The Chemistry of Penicillin, Eds. Clarke, Johnson
     and Robinson, Univ. Press, Princeton, N. J., U.S.A. (1949).
     Chapt. XIII.
159. J. T. Edsall, J. Chem. Phys., 4, 1 (1936) and later publica-
     tions.
160. S. E. Darmon and G. B. B. M. Sutherland, J. A. C. S., 69, 2074
     (1947).
161. W. T. Astbury, C. E. Dalgliesh, S. E. Darmon and G. B. B. M.
     Sutherland, Nature, 162, 596 (1948).
162. S. E. Darmon and G. B. B. M. Sutherland, Nature, 164, 440
     (1949).
163. E. J. Ambrose and A. Elliott, Proc. Roy. Soc. (London), A205,
     47 (1951); A206, 206 (1951); A208, 75 (1951).
164. G. B. B. M. Sutherland. In Advances in Protein Chemistry, 7,
     291 (1952).
165. Infrared and Raman Spectroscopy of Biological Molecules.   Ed.
     T. M. Theophanides, Proc. NATO Advanced Study Inst., D. Reidel
     Pub. Co., Dordrecht, Boston, London (1978).
166. R. N. Jones and F. Herling, J. Org. Chem., 19, 1252 (1954).
167. R. N. Jones and K. Dobriner. In Vitamins and Hormones, Ed. R.
     S. Harris and K. V. Thimann. Academic Press Inc., New Yori.
     Vol. VII (1949) pp. 293-363.
168. K. Dobriner, E. R. Katzenellenbogen and R. N. Jones, Infrared
     Absorption Spectra of Steroids -- An Atlas, Vol. I, Inter-
     science Pub. Inc. New York (1953).
169. G. Roberts, B. S. Gallagher and R. N. Jones, Infrared Absorp-
     tion Spectra of Steroids -- An Atlas. Vol. II, Interscience
     Pub., Inc., New York (1958).
170. R. B. Barnes, R. S. McDonald, V. Z. Williams and R. F.
     Kinnaird, J. Appl. Phys., 16, 77 (1945).
171. C. P. Rhoads. See Preface to Ref. [168].
172. R. B. Barnes and F. Matossi, Z. Physik., 76, 24 (1932).

173. W. J. H. Moll and H. C. Burger, Phil. Mag. (6), 50, 618, 626 (1925).
174. D. J. Pompeo and C. J. Penther, Rev. Sci. Instr., 13, 218 (1942); Bull. Am. Phys. Soc., 16, 9 (1941).
175. K. Dobriner, S. Lieberman, C. P. Rhoads, R. N. Jones, V. Z. Williams and R. B. Barnes, J. Biol. Chem., 172, 297 (1948).
176. F. Twyman, Chem. and Ind., 49, 578 (1930).
177. W. Brügel, Einführung in Die Ultrarotspektroskopie, Dietrich Steinkopff Verlag, Darmstadt (1954).
178. W. Brügel, An Introduction to Infrared Spectroscopy. Methuen & Co., London: John Wiley and Sons, Inc., New York (1962).
179. See pp. 30-67 of Ref. [35].
180. The Coblentz Society. Anal. Chem., 38, No. 9, 27A (1966): revised Anal. Chem., 47, No. 11, 945A (1975).
181. IUPAC, Pure and Appl. Chem., 50, 231 (1978).
182. L. J. Bellamy, The Infra-red Spectra of Complex Molecules. Methuen & Co. (London): John Wiley and Sons, Inc., New York (1954).
183. R. N. Jones and C. Sandorfy. The Application of Infrared and Raman Spectrometry to the Elucidation of Molecular Structure. In Technique of Organic Chemistry. Ed. A. Weissberger. Vol. IX. Chapter IV. (1956). pp. 247-580.
184. J. N. Shoolery, Dis. Faraday Soc., 19, 215 (1955).
185. H. S. Gutowsky, Dis. Faraday Soc., 19, 187 (1955).
186. R. N. Jones, J. B. DiGiorgio, J. J. Elliot and G. A. A. Nonnenmacher, J. Org. Chem., 30, 1822 (1965).
187. Jeanette G. Grasselli, Marcia K. Snavely and B. J. Bulkin, Chemical Applications of Raman Spectroscopy, John Wiley and Sons, New York, Chichester, Brisbane, Toronto (1981).
188. T. Takenaka, Advances in Colloid and Interface Science, 11, 291 (1979).
189. T. Miyazawa, T. Shimanouchi and S. Mizushima, J. Chem. Phys., 29, 611 (1958).
190. T. Miyazawa and E. R. Blout, J. A. C. S. 83, 742 (1961).
191. J. H. Schachtschneider, Tech. Report No. 57-65, Shell Development Co., Emeryville, Cal., U.S.A. (1966).
192. R. G. Snyder and J. H. Schachtschneider, Spectrochim. Acta, 21, 165 (1965).
193. R. N. Jones, Pure and Appl. Chem., 18, 303 (1969).
194. A. Savitzky and M. J. E. Golay, Anal. Chem., 36, 1627 (1964).
195. J. K. Kauppinen, D. J. Moffatt, H. H. Mantsch and D. G. Cameron, Appl. Spec., 35, 271 (1981).
196. J. K. Kauppinen, D. J. Moffatt, D. G. Cameron and H. H. Mantsch, Appl. Optics, 20, 1866 (1981).
197. R. N. Jones and K. Shimokoshi, Appl. Spec., 37, 59 (1983).
198. Pittsburgh Conference on Analytical Chemistry and Applied Spectroscopy. Program (1963) p. 36.
199. P. B. Fellgett, Ph.D. thesis, University of Cambridge (1951).
200. P. Jacquinot, J. O. S. A., 44, 761 (1954).
201. H. Rubens and R. W. Wood, Phil. Mag., 21, 249 (1911).
202. J. Connes, Rev. Opt. 40, 45, 116, 171, 249 (1961).
203. E. V. Loewenstein, Appl. Optics, 5, 845 (1966).
204. P. R. Griffiths, Chemical Infrared Fourier Transform Spectroscopy. John Wiley and Sons, Inc., New York (1975).

205. J. E. Bertie, in <u>Vibrational Spectra and Structure</u>, ed. J. R. Durig, Elsevier, New York (1985) Vol. 14.
206. H. A. Gebbie and G. A. Vanasse, Nature, <u>178</u>, 432 (1956).
207. J. W. Cooley and J. W. Tukey, Math. Comput., <u>19</u>, 297 (1965).
208. H. J. Buijs, D. J. W. Kendall, G. Vail and G. N. Berube, Soc. Photo-optical Instrumentation Engineers, <u>289</u>, 322 (1981).
209. Pittsburgh Conference on Analytical Chemistry and Applied Spectroscopy. Program (1976), p. 101.
210. A. S. Pine, Phil. Trans. Roy. Soc., (London), <u>A307</u>, 481 (1982).
211. J. E. Bertie, V. Benham and R. N. Jones, Appl. Spec., (in press).
212. A. S. Pine, J. O. S. A., <u>64</u>, 1683 (1974).

THE IMPORTANCE OF GROUP FREQUENCIES IN PRACTICAL ANALYSIS

Clara D. Craver

Chemir Laboratories
761 West Kirkham
Glendale, MO 63122

INTRODUCTION

Practical chemical problems are solved by infrared spectroscopy in
thousands of locations daily. These chemical problems include:

- Quality control on highly refined pharmaceuticals, chemical
  reagents and other closely controlled chemicals.

- Composition determination for process control in plant streams
  or environmental monitoring.

- Structure determination in polymers, and at polymer and filler
  or reinforcement interfaces in composites.

- Identification and tracking of chemical wastes.

- Chemical characterization of surfaces.

- Identification of unknown chemical substances, whether they
  arise from arson, pipe line deposits, gasoline contaminants,
  grease or oily contaminants, paper or paper chemicals, paint
  blemishes which occur during application or on a finished product.

- Materials characterization in product failure, answering
  questions such as:

    Were the right components used in the first place?

    Were impurities added in the raw materials either by
    misidentification at a plant or in shipping or by con-
    tamination in a barge, truck or a tank car?

- Forensic science.

- Contaminants critical in semiconductors at point of manufacture or in finished products ranging from solder flux to
  spilled coffee and instant tea in electronic systems.

   • Determining cause of product failure in liability cases for
      loss of adhesion. Examples are parquet floors that don't stay
      down; caulking that embrittles or peels out; conveyor belts
      that don't hold up to heavy service; plastics, which embrittle
      and crack prematurely; packaging materials, intended to be
      vapor barriers, that fail to protect the product; heat seal-
      ants for candy bars, food products or other industrial lines
      that rely upon adhesion by application of heat.

The list of practical applications goes on and on. It is the
versatility of IR that makes it the first method of analysis for so
many of these problems. Whenever the questions of: "What is it?",
"Is it there?", or "Are these materials the same?" need to be
answered about molecules or chemical substrates in broad, complex
mixtures or purified fractions, infrared spectroscopy is an
important part of the array of scientific tools to help answer
these chemical questions. Practical analysis can range from the
quick job, which can be answered from group-frequency assignment of
a few bands, to analytical research projects on complex mixtures.
The emphasis in this chapter is on how to go about using IR in
problem solving.

OBTAINING QUALITATIVE SPECTRA

Sample preparation depends upon the state of the sample, how
limited the sample size is, and the degree of quantitativeness
required. Solids, liquids and gases can be readily analyzed.
There are flow-through sample cells for continuous analysis,
specular reflection for coatings on metal, and diffuse reflection
techniques for highly opaque materials such as coal. The most
common sample preparation methods are described in excellent books
[1,2]. Detailed recommendations by ASTM Committees for general
samples, microsamples and for surfaces by internal reflection are
available [3] along with the annual updates [4]. The Coblentz
Society Board of Managers has published experimental details for
production of high-quality reference data [5].

The most common technique for liquids and semi-solids simply
requires making a thin layer of an unknown material, either sand-
wiched between transparent sample plates, commonly called "neat" or
"capillary", or as a deposited film or "smear" on a plate. Acceptable qualitative data can be quickly obtained by this method.
Inferior data are obtained if there are areas of light by-pass, or
if the sample is uneven in thickness. Crystalline solids are most
often run as pellets or mulls [1-5].

A most important sample preparation parameter is the use of the
right amount of sample to obtain a good spectrum which has bands in
optimum intensity ranges. For qualitative purposes, it is
generally considered optimum to have the strongest band in the
spectrum in the range of absorbance of 1.0 to 1.5, except for the
case of relatively nonpolar hydrocarbons, in which case better
qualitative analysis can usually be obtained by allowing the carbon
to hydrogen stretch to be essentially totally absorbing and to have

the $-CH_2-$ deformation band at 1450-1460 cm^{-1} at about 1.5 absorbance. Optimum qualitative interpretation of spectra can usually be
made if full scale display is achieved. This may require compensating out background or any light scattering that is otherwise
impossible to eliminate experimentally. Minimizing light scattering or broad energy loss is important for solid state samples, and
is especially bothersome in carbon-filled materials.

IDENTIFICATION OF SINGLE COMPOUNDS

Complete identification of a pure compound is possible by IR alone
if there is a reference spectrum of the compound. This is the
reason that reference spectrum libraries are so important in this
field. An IR analytical laboratory is severely limited in its
capabilities if it does not have ready access to large reference
data banks. Without reference spectra, complete identification
usually requires elemental analysis, NMR spectra and molecular
weight or mass spectrometry. Raman spectra may also be necessary
for highly symmetrical compounds.

There are several chemical structures that a beginner can learn to
identify with a high level of certainty in a few hours. Characteristic bands of some structures fall in distinctive regions of
the spectrum and are either exceptionally strong or have distinctive shapes. The first step is usually to establish whether any
polar structure is present, such as hydroxyl, carbonyl, nitrile,
phosphoryl, etc., and whether there is a large or small smount of
alkane structures and whether olefinic unsaturation or aromatic
rings are present. Alicyclic structures are more difficult to
specify by group frequency analysis.

Some of the background needed for IR interpretation is now commonly
included in undergraduate organic, analytical and instrumentation
courses. Special short courses and concentrated continuing education courses have also been available for thirty years and are
still very popular. A good course uses stepwise teaching of group
frequencies and lays a foundation of understanding of IR absorption
that permits logical assimilation of the many details that need to
be mastered. Figure 1 presents a useful orientation to broad
spectral regions and highlights the most distinctive groups. The
importance of reference spectra for amplifying this basic data is
demonstrated in Figs. 2 and 3. The spectra in Fig. 2 show straight
chain and branched chain alkanes, and the big difference that the
introduction of a hydroxyl group introduces. Figure 3 shows a
parallel example with toluene representing a typical aromatic
hydrocarbon with one substituent and the spectrum of phenol showing
the enormous spectral difference that results when the methyl group
of toluene is replaced by the OH group of phenol.

Further skill in interpretation soon permits deciding the most
likely kind of carbonyl, acid, ester, aldehyde, ketone, amide,
etc., and whether an -OH is a primary or branched alcohol or a
phenol, and whether these and other polar structures are attached
to a fully saturated carbon atom or may be attached directly to an

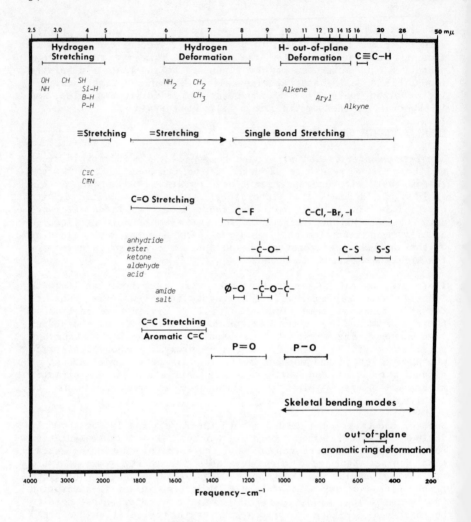

FIG. 1. A diagrammatic guide to IR group frequency regions.
Group frequencies are systematic, as shown by the clear distinction between hydrogen stretching and hydrogen deformation
regions above and by the narrow region of triple bond stretching. However, since both mass and bond strength affect the
frequencies, carbonyl frequencies overlap carbon-to-carbon
double bonds and phosphoryl bands overlap carbon-to-fluoride
single bonds. It is the overlap of distinctive bands from
different chemical structures that requires caution in
interpretation of IR spectra.

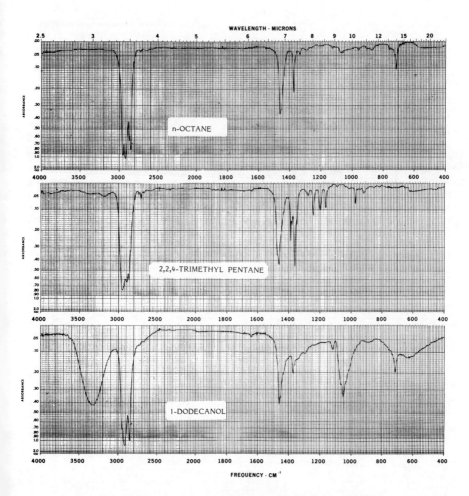

FIG. 2. Infrared spectra demonstrate the difference between straight chain and branched chain alkanes and the effects of introducing a primary alcohol structure to the alkane skeleton in dodecanol.

aromatic ring or other unsaturated group. Until skill in interpretation is acquired, it is desirable to have an elemental analysis, especially for halogens, which can have considerable impact on the frequency and strength of IR absorption of nearby structures and for nitrogen and sulfur, which may not always give distinctive infrared spectral characteristics. Elemental analysis is necessary for identifying the exact cation in salts, both organic and inorganic, and for specific identification of most heavy metal oxides and metallo-organics.

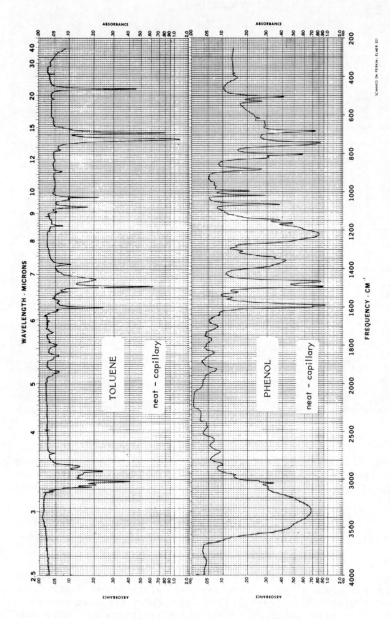

FIG. 3. The infrared spectrum of an aromatic hydrocarbon, toluene, is greatly changed by replacement of its methyl group by the OH group in phenol in which characteristic group frequencies of the OH structure and of the mono-substituted aromatic ring are clearly evident.

Guidance for beginners and spectral examples of major chemical classes are available in The Coblentz Society's Desk Book of Infrared Spectra [6]. The much larger IR reference library published by The Aldrich Chemical Company is also organized by chemical structure. Both of these publications are of considerable value for comparing an unknown to reference spectra after preliminary deductions about possible structures have been made. That way, support for the preliminary interpretation can be established or important differences between the postulated structure and the unknown spectrum are quickly detected. Good quality IR reference spectra are an important tool in the analytical laboratory. Group frequency summary charts should be used only as a preliminary guide or quick reference for spectral interpretation. Detailed books on infrared interpretation are essential aids for completing interpretation of an unknown spectrum. Multiple books are recommended because the emphasis is different in such classics as the compilations by Lionel Bellamy [7-9] and the detailed examples found in Colthup and Wiberley [10].

Not only are pure compound reference spectra most useful when they are grouped by chemical class, but polymer spectra are also more accessible when similar polymer structures are grouped together. The spectra in Fig. 4 are of common polymers and show the same kind of variation from alkane to aryl as was noted for compounds. A carbonyl compound, vinyl acetate, as a monomer, is readily identified as a homopolymer or copolymer by the spectra in Fig. 5, even at the level of 14% as shown in the top curve. Identification of a few percent PVAc is possible. Note that the ethylene monomer, as polyethylene, shows up as aliphatic CH stretching at 2920 and 2850 cm^{-1}, and at reasonable concentrations as long chain CH_2 at 720 cm^{-1} which shows that some block homopolymer is present.

All polymers are not this easy to identify. Examples of the need for extreme caution are given [11], and examination of spectra of different acrylic resins or phthalate-modified polymers brings respect for the difficulties involved in identifying some polymers, and especially polymer blends. The most useful reference books on polymer spectra are by Hummel and Scholl [12] and, for materials specifically related to coatings, the book by the Paint Federation is invaluable [13]. The Coblentz Society also publishes spectra of polymers, plasticizers, surfactants and other additives grouped by chemical class [6,14]. Haslam and Willis' well known [15] work combines chemical analysis with spectral data and Sadtler Laboratories [16] publishes extensive reference libraries.

The identification of a spectrum is always simplified if enough questions have been asked in the beginning to determine what kinds of answers might be expected. The more the analyst knows about the system from which the sample arose, the easier it is to put together the group frequency interpretations for chemical structures with suspect compounds and look them up in infrared libraries. This kind of sample, in which nearly all of the bands can be attributed to one compound or polymer, is the ideal sample spectrum for computer-assisted searching.

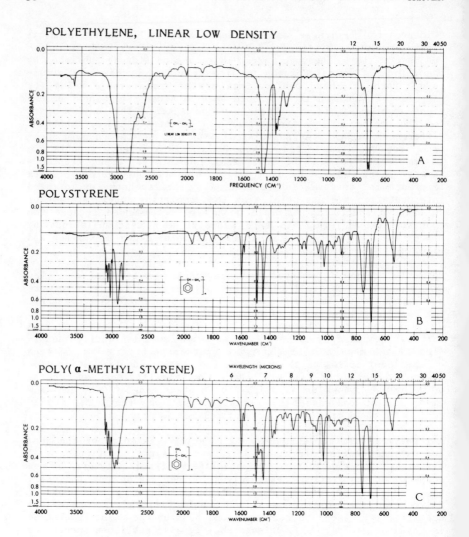

FIG. 4. Infrared spectra of polymers are characterized by the same group frequencies as lower molecular weight compounds. For example, in Chart A, polyethylene, which is principally straight chain or lightly branched alkane, is spectrally similar to n-octane, in Fig. 2. Correspondingly, the spectra of polystyrene and poly α-methyl styrene, B and C, parallel the spectrum of toluene, see Fig. 3, in major group frequency bands.

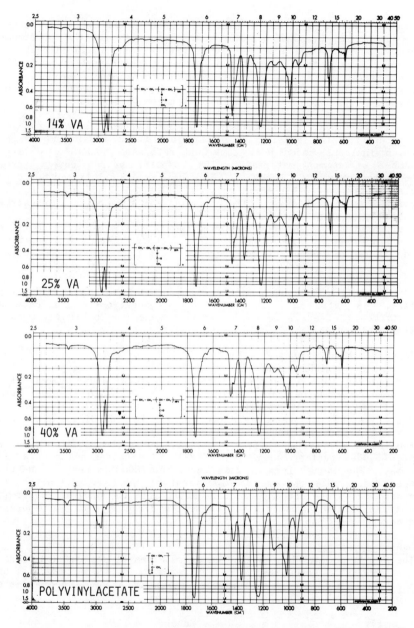

FIG. 5. Infrared spectra of polyvinyl acetate and its copolymers with ethylene. Observe that the characteristic spectrum of polyvinyl acetate is maintained even in the lower percentage copolymer, 14% VA. Polyethylene is evident by the CH stretching bands at 2900 and 2850, the $-CH_2-$ deformation at 1460 and the $-CH_2-$ chain rocking at 720 cm$-^1$.

Computer search systems of infrared libraries may be on dedicated
computers, e.g., the ones which are integral parts of spectrometers, or they may access a central data bank as in time-shared
commercial systems. Time-shared search systems have been available
since 1970, and dedicated systems with spectrometers are becoming
increasingly available. These change and grow continuously and the
examples cited here are not intended to be comprehensive.

The Perkin-Elmer Corporation system supplies both a spectrum interpretation for possible structure units and a library search,
which includes coded peak data and reference books of a large
number of common materials. The computer search covers three
Coblentz Society books: the Desk Book of Infrared Spectra,
Plasticizers and Other Additives, and Halogenated Hydrocarbons; The
Paint Federation book on Infrared Spectra in the Coatings Industry,
and polymer data from Hummel [12]. The Nicolet system offers The
Aldrich Library, the Georgia Crime Lab file, EPA Vapor Phase files
and Hummel data. Nicolet also offers a choice of search algorithms
and is rapidly acquiring an increased data base through collaborative programs generating reference data. Digilab® spectrometers
may provide search capability for the Sadtler IR library and a
peak-listing by frequency interval, or Spec-Finder® type of Sadtler
file, is also available on other spectrometers. Beckman
Instruments offers a search of spectra generated by that company,
plus public vapor phase and forensic files.

Spectral search systems which are not instrument-specific are
available by modem connection to telephone lines from practically
any terminal and, in most U.S. areas, by a local phone number. The
principal system in Canada is available from the Canadian Institute
for Science and Technology [17], and the principal European and
U.S. time-sharing system is IRGO©, available from Singer Technical
Services, Inc., through Chemir Laboratories [18]. The search
routine is similar in these systems and covers the public data base
of 145,000 spectra coded by ASTM E-13 plus subsequent data addi-
tions. The manual form of data input for IRGO© is most frequently
used and permits the user a wide range of choices. The automatic
system has preselected parameters built into the program to permit
a technician or nonspecialist to search directly from a computer-
generated peak table.

Manual Input
The system prompts the user to enter a series of search parameters.
Most analysts decide how to search a spectrum before they dial up
the computer. After the computer log-in is accomplished and the
program called up, the dialog is as follows:

Prompt: Enter:

Sample I.D.   Description, name, or number of sample

PHYSICAL/CHEMICAL
DESCRIPTION?   "yes" or "no". Most often "no", but if "yes',
                       a subroutine appears in which elements and

known structural features may be entered as either "yes" or "no" for each one present or not present.

MICRONS OR
WAVENUMBERS?   An "m" or "w" identifies the input for the computer.

(The following selections are all optional)

NO-DATA AREAS   Offers the opportunity to enter a region or single position where there is uncertainty in the data. This is important for discounting regions obscured by solvents or known components, and for avoiding uncertain entries such as weak bands or side-bands. The proper use of no-data areas greatly sharpens the specificity of a search.

BLACK AREAS   A bottomed-out band can be entered as a region. Even more useful, a series of multiple closely adjacent bands can be entered as a group. This frequently avoids high statistical mismatch from differences in peak-picking routines or from sample state effects on band shapes and multiplicity of bands from polymorphs.

MANDATORY BLANKS   Spectral regions, or points, where there are no absorption bands.

MANDATORY PEAKS   Usually entered for at least one principal distinctive band. Entry of more mandatory bands significantly shortens the calculation, but could lead to loss of an answer if the sample is an unsuspected mixture and if the mandatory peaks selected belong to different compounds.

PRIMARY PEAKS   All remaining peaks for which there is confidence that they have been coded in the data base. These are rated high in the calculation.

SECONDARY PEAKS   Peaks expected to be coded, but not as certain as primary peaks. Usually these are weak-to-medium bands which would not necessarily have been observed stronger than the coding threshold for a weak spectrum. These are weighted less than "primary peaks".

There are additional inputs for whether the entire public file is to be searched, or "Standard American Files", or separate spectral catalogs. There is also an option to adjust the band position tolerance which is to be accepted as a hit. This is an important selection option for spectra wih broad, irregular bands. By these selected input parameters, a spectroscopist can combine knowledge of the sample history, the data base, and factors which affect band locations to make a highly effective search input.

Automated Input

For a routine search from peak tables, most of the above parameters are preselected. A format file for any particular spectrometer's peak-finding system is set up and other parameters are preselected based on peak intensity in normalized spectra and, to an important but lesser degree, on spectral regions. For example, IRGO© typically selects the strongest band below 2000 cm^{-1} as the only mandatory peak. For obvious reasons, the common 1460-1450 cm^{-1} region is excluded from this automatic selection as a mandatory peak. Other significant peak input variables are also preselected. The user is offered an opportunity to make a selection of no-data areas to cover circumstances such as retained solvent bands which can be known to the chemist, but cannot be programmed into the search routine. A typical search for a compound by the routine Perkin-Elmer/IRGO© interface is given in Fig. 6.

The success of computerized spectral searches with such a sophisticated program is limited only by the quality and quantity of reference data files available. As pointed out earlier, computer search systems work best on pure or nearly pure compounds. However, it is possible to identify components of a simple mixture if each component has enough distinctive bands, and common mixtures can be found as easily as if they were pure compounds if a reference spectrum of the mixture is entered into the data base.

IDENTIFICATION OF MIXTURES

Commercial mixtures or products require a different approach from pure compounds. It is desirable to obtain a spectrum of the total sample, but it may serve only to establish the complexity of the sample and to guide subsequent work. The spectrum may be dominated by additives and fillers. The most common ones usually obscure too much of the spectrum to permit identification of other important ingredients. Ubiquitous among these are various forms of silica and silicates such as talc and clay, carbonates and heavy metal oxides. Organic components, which may be an important part of the unknown or which may be adventitious contaminants, are commonly silicones, phthalates and a wide assortment of aliphatic esters including dibasic acid esters. Contamination arises from the most unsuspected sources. An analyst may be naturally cautious about using plastic sample vials, funnels, etc. and still be misled by contaminants from organics on "disposable" glass or from "chromatographically pure" solvents. Whenever a small sample is being analyzed, or a larger sample is being separated into small fractions, it seems easier to obtain spectra of impurities than of the sample. The tried and true analytical practice of running complete blanks is essential for reliable analytical spectroscopy.

Polymers

The use of plastics in consumer products creates many needs for analyses. Development work to establish a composition of polymers

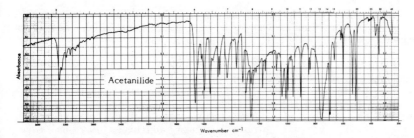

PHYSICAL/CHEMICAL DESCRIPTION OF UNKNOWN?   N (CR)
YOU CAN DECLARE NO-DATA AREAS. ALL OTHER PEAK DATA ARE TAKEN FROM TABLE.
DO YOU WISH TO ENTER NO-DATA AREAS?   N (or Y)

MANDATORY PEAKS   1672
  PRIMARY PEAKS   1596 1554 1536 1498 1485 1432 1375 1321 1262 755 694
SECONDARY PEAKS   1038 1011   960   906
DO YOU WISH TO RECEIVE IRGO OUTPUT ON A DISK FILE?   Y or N
DO YOU WISH TO SEARCH THE ENTIRE PUBLIC FILE?   Y or N (CR)
ENTER OTHER FILE TO SEARCH   (CR)

      (Final response with CR only, i.e., no new file entry, initiates search)

S#	SCORE	FILE	NO.	
1	884	EA *(Literature)*	23657	Acetanilide
2	769	GA *(Coblentz)*	1688	"
3	730	MCA *(Manuf. Chem.)*	4	
4	576	DA *(NRC-NBS)*	691	"
5	538	KA *(Aldrich)*	7845	"
6	519	CA *(Sadtler)*	19805	"
7	500	CA *(Sadtler)*	34648	Not Acetanilide

FIG. 6. Example of an automatic infrared peak search in which
peaks to be searched as mandatory, primary and secondary are
selected through a format table. The output score is a ranking relative to a value of 1000 for a perfect data match. The
File identification and number correspond to the ASTM system
for all published files.

and additives which will yield a desired product may be done
empirically and with little analytical data. In fact, only the
most experienced or foresighted manufacturers retain samples of the
original ingredients as a basis for future comparison. Sources of
supply change, or blenders switch raw ingredients to substitute
something conveniently on-hand or less expensive, and trouble sets
in, usually as cracking, crazing, failure to get adhesion, or
deformation of the product. It is a lucky end-manufacturer or
assembler of components who detects this problem before there are a
lot of faulty products in the warehouse, or worse yet, in the
stores or in the customers' hands. The escalation of replacement
costs is considerable as a plastic product incurs the add-on costs

of assembly and distribution, and becomes a serious business risk in labor-intensive industries such as construction or in expensive products such as automobiles.

Some of these formulation problems are difficult to trace because the product itself may be marginal. That is, it may be pushing the limits of the innate properties of the polymer being used. This is most apt to happen in the more inexpensive polymers: polyethylene, polypropylene, polystyrene, etc. For these materials, molecular weight control, thermal properties and additives may be more important than the small chemical structure differences that can be determined from a qualitative infrared spectrum.

On the other hand, some classes of polymers, for example, ABS (acrylonitrile/butadiene/styrene terpolymers) show widely different properties as a function of the relative amounts of the three monomers. A recent example was a molded plastic cover which had a long history of reliable performance, but the assembly plant started observing a splitting at the rounded corners during assembly, and the problem soon became essentially continuous with a large accumulation of molded covers with cracked corners. Infrared spectra showed the polymer to be ABS, but with a much lower content of butadiene and much higher styrene than the resin product specified. This kind of polymer comparison is quick and easy and usually does not require precise quantitative data. See Fig. 7 in which the trans olefin band of butadiene at 970 cm^{-1} is weaker than the 760 cm^{-1} styrene band in the failed product, curve A (----), and stronger in intensity in the originally specified resin, curve B (——). This spectral difference translates to the difference between 13% butadiene and 37% butadiene when a quantitative analysis is performed. Further results based on solvent fractionation of the plastic in the product showed that a resin blend had been used by the supplier instead of the specified terpolymer.

The analysis of a blended product relies upon solvent extraction to separate the components into identifiable fractions. A combination of weight percent yields from the extraction, and identification of the fractions by IR spectra gives a good quantitative analysis if the fractionation gives a clean separation, or only semi-quantitative data if the fractions are impure. The spectra in Fig. 8 show a caulking compound and an additive separated by heptane extraction. The bands of "Mesamoll"® are distinctive and it is possible that an analyst familiar with this kind of formulation could identify it without fractionation. In fact, an experienced spectroscopist could discount the 1730, 1590, 1530 and 1100 bands as probably arising from entirely different materials and make a successful computer search identification based on the sharper bands remaining in the spectrum. The identity of the other components would be elusive without fractionation. See Fig. 9. A methyl ethyl ketone extract showed so many different structures that it was suspected that more than one other major organic constituent was present. This extract was separable into benzene soluble and benzene insoluble materials which gave spectra characteristic of a polyetherurethane, high in polyether content, and of

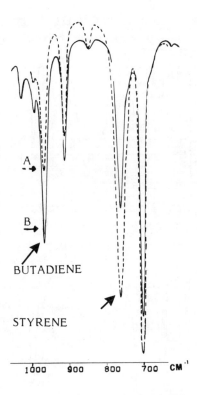

A
B
BUTADIENE

STYRENE

1000   900   800   700   CM⁻¹

FIG. 7. A portion of the spectra of ABS terpolymers showing butadiene and styrene characteristics. The difference in ratio of these components greatly effects properties.

polyvinyl chloride modified with a carbonyl compound. The presence of a urethane was predictable from the total spectrum, but the high level of polyether in it could not be confirmed without solvent fractionation. Polyvinyl chloride was obscured by other components including fillers which were removed by the extraction.

Complex mixtures such as crude oil, fuel oil, lubricating oil, asphalt and creosote may contain so many compounds that fractionation followed by analysis of so many fractions is prohibitive as a way of identifying the material. A combination of group frequency interpretation to establish the class of material and fingerprinting by absorption peaks characteristic of substructures within the mixture can give conclusive identification for some samples, and useful data in the form of "probable" or "consistent with" analyses in other cases.

The example in Fig. 10 of creosote as a suspected contaminant in a city sewer gives clearly conclusive identification. Spectra of spilled fuel oils or crude oils are generally less conclusive, but

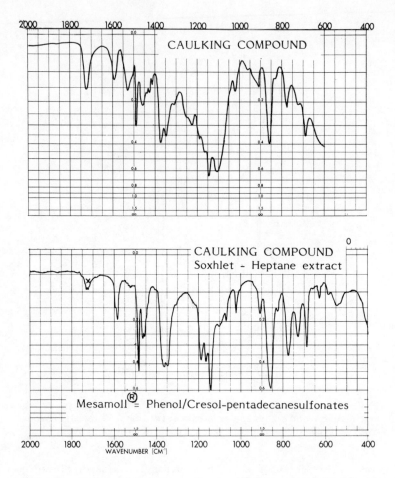

FIG. 8. Infrared spectra of a caulking compound and a component extracted from it.

very useful along with other data in monitoring spills in the
world's waterways. The careful attention that must be paid to
sample preparation technique for these comparisons of closely
similar materials are described in ASTM Methods D3414, on Comparison of Waterborne Petroleum Oils by Infrared Spectroscopy, and
examples of applications from the Santa Barbara Channel to the
Mediterranean are given in a report on the oceans [19].

The Coblentz Society reference book on regulated chemicals [20]
contains useful spectra of petroleum solvents, gasolines, kerosene
and crude oils. It is important to know the boiling range of
petroleum products as well as their IR spectra to characterize them
accurately. This is especially true for the higher molecular
weight fractions used for fuel oils or lubricating oils.

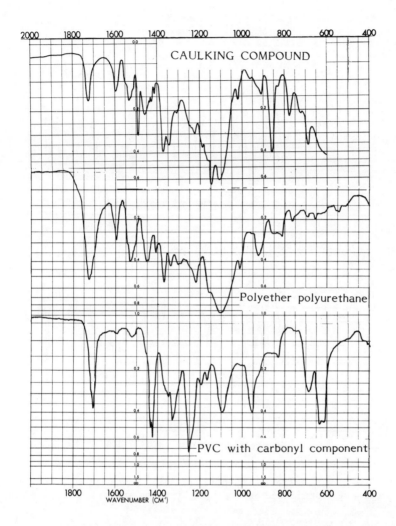

FIG. 9. Infrared spectrum of caulking compound and of two components isolated by benzene and methyl ethyl ketone extraction.

Characterization of more closely similar broad mixtures depends upon difference, or subtraction spectra in which the structures in common are cancelled and the resulting fingerprint difference is used for identification. Asphalts can be broadly characterized this way. The spectral differences which characterize asphalts arise from the amounts of aliphatic groups of which the most readily detectable are methyl and straight chain structures, and aromatic bands characteristic of ring substitution. These are small spectral differences as is demonstrated in the spectra of a

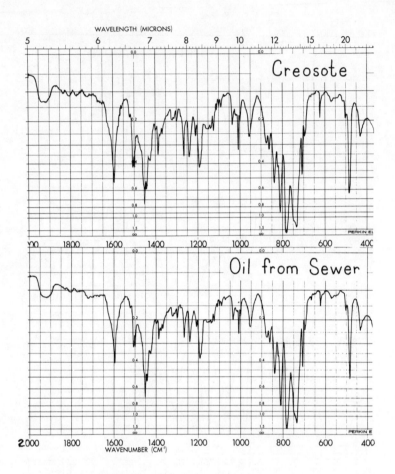

FIG. 10. Infrared spectra of oil drops from a sewer are clearly identified as creosote by these spectra.

wide range of asphalts in Fig. 11. More distinctive difference spectra, in which a highly aromatic California asphalt is subtracted from other more naphthenic and paraffinic asphalts, are shown in Fig. 12. The spectra show qualitative differences that serve much better as fingerprints than the spectra of the total asphalts in Fig. 11. Quantitative data on individual bands permit more complete characterization of asphalts [21].

Paper products often need to be analyzed in situ to establish composition of the paper and of additives. This kind of analysis can be done by internal reflection if surface characterization is required. However, transmission spectra are necessary to analyze the paper as a whole. Transmission through paper can be achieved by coating it with oils which are high in refractive index (Fig.

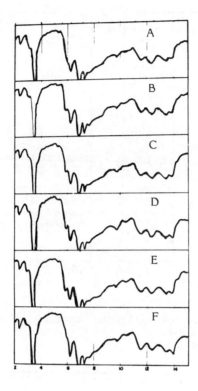

FIG. 11. Infrared spectra from 5000-670 cm-1 of asphalt
films. A-California, B-Venezuela, C-Saskatchewan, D-Arabian,
E-West Texas, F-Midcontinent.

13). Then it is possible to determine the groundwood content of a
Kraft/News blend (Fig. 14), and to measure additives in situ [22].

Excellent reviews of the applications of infrared spectroscopy to
commercial products are covered in alternate years in the April
Review edition of Analytical Chemistry, published by the American
Chemical Society. The references cited there help applied spectroscopists keep abreast of the diverse applications of this
versatile analytical tool.

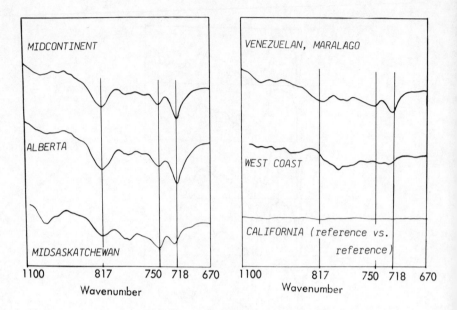

FIG. 12. Infrared difference spectra of asphalts from 10% solutions in 1.0 mm cells. Highly aromatic West Coast asphalt used as reference for all spectra.

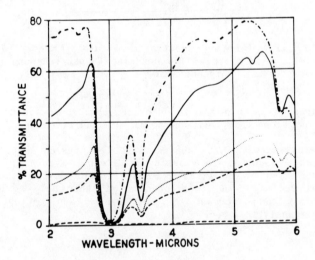

FIG. 13. Infrared transmissivity of paper coated with liquids of different refractive indices. 50/50 Kraft/News - 0.003 g/cm², --- no coating; ·--- R.I. 1.3995 - Fluorolube S-30; —— R.I. 1.4506 - Polybromotrifluoroethylene; ══ R.I. 1.5018 - Tetrachloroethylene; ·—·—·— R.I. 1.5490 - Dibromodichloromethane.

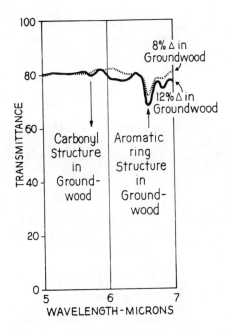

FIG. 14. Composition differences of paper shown by IR subtraction spectra with paper richer in groundwood in the sample
beam of a double-beam spectrometer.

REFERENCES

1. W. J. Potts, Jr., Chemical Infrared Spectroscopy, Vol. 1,
   Wiley-Interscience, NY, 1962.
2. A. L. Smith, Applied Infrared Spectroscopy, John Wiley and
   Sons, NY, 1979.
3. Manual of Practices in Molecular Spectroscopy, Fourth Edition,
   Committee E-13, American Society for Testing & Materials,
   Philadelphia, PA, 1979.
4. Annual Book of ASTM Standards, Volume 14.01/Molecular and Mass
   Spectroscopy; Chromatography; Resinography; Temperature
   Measurement; Microscopy; Computerized Systems; 980 pages; 73
   Standards, American Society for Testing & Materials,
   Philadelphia, PA, 1985.
5. The Coblentz Society, Board of Managers, Analytical Chemistry,
   47, 945A, 1975.
6. The Coblentz Society, Desk Book of Infrared Spectra, 2nd
   Edition, C. D. Craver, Editor, P. O. Box 9952, Kirkwood, MO
   63122, 1982.
7. L. J. Bellamy, Advances in Infrared Group Frequencies, Methuen
   & Co., Ltd., London, 1968.

8. L. J. Bellamy, The Infrared Spectra of Complex Molecules, 3rd Edition, John Wiley & Sons, NY, 1975.
9. L. J. Bellamy, The Infrared Spectra of Complex Molecules, Vol. 2, 2nd Edition, Chapman and Hall, London, 1980.
10. N. Colthup, L. Daly and S. Wiberley, Introduction to Infrared and Raman Spectroscopy, 2nd Edition, Academic Press, NY, 1975.
11. G. Brame, Jr. and G. Grasselli, Infrared and Raman Spectroscopy, Part C, Chapter on "Surfaces" by C. D. Craver, Marcel Dekker, Inc., NY, 1977.
12. D. O. Hummel and F. Scholl, Atlas of Polymer & Plastic Analysis, 2nd Edition, Vol. 1, Polymers, 1978; (Vol. 2, Plastics, Fibers, Rubbers, Resins; Vol. 3, Additives and Processing Aids) Carl Hanser, Verlag, Munich, Verlag Chemie International, New York.
13. IR Spectra Comm. of the Chicago Soc. for Paint Tech., An Infrared Spectroscopy Atlas for the Coatings Industry, Fed. of Soc. for Paint Tech., 1315 Walnut St., Philadelphia, PA, 1980.
14. C. D. Craver, Ed., Infrared Spectra of Plasticizers and Other Additives, 2nd Edition, The Coblentz Society, P. O. Box 9952, Kirkwood, MO 63122, 1980.
15. J. Haslam, H. A. Willis and D. C. M. Squirrel, Identification & Analysis of Plastics, 2nd Ed., Iliffe Publishers, London, and Van Nostrand, Princeton, NJ, 1972.
16. Catalogs and Atlases of Infrared Spectra, Sadtler Laboratories, 3316 Spring Garden St., Philadelphia, PA 19104, 1985.
17. SPIR, Canadian Institute for Scientific & Technical Information, National Research Council, Montreal Road Laboratory, M-55, Montreal and Blair Road, Ottawa, Ontario, Canada, K1A 052.
18. IRGO, Chemir Laboratories, 761 West Kirkham, Glendale, MO 63122.
19. P. D. Wilmot, A. Slingerland, Technology Assessment and the Oceans, "Oil Spills and Natural Seeps - Status of Analytical Methods for Distinguishing Hydrocarbonaceous Materials in the Marine Environment", C. D. Craver, IPC Science and Tech. Press Ltd., London, 1977.
20. The Coblentz Society, Inc., Infrared Spectra of Regulated and Major Industrial Chemicals, C. D. Craver, Editor, P. O. Box 9952, Kirkwood, MO 63122, 1983.
21. C. D. Smith-Craver, C. C. Schuetz and R. S. Hodgson, I&EC Product Research and Development, 5, 153, 1966.
22. C. D. Smith-Craver and J. K. Wise, Analytical Chemistry, 39, 1698 (1967).

INFRARED GROUP FREQUENCIES FOR STRUCTURE DETERMINATION IN ORGANO-SILICON COMPOUNDS

A. Lee Smith

Dow Corning Corporation
Midland, MI 48640

INTRODUCTION

Organosilicon molecules show exceptionally good correlation of their infrared group frequencies with molecular structure. The wavelength range over which a substituent absorbs is usually small, and the position of an absorption band within that range can often be accurately predicted, either by geometric factors, or through correlation with parameters that are characteristic of the substituents. In this paper we discuss several such absorptions: SiH, SiOSi, SiVinyl, and SiOAlkyl.

CORRELATIONS WITH ELECTRONIC EFFECTS

It is appropriate first to touch briefly on the characteristics of the Si atom that makes it such an interesting subject for IR studies. Its mass is greater than that of most substituents (except Cl and Br), in contrast to carbon, which is of equal or lesser mass than all its common substituents except hydrogen. Thus, mass effects are relatively less important for Si. It is somewhat more electropositive than carbon, so it forms strongly polar bonds with the more electronegative elements. Its atomic radius is substantially larger than that of carbon, so its bonds are longer. It is also more polarizable. Bonds to oxygen seem to be less rigorously directional and more flexible and, at the same time, stronger than expected. Only recently has a true double bond involving silicon been demonstrated, and it is extremely unstable. Since Si is in the second row of the periodic table, it has d-orbitals which can be used in bonding to other atoms. Proof that d-orbitals are actually used is elusive, however, and alternative explanations are possible for every case in which $d\pi-p\pi$ bonds have been alleged to exist.

The SiH stretching frequency is convenient to study; it is virtually insensitive to mass changes of the other Si substituents; it is usually free of interference from other absorptions; and its frequency and intensity can be accurately measured. It can therefore serve as a sensitive probe into the electronic structure of the Si atom. In 1959, N. C. Angelotti and I published a paper [1] in which we showed that the position of the SiH band was precisely predictable, and developed a set of substituent parameters which, when summed, predicted the wavenumber of the SiH stretch

$$\nu(\text{SiH}) = \sum_{i}^{3} E_{i} \tag{1}$$

Sir Harold Thompson subsequently published a suggestion [2] that the use of Taft's σ^*, a popular measure of the inductive effect in organic compounds, was more appropriate than E_i even though the correlation was not as good. The rationale was that a single inductive parameter scale was preferable to multiple parameter scales. In the years following, a large number of publications have appeared, particularly from Russia, in which both frequency and integrated intensities of many different substitutents were measured and correlated with σ^* [3].

In the last ten years or so, there has been a great deal of interest in amorphous silicon (a-Si) by the semiconductor industry. Amorphous Si contains H and (sometimes) F or O. These elements disrupt the crystal lattice and are responsible for the amorphous nature of the solid. The SiH vibration has been intensively studied in order to obtain information on the local structure of the solid. Equations similar to Eq. (1) have been proposed to relate the SiH stretching frequency to the nature of the R groups for the moiety $R_1R_2R_3$ SiH (R = Si, H, O, F, etc.). The "stability ratio" electronegativity of the substituent [4] was used for this correlation [5]. Assignments, however, are still somewhat in dispute even though the SiH, SiH_2, and SiH_3 deformations have been studied as well as the stretching modes. Specifically, an absorption at 2000 cm^{-1} is attributed to SiH, 2100 to SiH_2, and 2150 to SiH_3 by Lucovsky and others [5-7], whereas another school of thought [8] suggests that the 2000 and 2100 cm^{-1} absorptions both arise from SiH_x groups within the amorphous network and on the inner surface of voids, respectively. These authors point out that Raman spectra of well characterized polysilanes show a maximum frequency difference of 20 cm^{-1} between SiH and SiH_2 stretch modes.

We do not take sides in this controversy, but it is of interest to observe that in silanes, substitution of a H for an Me_3Si- group raises the SiH frequency by about 40 cm^{-1}. A shift of 100 cm^{-1} seems much too large. Considering all the possible combinations of Si, H, F and O, and the lack of authentic model compounds, it is not surprising that there is some difficulty in arriving at convincing assignments.

Returning to our comments on substituted silanes and siloxanes, we pointed out in our original paper that siloxane substituents, in contrast to alkyl and aryl substituents, gave a non-linear change in the SiH frequency with the number of substituents. We and others have subsequently found that certain substituents, namely Cl, F, OMe, and NMe_2 also give non-additive shifts. These groups also show anomalous NMR shifts of the SiH protons in silanes [9], and are precisely the same groups that are postulated to form $d\pi-p\pi$

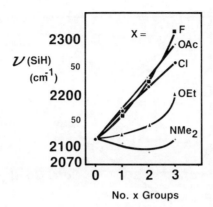

FIG. 1. Effect of increasing substitution on the SiH stretching frequency.

dative double bonds to Si. It is tempting to conclude that such bonding is responsible for the apparently anomalous behavior of these groups, and this explanation has been embraced by Egorchin and others [10].

Let us observe at this point that we believe our IR evidence neither supports nor refutes the concept of increased $d\pi$–$p\pi$ bonding in halogen, oxygen, and nitrogen substituents. The reasons are as follows:

(1) The curvature of plots for F, OMe, and OEt is all in the same direction, i.e, increasing effect with increasing substitution, suggesting that successive substituents become more electronegative in character. However, Cl shows just the <u>opposite</u> effect, having a small curvature in the opposite direction. The AcO group, which should interact as strongly as, say, a MeO group, gives virtually a straight line plot. The NMe$_2$ group gives a pronounced effect, first decreasing and then increasing in electronegativity as the number of groups increases. These effects are shown in Fig. 1. Thus, we see little consistency in the behavior of substituents that should presumably all interact in the same way.

(2) In organic molecules, the CH stretching frequencies do not follow a simple linear relationship with substituents, and no $d\pi$–$p\pi$ bonding can occur with the carbon atom. Possibly the small size and low polarizability of the carbon atom compared with other Group IV elements leads to a greater interaction of the substituents (or their lone pairs) with the CH bond, with the resulting steric and field effects obliterating all but gross trends.

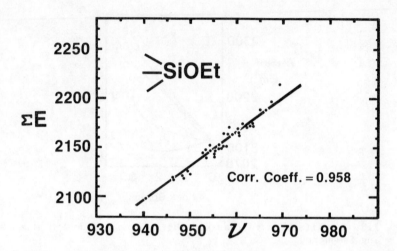

FIG. 2. Frequency of the SiO stretch for the SiOEt moiety versus the sum of inductive parameters for the other substituents.

Some comments are appropriate at this point. First, it is entirely possible that $d\pi$–$p\pi$ interaction in Si compounds is indeed present as suggested by MO calculations, but it may not be large enough to influence the IR absorptions [11,12]. Second, in carrying out correlations of frequencies with substituent parameters, one should use the widest possible variety of substituent types. Many authors have confined their studies to a relatively small range of substituents. Third, there is no reason to expect Taft's σ^* inductive parameter, which was developed with and for carbon compounds, to hold for Si except in a rather superficial way. Indeed, several other inductive parameters are also in use for carbon compounds such as Brownlee and Topsom's σ_I [13], the pK_a of substituted acetic acids, etc. Thus, it appears entirely reasonable to expect that the $\nu(SiH)$ values themselves give the best set of inductive parameters for use with silicon substituents. On that premise, we will now examine some additional correlations with IR frequencies.

SiOEt
Bands characteristic of the SiOEt group are found at 1165 ± 4 cm^{-1}, 1105 ± 4 cm^{-1}, 1078 ± 4 cm^{-1}, and 955 ± 15 cm^{-1}. The latter absorption is believed to be the SiO stretching vibration, and it is a useful group frequency in that it correlates rather well with the sum of the inductive parameters for the other substituents (Fig. 2 and Eq. 2):

$$\nu(EtOSi) = 407.1 + 0.255 \ \Sigma \ E_i \ \text{cm}^{-1} \qquad (2)$$

SiOiPr
Fewer data points are available for this substituent, and relative errors are larger, but the same trend of higher frequencies

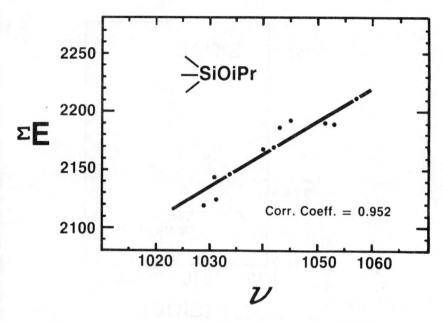

FIG. 3.   Frequency of the SiO stretch for the SiOiPr moiety versus inductive parameters of the other substituents.

accompanying higher inductive effects of substituents is seen (Fig. 3).

SiOAc
Group frequencies for the acetoxy group are well known; both the 945 ± 10 cm^{-1} absorption and (surprisingly!) the C=O stretch at 1740 ± 30 cm^{-1} show a correlation with the inductive parameters (Figs. 4 and 5).

The large shift of the carbonyl stretching vibration is unexpected, and clearly reflects a very significant interaction between Si and multiple oxygen substituents, which increases the electronegativity of the Si atom and thereby competes more effectively with the carbonyl oxygen for electrons. The net result is an increase in the double bond character of the C=O bond and an increase in its frequency of vibration.

SiOH
Data for the SiO stretch vibration in silanols are shown in Fig. 6 and show the same trends as discussed above. The exact frequencies depend on whether or not the SiOH groups are associated or not; unassociated OH shows a greater scatter of the points.

SiOPh
Only a few points are available for this group, but they show the same good correlation as do alkoxy compounds (Fig. 7).

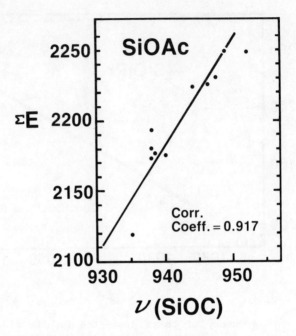

FIG. 4.   Frequency of the SiO stretch in SiOAc compounds versus inductive parameters.

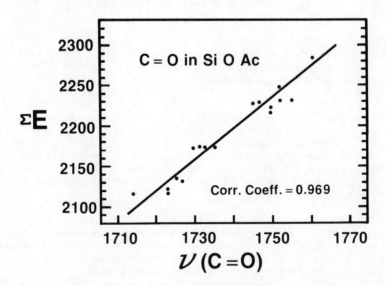

FIG. 5.   Frequency of the C=O stretch in SiOAc compounds versus inductive parameters for the other substituents.

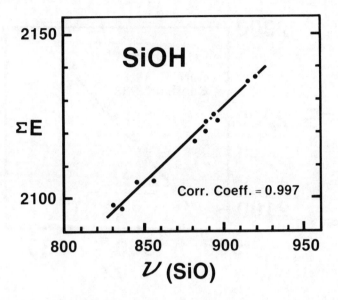

FIG. 6. Frequency of the SiO stretch in silanols versus inductive parameters.

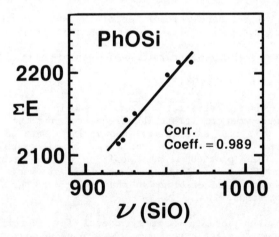

FIG. 7. Frequency of the SiO stretch in SiOPh compounds versus inductive parameters.

Si-Vinyl

The out-of-plane hydrogen deformation frequencies of substituted olefins $GCH=CH_2$ have been correlated with inductive and resonance parameters characteristic of the substituent G [14]. The CH_2 "twist" mode correlates with the inductive effect of G (as measured by the pKa of the corresponding acid GCH_2COOH), and the CH_2 "wag"

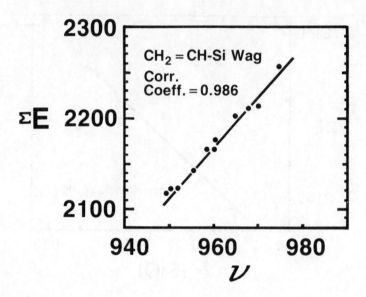

FIG. 8. Frequency of the CH_2=CH-Si wag versus inductive parameters in CH_2=CHG compounds where G = SiX_3.

correlates with resonance properties of the substituent. We have taken G = SiX_3 and plotted E_i for G vs the twist frequencies of the GCH=CH_2 group. Only a small wavenumber range is involved ($\sim$20 cm^{-1}) but the correlation appears to be reasonable (Fig. 8).

CORRELATIONS WITH GEOMETRIC EFFECTS

For cyclosiloxanes, geometric factors predominate in determining the SiOSi antisymmetric stretching frequency, and the correlation of frequency with ring size is well known [15]. Data on cyclics in the vapor phase, as they elute from a gas chromatograph, are shown in Fig. 9. Also plotted is the position of the methyl rocking frequency near 800 cm^{-1}. Although some overlap in range occurs, it is usually possible to distinguish the cyclics by the frequencies of their SiOSi bonds.

Such correlations become extremely useful for identification of cage and ladder siloxane structures. We are using a combination of gas chromatography (CG), mass spectroscopy (MS), and Fourier transform infrared (FTIR) spectroscopy to examine complex mixtures of condensed siloxane structures. The GC retention time gives some clues as to the structures and groups present. Also, some literature data are available; Alexander and Garzo [16] have given GC retention times for a large number of condensed siloxane structures. Condensed structures occurring in aqueous silicate solutions have been identified by NMR [17]. GC/MS gives essential molecular weight information, but usually cannot identify specific

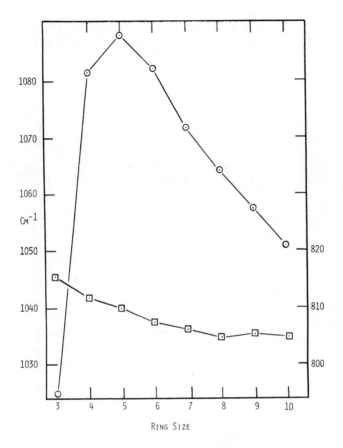

FIG. 9. Correlation between the ring size of cyclosiloxanes in the vapor phase and the SiOSi antisymmetric stretching frequencies. Also shown in the position of the methyl rocking frequency near 800 cm-[1].

isomeric structures. GC/IR uses group frequency correlations to establish which structures can or cannot be present.

Thus, the combination of GC, IR, and MS gives the information necessary to identify specific structures.

A number of novel ladder/cage siloxanes have been found in the reaction products from the base-catalyzed rearrangement of poly-dimethylsiloxanes. A Gram-Schmidt reconstruction of the chromatogram from such a mixture is shown in Fig. 10. Some typical spectra from the same run will now be shown.

Figure 11 shows file 443. A trimer ring is seen at 1032 and a tetramer ring at 1090 cm-[1]. There is no Me_3Si and no SiOH. The GC

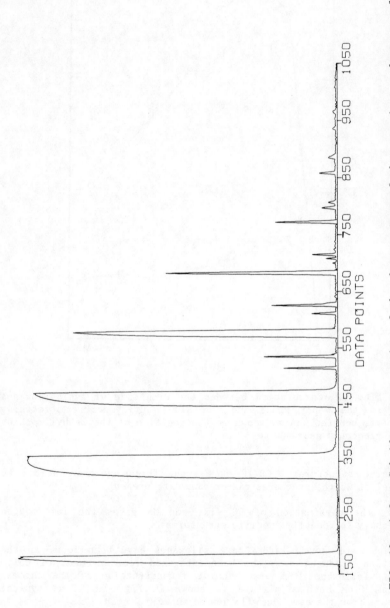

FIG. 10. A Gram-Schmidt reconstruction of the chromatogram of the reaction products resulting from a base-catalyzed rearrangement of polydimethylsiloxanes.

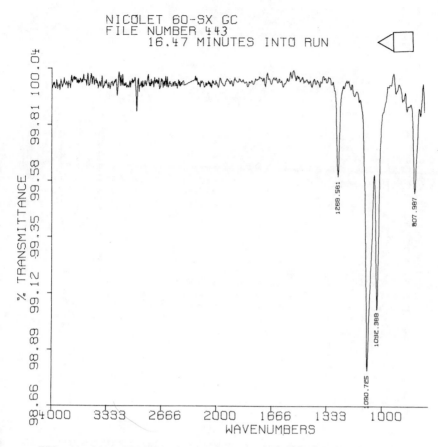

FIG. 11. A typical spectrum showing file 443.

retention time and MS data are consistent with an Si_5 compound.
The structure is the fused ring system shown in Table I.

Figure 12 shows file no. 669. The band at 1084 cm^{-1} suggest a
cyclic tetramer or pentamer; the 841 cm^{-1} band indicates -OSiMe$_3$,
and the 1051 cm^{-1} siloxane band suggests a linear polymer. Thus,
we must have a linear "tail" attached to a cyclic with the 1097
cm^{-1} band originating from the monosubstituted Si. This is an Si_6
molecule, and must have either the structure indicated, or an Si_5
ring with a tail.

All in all, 26 structures in the Si_3 to Si_9 range were identified,
some of which are shown in Table 1. Not all of them are completely
unambiguous, but the IR, MS, and GC data are all consistent with
the structures shown. It is worth noting that some of the minor
constituents that were identified are present to the extent of only
a few parts per million in the original sample.

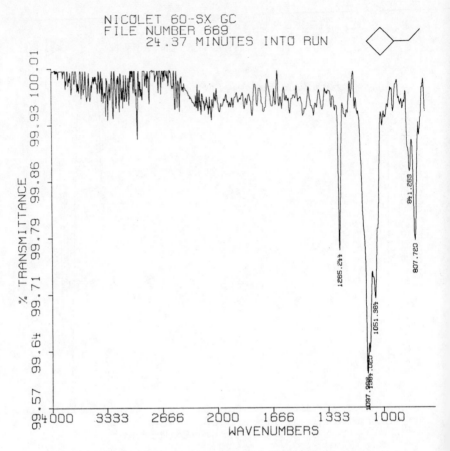

FIG. 12.   A typical spectrum showing file 669.

In conclusion, infrared group frequencies are an indispensible aid
in characterizing silicon-containing molecules; and infrared, when
used in combination with other analytical techniques, plays a key
role in extending our knowledge of organosilicon chemistry.

I would like to thank N. C. Angelotti, R. J. Robinson, and J. A.
Moore of Dow Corning, and Fred Walder of Nicolet Scientific Instruments for their help in obtaining the data.

EXPERIMENTAL

Spectra were obtained on dilute solutions in CCl_4 (4000-1370 cm^{-1})
and CS_2 (1370-600 cm^{-1}), using 0.1 mm thick cells with a Perkin
Elmer Model 521 IR spectrometer. Each individual run was carefully
calibrated by simultaneously running a calibrating gas (NH_3, CO,
etc., as appropriate) on each side of the absorption to be
measured.   GC/IR spectra were obtained at 225°C using a Nicolet

TABLE 1.   Structures in the Si_3 to Si_8 Range Identified from the Reaction Rearrangement of Polydimethylsiloxane.
$D = Me_2SiO$; $M = Me_3SiO_{1/2}$; $T = MeSiO_{3/2}$.

File No.	Identification	Structure
182	D_3	
361	D_4	
433	T_2D_3	
443	T_2D_3	
472	D_5	
515	T_2D_4	
535	T_2D_4	
580	D_6	
611	T_2D_5	
626	T_2D_5	
642	T_2D_5	
657	T_2D_5	
669	MD_5T	
683	D_7	
700	T_2D_6	
708	T_2D_6	
716	T_2D_6	

60-SX spectrometer (8 cm^{-1} resolution) coupled to a Hewlett-Packard gas chromatograph equipped with a capillary column.

REFERENCES

1.   A. L. Smith and N. C. Angelotti, Spectrochim. Acta, 14, 412 (1959).
2.   H. W. Thompson, Spectrochim. Acta, 16, 238 (1960).
3.   See, for example, M. G. Voronkov, T. V. Kaskik, and N. I. Shergina, Dokl. Akad. Nauk SSSR, 232 (4), 817 (1977).

4.   R. T. Sanderson, <u>Chemical Periodicity</u>, Reinhold, New York (1960).

5.   G. Lucovsky, Solid State Comm., <u>29</u>, 571 (1979).

6.   G. Lucovsky, J. Vac. Sci. Technol., <u>16</u>, 1225 (1979).

7.   P. John, I. M. Odeh, M. J. K. Thomas, M. J. Tricker, and J. I. B. Wilson, Phys. Stat. Sol. (b), <u>105</u>, 499 (1981).

8.   W. Wagner and W. Beyer, Solid State Comm., <u>48</u>, 585 (1983).

9.   H. J. Campbell-Ferguson, E. A. V. Ebsworth, A. G. MacDiarmid, and T. Yoshika, J. Phys. Chem., <u>71</u>, 723 (1967).

10. A. N. Egorochin, S. Ya. Khorschev, N. S. Vyazankin, T. I. Chernysheva, and O. V. Kuzmin, Izv. Akad. Nauk SSSR, Ser. Khim. <u>1971</u> (3), 544; p. 478, Consultant's Bureau Translation.

11. H. Oberhammer and J. E. Boggs, J. Amer. Chem. Soc., <u>102</u>, 7241 (1980).

12. T. Veszpremi and J. Nagy, J. Organometal. Chem., <u>255</u>, 41 (1983).

13. R. T. C. Brownlee and R. D. Topsom, Spectrochim. Acta, <u>31A</u>, 1677 (1975).

14. W. J. Potts, Jr., and R. A. Nyquist, Spectrochim. Acta, <u>15</u>, 679 (1959).

15. C. W. Young, P. C. Servais, C. C. Currie, and M. J. Hunter, J. Amer. Chem. Soc., <u>70</u>, 3758 (1948).

16. G. Alexander and G. Garzo, Chromatographia, <u>7</u>, 190 (1974).

17. R. K. Harris and M. J. O'Connor, J. Mag. Reson., <u>57</u>, (1984).

THE ROUTINE USE OF FT-IR AND RAMAN SPECTROSCOPY FOR SOLVING NON-ROUTINE PROBLEMS IN AN INDUSTRIAL LABORATORY

M. Mehicic, M. A. Hazle, J. R. Mooney, and J. G. Grasselli

The Standard Oil Company (Ohio)
Research & Development Center
Cleveland, Ohio 44128

INTRODUCTION

In the industrial laboratory today, it is exciting to see the explosive growth of sophisticated methods and instruments which have been developed to attack a variety of problems of ever-increasing complexity. Every "non-routine" problem brings opportunities for innovative solutions. Infrared spectroscopy and, to a lesser but equally important extent, Raman spectroscopy have long been the stalwarts of the instrumental methods for such problem solving.

Today in our laboratories, Fourier transform infrared spectroscopy and Raman spectroscopy are an indispensable pair of reliable spectroscopic techniques. Both instruments are computerized, so powerful data processing methods are also available to enhance the speed and sensitivity of the analyses performed. They are routinely used to address problems which impact significantly on R&D successes and corporate operations.

Applications in catalysis are extensive, from structural analysis of the solid itself to the characterization of surface adsorption sites, in-situ analysis of adsorbed species at moderate conditions as well as at high temperatures and pressures, isotope substitution studies, etc. In recent years these infrared and Raman studies have been greatly aided by the development of many excellent sample handling techniques such as DRIFT and FT-PAS (infrared) and optical fibers for remote detection (Raman), as well as the very rapid advancements in the use of combined techniques such as GC/FT-IR and TGA/FT-IR. Other areas of application for infrared and Raman spectroscopy include polymers, ceramics and materials from alternate energy sources or biotechnology.

FT-IR is now widely accepted as an essential problem-solving technique in the industrial laboratory [1-3]. The literature abounds with elegant examples of its use. Many problems that were previously never amenable to solution by IR spectroscopy because of energy limitations or time constraints are now routinely addressed.

Raman spectroscopy is not usually recognized as a viable partner to IR in a practical sense in the industrial laboratory, although many books and papers have tried to dispel that view [4-9]. At the XXI Colloquium Spectroscopicum Internationale and the sixth International Conference on Atomic Spectroscopy in Cambridge, England, 1979, we predicted that wider spread use of Raman spectroscopy would soon be realized, especially since microscope and optical multichannel detection devices had become available. We would like to measure that prediction and give another update on the industrial applications of Raman spectroscopy in this paper. It seems clear that even though the rapid proliferation of applications of Raman spectroscopy has already encompassed and impacted many fields of chemistry, we are on the verge of another era of explosive growth in the next decade. We will illustrate some areas where Raman spectroscopy offers unique advantages and others where it is a partner to infrared spectroscopy.

DISCUSSION

Amorphous Silicon Films
Thin films are of great importance in many areas of science and technology. Especially prominent are amorphous silicon (a-Si) films which are being developed for solar energy devices. The electrical and optical properties of these films depend to a great extent upon dangling silicon bonds, and on the amount of hydrogen (or other suitable element) available to saturate these bonds. The films contain between 1 to 50 atomic % of bonded hydrogen which can be present as H_2, SiH, SiH_2 or SiH_3. Many methods have been applied to the study of these materials including Induced Nuclear Reaction Spectroscopy, Secondary Ion Mass Spectroscopy, ^1H NMR and ESR. These methods generally suffer from two problems, lack of commonly available equipment and difficulty of use. Transmission IR studies were first suggested for use in quantitating the hydrogen in amorphous silicon by Brodsky [10]. Others have subsequently investigated the amorphous silicon [11-13] and the assignments listed in Table 1 are those accepted for the system [13].

The infrared transmission method is normally carried out on specially prepared films that are deposited on crystalline silicon simultaneously with normal deposition on a steel substrate. Crystalline silicon has a strong phonon band at 607 cm-1 which interferes with the analytical band for Si-H at 640 cm-1 used for quantitation. Historically this interference was eliminated by using a carefully matched reference sample in a double beam instrument. This is more conveniently done now by subtraction on an FT-IR. One of the problems of the single beam FT method is large interference fringes due to the crystalline silicon substrate. These may be removed by the method suggested by Hirschfeld [14-15] or, more simply, by taking the spectrum at a resolution such that the secondary interferogram is not sampled (typically 4 cm-1). The band area of the 640 cm-1 band is integrated from 740-540 cm-1. The band is calibrated over the concentration range of interest using samples whose absolute hydrogen concentrations are obtained by Induced Nuclear Reaction Spectroscopy. Film

TABLE 1. IR Assignments for the Amorphous Silicon System

Group	Frequency (cm^{-1})	Assignment
SiH	2000	stretch
	640	bend
SiH_2	2090	stretch
	890	scissors
	845	wag
	640	rock
SiH_3	2140	stretch
	907	degenerate deformation
	862	symmetric deformation
	640	rock
SiH_3	2140	stretch
	907	degenerate deformation
	862	symmetric deformation
	640	rock
a-Si-O-Si	980	stretch
c-Si-O-Si	1130	stretch
Si-Si lattice	607	phonon band
SiC	605	

thickness measurements are done either mechanically or by using interference fringes in the near IR/Visible region. A typical spectrum used for the quantitation is shown in Fig. 1.

A clear limitation of this method is that it must be run on specially prepared samples. It would be far more convenient to measure spectra on samples from normal preparation on steel surfaces. We investigated the use of both diffuse reflectance and specular reflectance spectra to examine films prepared on steel. Figure 2 shows a specular reflectance spectrum obtained from an amorphous silicon film on steel. One difficulty in this procedure is that because of the film thicknesses, fringes are observed in the spectrum. In this case, Hirschfeld's method will not work because the secondary interferogram lies nearly coincident with the primary interferogram. The fringes can be removed by curve fitting and subtraction of the calculated fringe curve. This yields an excellent spectrum (Fig. 2, bottom) that can be used qualitatively or semi-quantitatively using band ratios.

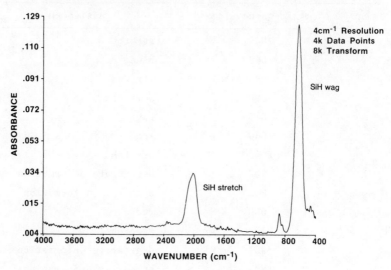

FIG. 1. FT-IR spectrum of amorphous silicon after subtraction of crystalline silicon wafer substrate.

Another problem of significance in amorphous silicon films is the extent of crystallinity present. This problem can be conveniently examined using Raman spectroscopy; Fig. 3A shows the Raman spectrum of crystalline silicon and Fig. 3B amorphous silicon prepared on crystalline silicon. The sharp phonon band of crystalline silicon at 520 cm^{-1} is not observable through the amorphous silicon film. But because of the excellent spatial resolution of the Raman technique, crystallinity can be checked in localized areas. Figure 4 shows the spectra obtained from an a-Si film that had been subjected to laser annealing. Clearly, areas of amorphous material, mixed crystalline, and amorphous material and pure crystalline material can be readily detected [16].

In-situ Raman
In-situ experiments are increasingly important in many research areas today where it is important to obtain data under dynamic and real world conditions for a better understanding of kinetics and mechanisms of important processes. Such experiments have been especially useful in the field of heterogeneous catalysis for evaluating surface and bulk properties of the inorganic solid catalysts.

Of all the molecular and surface spectroscopies, Raman is the easiest for in-situ work. This is so because of the ease of getting visible laser radiation in and out of the cell, which can be made of glass or quartz. But Raman suffers from the weakness of

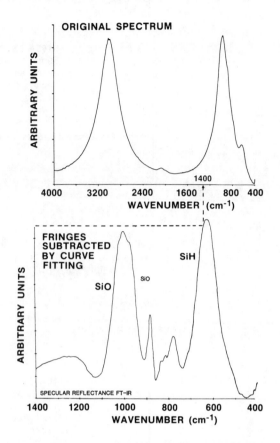

FIG. 2. FT-IR spectrum of amorphous silicon on stainless steel. Top, original spectrum; bottom, after fringes are removed.

the scattering effect and this can be a significant problem with poor scatterers, such as the majority of solid catalyst materials. An in-situ cell also will reduce the observed signal. Since most materials are sensitive to laser radiation, one cannot simply increase the power of the incident radiation. The ways to cope with this problem include: sample spinning (or rapid beam scanning), wide slits (or wide collection apertures such as a Cassagrain collector), cooled photomultipliers, photon counting, optical multichannel detection (OMA), and computerized data acqui- sition with signal averaging.

Figure 5 illustrates a very simple set-up used for study of hetero- geneous catalysis (solid-gas interface) [5-6]. It consists of a quartz tube in a ceramic furnace which can be heated to 650-700°C. Most of the radiation backscattered from the catalyst is collected by a 45° mirror for spectral analysis.

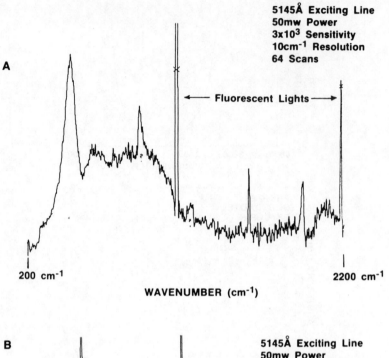

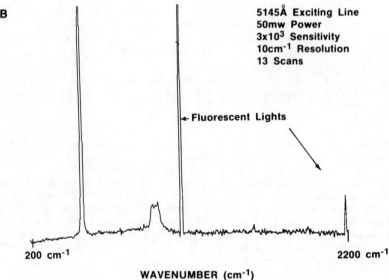

FIG. 3. Raman spectra of silicon wafers: (A) amorphous silicon on crystalline silicon substrate; (B) crystalline silicon.

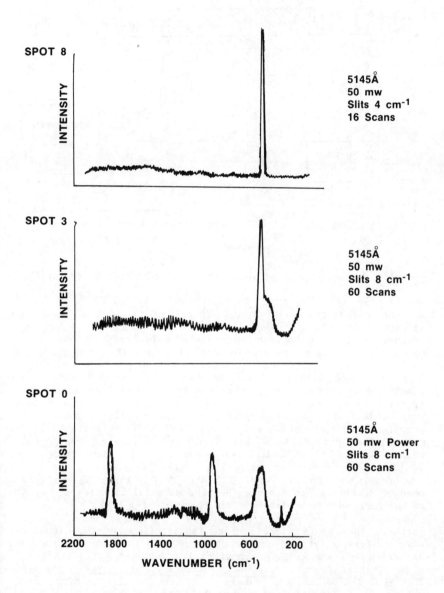

FIG. 4. Raman spectra of individual spots from laser annealing of amorphous silicon, 1300 Å thick, on glass (from Ref.
[16], with permission).

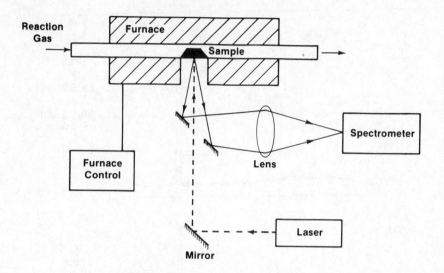

FIG. 5. In-situ Raman experiment - schematic of heated cell.

Bismuth molybdates (α-$Bi_2Mo_3O_{12}$. β-$Bi_2Mo_2O_9$, and γ-Bi_2MoO_6) and related compounds have been extensively studied for many years because of their importance in selective oxidation and ammoxidation of olefins. Raman spectra of the stoichiometric compounds are shown in Fig. 6. Disproportionation of β-$Bi_2Mo_2O_9$ in a redox cycle into the more stable α- and γ- forms was postulated based on kinetic measurements [17]. In-situ Raman study provides direct spectroscopic evidence for this process. Figure 7 clearly indicates that β-$Bi_2Mo_2O_9$ disproportionates under cyclic reduction with propylene and reoxidation with air at 430°C. This restructuring of single phase β-$Bi_2Mo_2O_9$ is a slow process. Although β-$Bi_2Mo_2O_9$ is metastable up to 500°C it disproportionates at lower temperature due to the presence of α- and γ- forms that act as nucleation centers. Under similar conditions α-$Bi_2Mo_3O_{12}$ and γ-Bi_2MoO_6 do not change. After disproportionation, the material kinetically behaves as the γ-Bi_2MoO_6 form via a surface restructuring [18].

Further, one can use $^{18}O_2$ isotope substitution to study functionally different lattice oxygens in such catalysts [19]. Propylene oxidation and ammoxidation requires two distinct surface sites in close spatial proximity for α-H abstraction and oxygen insertion. The reoxidation of the catalyst occurs by dissociative chemisorption of O_2. Oxide ions fill anion vacancies. Oxygen adsorption occurs on sites that are spatially and structurally different from the α-H abstraction and oxygen insertion sites [20]. Comparison studies of propylene and butene reactions over bismuth molybdate catalyst indicate the existence of two types of lattice oxygens [21-22].

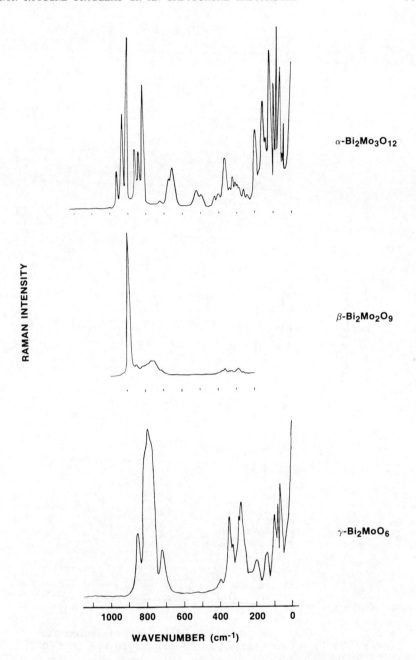

FIG. 6. Raman spectra of bismuth molybdate phases (from Ref. [17], with permission).

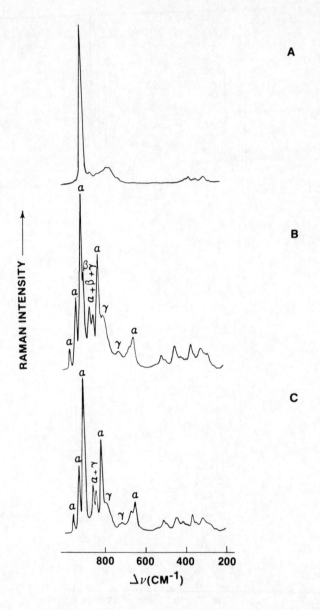

FIG. 7. Raman spectra of $Bi_2Mo_2O_9$ disproportionation (from Ref. [17], with permission). (A) $Bi_2Mo_2O_9$ at 500°C; (B) $Bi_2Mo_2O_9$ reduced with propylene and reoxidized in air at 430°C for 0.5 hrs.; (C) Reoxidized at 430°C for 24 hrs.

γ-Bi_2MoO_6 was used to identify the sites responsible for different catalyst functions. In principle one should be able to see the reduction of integrated intensity of the vibrational band involving the particular oxygen. This is not straightforward, however, because the reduced catalyst is a very poor scatterer, and good quantitative data are not readily available. A better way is to reoxidize with $^{18}O_2$ (such as in Ueda's kinetic studies) [23]. Isotope-shifted bands would then identify which lattice oxygen is involved in a particular function. The Mo-O stretching bands in the 800 cm^{-1} region were used for the study. The lower frequency region contains Bi-O stretches which would be of great interest, but they were interfered with by a number of bending modes. They could not be sorted out to yield good interpretations.

It has already been stated that oxidation of propylene requires α-H abstraction and oxygen insertion. On the other hand, oxidation of 1-butene to butadiene obviously requires only α-H abstraction. It has also been shown [24] that oxidation of methanol to formaldehyde occurs on Mo-O centers in molybdates. These three reduction gases were therefore well suited for identification of characteristic sites.

The results of $^{18}O_2$ studies are summarized in Table 2.

TABLE 2. Raman Spectra of Partially Reduced γ-Bi_2MoO_6 Reoxidized with ^{18}O-Oxygen

		Major Band Positions (cm^{-1}) after:					
Reductant	Initial	First Cycle	Second Cycle	Third Cycle	Fourth Cycle	Fifth Cycle	Sixth Cycle
Propylene	844	835	832	830	830	830	830
	803	792	792	790	790	786	785
	725	715	715	713	712	710	708
1-Butene	844	841	841	841	840	840	840
	803	801	802	802	798	798	797
	725	724	724	725	722	722	723
Methanol	844	839	839	838	836	832	831
	803	804	801	802	799	787	787
	725	721	720	719	717	709	709

1-Butene does not show shifts in frequency indicating that Mo-O oxygens do not participate in α-H abstraction. The shifts with propylene and methanol indicate involvement of Mo-O oxygens in oxygen insertion. From this work it can be concluded that Bi-O centers are responsible for α-H abstraction and Mo-O centers for oxygen insertion. Single crystal XRD [25] and neutron diffraction [26] indicate three types of oxygen: Bi-O-Bi in Bi_2O_3 layers; Mo-O-Mo in molybdenum polyhedra layers; and Bi-O-Mo bonds that bridge two layers. Keeping in mind that α-H abstraction and oxygen insertion sites should be spatially close, α-H abstraction is assigned to Bi-O-Mo oxygens, while oxygen insertion is assigned to Mo-O-Mo oxygens. The third function, dissociative chemisorption of O_2 and reoxidation of the catalyst by diffusion of oxide ions to lattice vacancies, could be provided by Bi-O-Bi sites where two electron pairs are available for O_2 reduction. It should be pointed out that the ability of Bi_2O_3 oxygen electrolytes to dissociatively chemisorb O_2 is documented [27]. This can be shown in schematic fashion as follows:

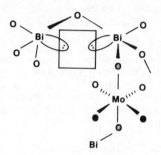

Evolved Gas Analysis/FT-IR

The term "evolved gas analysis" covers a wide range of applications such as the analysis of reactor effluent, engine exhaust, or the effluent gas from an instrument such as a Thermal Gravimetric Analyzer. In the past, it was necessary to bring the evolved gases to the analyzing instrument - in this case an FT-IR - either by bringing the gas sources to the instrument lab or as samples in gas bombs. Lephardt [28] first described such methods. We have used a novel approach to this problem by mounting a small, rugged FT-IR (a Nicolet 5MX) and its accessories on a lab cart. This allows us to wheel the mobile FT-IR to various locations at our Research Center and provide on-line gas analysis. In an early experiment, we took the evolved gas from a pilot reactor and directed it via stainless steel piping to a heated stainless steel infrared cell having 6 cm × 6 mm internal dimensions. The spectrum of the effluent from the reactor showed peaks due to carbon dioxide, carbon monoxide, carbon oxysulfide, methane and hydrogen sulfide. These were readily identifiable, as the spectra had high signal to noise. The spectrum also indicated the possible presence of methane thiol which was confirmed later by MS analysis of a portion of the effluent.

A similar experimental set-up has been used for TGA/FT-IR evolved gas analysis [29]. The FT-IR was placed close to a DuPont 990 TGA and the evolved gas passed through a heated transfer line of minimum length to the heated gas cell.

Catalysts used to synthesize alcohols from CO/H_2 were analyzed by TGA/FT-IR to help determine the reason for their decline in activity. Figure 8 shows the TGA curves for fresh versus used catalyst.

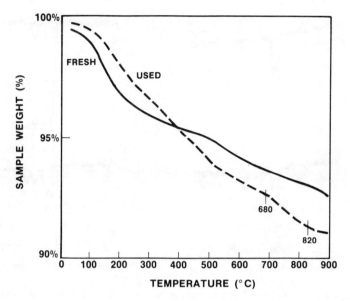

FIG. 8. Thermal gravimetric analysis of alcohol synthesis catalysts (from Ref. [30], with permission).

From 400°C to 800°C, used catalyst shows about 1% more weight loss than a fresh sample. Profiling the evolved volatiles through this temperature region using FT-IR showed that the used sample released significantly more CO_2 than the fresh (Fig. 9). With some supporting data, it was conclusively shown that the activity decrease is due to formation of carbonates on the catalyst surface. TGA/FT-IR was further used to quantify the respective amounts of carbon deposited on both catalysts as CO_2 and CO. A calibration curve was prepared by decomposing various weights of calcium carbonate in the TGA and integrating the CO_2 absorbances generated by these weights. Using this calibration curve for CO_2 analysis and the TGA weight loss data for CO analysis, carbon levels calculated from TGA/FT-IR and elemental analsis were found to be in good agreement [30].

Problem Solving
It is often the ability to solve problems quickly, especially ones involving the identification of deposits, stains, inclusions, etc., that make spectroscopic instruments invaluable in the industrial laboratory.

The heating element of a Leco total sulfur analyzer failed after formation of a yellow/green/brown deposit close to the edge of the silica insulation. The element is made of molybdenum disilicide, $MoSi_2$, and the furnace operates at a temperature of ~2500°F. The FT-IR spectrum of the brown deposit shows primarily inorganic silicate or SiO_2, while the yellow/green deposit indicates mostly MoO_3 plus the possibility of an inorganic molybdate (Fig. 10).

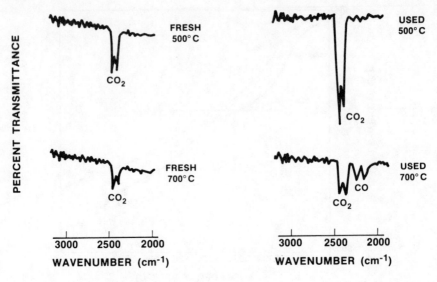

FIG. 9. Evolved gas analysis, FT-IR, of alcohol synthesis catalysts (from Ref. [30], with permission).

This inorganic molybdate, was positively identified by Raman spectroscopy as ferric molybdate $Fe_2(MoO_4)_3$ (Fig. 11). The presence of MoO_3 in the yellow deposit was also confirmed. It was obvious that failure of the heating element was caused by oxidation, and both the infrared and the Raman spectra were necessary to obtain the complete identification of the deposit.

Another typical type of problem in our laboratory involves staining of metal sheets during processing. A sample of rolled and annealed brass sheet was submitted for analysis of the dark stains. A dispute over the cause of the stains involved the possibility of an improper rolling operation, which would tend to result in surface oxidation, or a problem with the rolling lubricant or tarish inhibitor, which would tend to produce carbonaceous deposits. Figure 12 shows the Raman spectra of the stained material compared with an unstained portion of the brass. The peaks at 1355 cm^{-1} and 1575 cm^{-1} are indicative of graphitic material [31]. The ratio of these two bands suggests that the crystallite size in the direction of the graphitic plane is around 56 Å. There is no evidence of Cu_2O or ZnO. Raman spectroscopy was thus able to eliminate the rolling operation as the culprit.

However, in another instance where black stains were appearing on copper sheet metal rolls, a different conclusion was reached utilizing Auger and FT-IR spectroscopies. Auger spectra were obtained on both the stained and clean sections of the copper sheet. The stained copper showed a very strong carbon peak along with smaller peaks due to sulfur, chlorine, oxygen and nitrogen (Fig. 13). Comparing these results with the non-stained area,

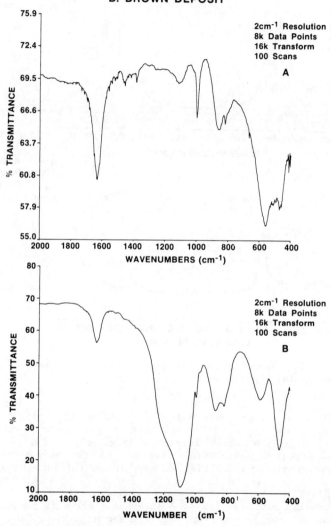

FIG. 10. FT-IR spectra of deposits on failed heating element:
(A) yellow; (B) brown.

there is a significant reduction in the amount of carbon in the
clean copper. The sulfur and chlorine peaks are also much smaller
relative to the largest copper peak.

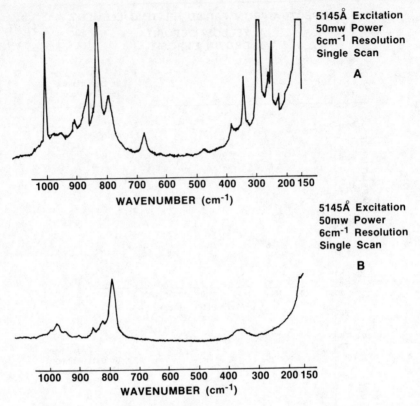

FIG. 11. Raman spectra of deposits on failed heating element:
(A) yellow; (B) brown.

The origin of these peaks was elucidated by examining the FT-IR
spectrum of the stain, using specular reflectance with subtraction
of a clean copper sheet reference (Fig. 14). The IR spectrum
shows hydrocarbon absorptions typical of the sulfur, nitrogen,
chlorine containing industrial oil used in the rolling operation.

Micro Raman
Fine focusing and scanning of a laser beam makes it possible to do
high spatial resolution Raman and a scanning/mapping Raman experiment (MOLE). Microscopes, generally equipped with TV monitors, are
available in both dispersive slit instruments and OMA-based instruments. Applications of micro Raman are well documented in the
literature [32-39].

Most heterogeneous catalysts have inhomogeneities (phase separation). They can create hot spots on the surface of a catalyst
particle, seriously affecting the lifetime and performance. Figure
15 illustrates the Raman spectra on a single ammoxidation catalyst
particle (10 μm sphere), and of a 5 × 10 μm irregular particle from

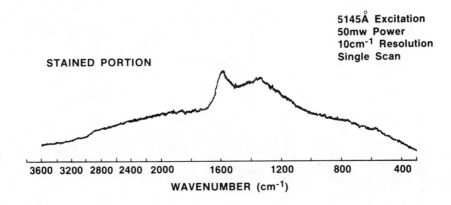

STAINED PORTION

5145Å Excitation
50mw Power
10cm⁻¹ Resolution
Single Scan

WAVENUMBER (cm⁻¹)

5145Å Excitation
50mw Power
10cm⁻¹ Resolution
Single Scan

UNSTAINED PORTION

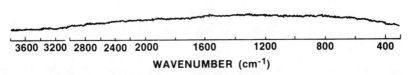

WAVENUMBER (cm⁻¹)

FIG. 12.   Raman spectra of stain on brass sheet.

the same catalyst batch. Differences in the amount of $Fe_2(MoO_4)_3$ (780 cm-^{1}) relative to $M^{2+}MoO_4$ and Bi-molybdates (900-1000 cm-^{1} region) can be easily discerned.

Optical Multichannel Analysis (OMA) Raman Spectrophotometry
The use of a diode array detector in optical multichannel analyses brings a new dimension to Raman spectroscopy. OMA instruments are characterized by fast data collection (current state-of-the-art is 10 milliseconds/scan) and improvement in S/N by the use of multi-plexing and signal averaging, which give similar results to FT-IR and FT-NMR. OMA instruments have high energy throughput, working with wide apertures typically in a Cassagrain configuration. Because of this, however, OMA spectrometers are more susceptible to stray light problems than conventional instruments. For samples where data close to the Rayleigh line is important, it is better to use a conventional instrument with slits. This is also true for high resolution work. Therefore, in our

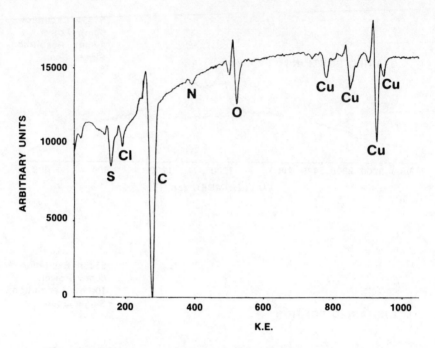

FIG. 13. Auger spectrum of stain on copper sheet.

opinion, the well-equipped lab should have both OMA and conventional spectrophotometers.

An example of fast acquisition is the spectrum of high density polyethylene acquired in 0.01 sec. (Fig. 16). Implications for sample throughput and kinetic studies are obvious.

Figure 17 illustrates the possibilities of an OMA Raman spectrometer for kinetic studies. The four spectra show the build-up of pyridine on a silver electrode immersed in 0.01 molar pyridine in water. Each spectrum took 5 sec. to record in 8 sec. intervals. They indicate different active adsorption sites on the surface.

Even more attractive could be the use of optical fibers with an OMA spectrometer for remote detection. This can have applications in studies of large and odd-shaped samples (automotive engines, refractory furnaces, etc.), catalytic reactor in-situ work, high-pressure reactors, and in many other areas. A spectrum of sulfur powder recorded using a 7 ft. optical fiber (and it could be much longer) was easily obtained. Total acquisition was 12.2 sec. (20 spectra accumulated, each 0.61 sec.).

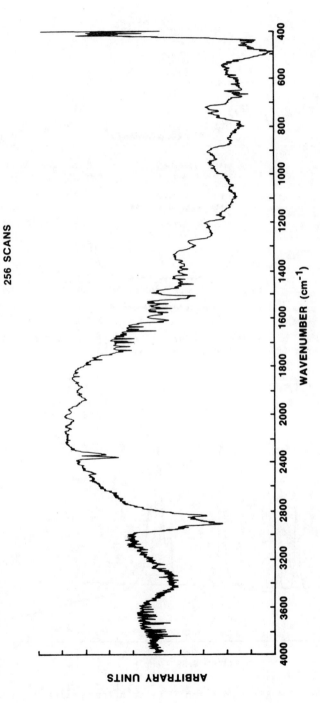

FIG. 14. FT-IR spectrum of stain on copper sheet – specular reflectance with subtraction of clean copper sheet reference.

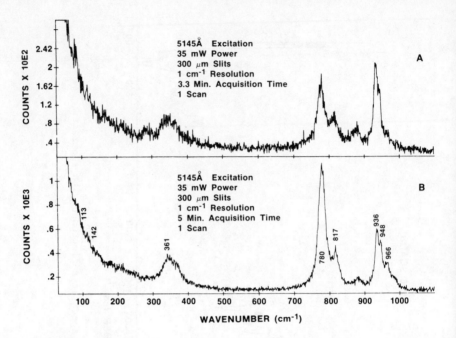

FIG. 15. Micro Raman spectra of catalyst particles: (A) 10 μm sphere; (B) 5 x 10 μm irregular particle.

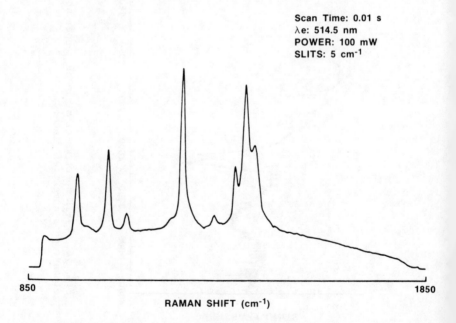

FIG. 16. OMA Raman spectrum of polyethylene pellet.

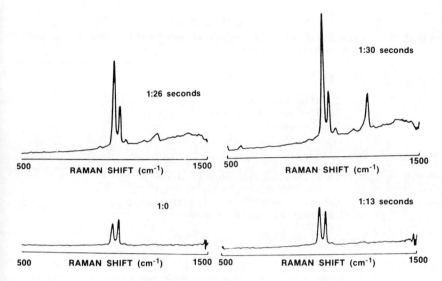

FIG. 17. OMA Raman spectra of pyridine adsorbed on silver.

CONCLUSIONS

It is clear that Fourier transform infrared spectroscopy and Raman spectroscopy are a true partnership and provide a powerful capability in any industrial laboratory for solving non-routine problems. We hve tried to illustrate complementary and unique applications for both. Our predictions for even wider spread use of Raman have been realized, especially since microscope and optical multichannel devices have become available. Table 3 shows some future directions for use of Infrared and Raman Spectroscopy.

Who knows what the next decade will show for this dynamic duo of vibrational spectroscopy.

ACKNOWLEDGMENTS

The authors gratefully acknowledge Dorothy Lukco for the IR analyses of amorphous silicon films; David Compton and Marty Mittleman for the evolved gas/FT-IR analysis; Linda Glaeser, Jim Brazdil, and Bob Grasselli for aid and discussion of the in-situ heterogeneous catalysis work; and Dave Surman for contributions in the Auger analysis, and Don Gerard (British Petroleum) for the OMA studies.

TABLE 3. Current and Expected Areas of Application for Raman and
          Infrared Spectroscopy

A. Catalysis

    1. In-situ studies
    2. Time resolved studies
    3. Non-homogeneous solids

B. Corrosion

    1. In-situ time resolved studies
    2. Aqueous studies (Raman primarily)
    3. Combination with surface spectroscopies
    4. IR/Raman microspectroscopy

C. Photovoltaics and Amorphous Materials

    1. Sensitive materials (rapid runs)
    2. Weak scatterers/low S/N problems (long acquisition)
    3. Carbon fibers, conductive polymers, capacitors

D. Optical Fibers/Diffuse-Specular Reflectance

    1. Ceramic inhomogeneities
    2. Crankshafts
    3. In-situ high pressure
    4. In-situ catalytic reactors

REFERENCES

1. P. R. Griffiths, Transform Techniques in Chemistry, (Plenum
    Press, New York, 1975).
2. J. G. Grasselli and L. E. Wolfram, Appl. Optics, 17, 1386
    (1978).
3. Fourier Transform Infrared Spectroscopy, T. Theophanides, ed.
    (Reidel Publishing Co., Holland, in press).
4. D. A. Long, Raman Spectroscopy, (McGraw-Hill, New York, 1977).
5. J. G. Grasselli, M. A. S. Hazle and L. E. Wolfram, Ch. 14 in
    Molecular Spectroscopy, A. R. West, ed. (Heyden & Son Ltd.,
    1977).
6. J. G. Grasselli, M. A. S. Hazle, J. R. Mooney and M. Mehicic,
    Ch. 7, in Proceedings, XXI Colloquium Spectroscopicum Inter-
    nationale, 8th International Conference on Atomic Spectros-
    copy, G. F. Kirkbright, ed. (Heyden & Son Ltd., 1979).
7. J. G. Grasselli, M. K. Snavely and B. J. Bulkin, Chemical
    Applications of Raman Spectroscopy, (Wiley-Interscience, New
    York, 1981).
8. D. L. Gerrard, Eur. Spectrosc. News, 41, 45 (1982).

9. F. R. Dollish, W. G. Fateley and F. F. Bentley, Characteristic Raman Frequencies of Organic Compounds, (Wiley-Interscience, NY, 1974).
10. M. H. Brodsky, M. Cardona and J. Cuomo, Phys. Rev. B, 16, 3556 (1977).
11. P. John, I. M. Odeh, M. J. K. Thomas, M. J. Tricker and J. I. B. Wilson, Phys. Status Solid, 105, 499 (1981).
12. P. John, I. M. Odeh, M. J. K. Thomas, M. J. Tricker, J. I. B. Wilson, J. B. A. England and D. Newton, J. Phys. Chem., 14, 309 (1981).
13. G. Lucovsky, Solar Cells, 2, 431 (1982).
14. A. Baghadi and R. A. Forman, Appl. Spectrosc., 36, 319 (1982).
15. T. Hirschfeld and A. W. Mantz, Appl. Spectrosc., 30, 552 (1976).
16. J. Moulet, M. S. Thesis, Cleveland State University (1984).
17. R. K. Grasselli, J. D. Burrington and J. F. Brazdil, Disc. Chem. Soc., 72, 203 (1981).
18. J. F. Brazdil, M. Mehicic, L. C. Glaeser, M. A. S. Hazle and R. K. Grasselli, in Proceedings of Symposium on the New Surface Science in Catalysis, ACS National Meeting, Philadelphia, August 1984.
19. G. W. Keulks, J. Catal., 19, 232 (1970).
20. J. D. Burrington, C. Kartisek and R. K. Grasselli, J. Catal., 81, 489 (1983).
21. I. Matsuura and G. C. A. Schuit, J. Catal., 20, 19 (1971).
22. T. Ono, T. Hironaka and Y. Kubokawa, Bull. Univ. Osaka Prefect., Ser. A, 30, 21 (1981).
23. W. Ueda, Y. Moro-Oka and T. Ikawa, J. Chem. Soc., Faraday Trans. 1, 78, 495 (1982).
24. C. J. Machiels and A. W. Sleight, in Proceedings of the Fourth International Conference on Chemistry and Uses of Molybdenum, p. 411 (1982).
25. A. F. VandenElzen and G. D. Rieck, Acta Cryst., B29, 2436 (1973).
26. R. G. Teller, J. F. Brazdil and R. K. Grasselli, Acta Cryst., in press.
27. M. J. Verkerk, M. W. Hammink and A. J. Burggraaf, J. Electrochem. Soc., 70 (1983).
28. J. O. Lephardt, Appl. Spectrosc. Rev., 18:2, 265 (1982-83).
29. D. A. C. Compton, J. G. Grasselli and M. L. Mittleman (The Standard Oil Company, Cleveland, OH), "Use of a Small FT-IR Spectrometer as a Mobile Detector for Fluid Streams", 1983 International Conference on Fourier Transform Infrared Spectroscopy, University of Durham, U.K.
30. M. Mittleman, P. Engler, and D. A. C. Compton, "Interfacing a TGA with an Inexpensive, Mobil, FT-IR", in Proceedings of the 13th North American Thermal Analysis Conference, Philadelphia, PA, (1984), in press.
31. F. Tuinstra and J. L. Koenig, J. Chem. Phys., 53, 1126 (1976).
32. M. Delhaye and P. Dhamelincourt, J. Raman Spectrosc., 3, 33 (1975).
33. G. J. Tosasco and E. C. Etz, Research and Devel., 28, 20 (1977).
34. F. Adar, Microbeam Analysis, p. 67-72 (1981).

35. M. E. Anderson and R. Z. Muggli, Anal. Chem., 53, 1772 (1981).
36. J. L. Chao, Appl. Spectrosc., 35, 281 (1981).
37. F. J. Purcell and E. S. Etz, Microbeam Analysis, p. 301-306 (1982).
38. J. Barbillat, P. Dhamelincourt and M. Delhaye, in Proceedings 10th International Congress on X-ray Optics and Microanalysis, p. C2-255, Les'editions de physique, Les Ulis (1984).
39. P. Dhamelincourt, M. Delhaye and F. Wallart, "Industrial Applications of Raman Microprobing Techniques", in Proceedings IXth International Conference on Raman Spectroscopy, Tokyo, Japan, p. 260 (1984).

THE QUANTITATIVE ANALYSIS OF COMPLEX, MULTICOMPONENT MIXTURES BY FT-IR; THE ANALYSIS OF MINERALS AND OF INTERACTING ORGANIC BLENDS

James M. Brown and James J. Elliott

Exxon Research and Engineering Company
Analytical Division
Clinton Township, Route 22 East
Annandale, NJ 08801

INTRODUCTION

Within recent years the number of quantitative applications of mid-infrared spectroscopy has increased dramatically. This increase, which is illustrated by the approximately threefold increase in the number of papers dealing with quantitative applications of mid-infrared between the 1979 and 1984 FACSS meetings, has resulted largely from improvements in the computer technology and software which are available for infrared applications. In particular, the development of software for spectral fitting, spectral stripping, least-square regression, cross-correlation and factor analysis have greatly enhanced the applicability of infrared spectroscopy for quantitative analyses of multicomponent systems. This paper will discuss two such analyses.

For the purposes of this discussion, we will divide the complex mixtures we wish to analyze into two categories; interacting systems and non-interacting systems. The reason for making this distinction is that the methodology employed in each case is quite different. For non-interacting systems, the spectrum of a mixture can be analysed in terms of a linear combination of the spectra of its individual components. In interacting systems, intercomponent interactions are expected to alter the intensity and/or the frequency of absorptions. The spectrum of the mixture is no longer a simple linear combination of those of the pure components, and calibrations must be developed from the spectra of mixtures. This paper will discuss examples of the analyses of both types of systems as they are currently being carried out in our laboratories. For non-interacting systems, we will use as example the quantitative analysis of minerals by FT-IR and its application to oil shale research. For interacting systems, we will discuss the analysis of multicomponent organic blends which has applications in chemical plant quality control.

NON-INTERACTING SYSTEMS: MINERALS

The quantitative analysis of minerals provides a nearly-perfect example of the non-interacting systems. The spectrum of a specific mineral component in a complex mixture is expected to be minimally effected by the other mineral components. The overall spectrum of the mixture is thus expected to be a simple linear combination of the spectra of the individual components. The analysis problem thus reduces to finding the coefficients of the linear combination which are in fact the concentrations of the individual components. These coefficients can be determined using spectral fitting techniques as will be discussed later. Infrared is particularly well suited for this type of application since the spectra of individual mineral phases are generally distinctive, allowing for their identification and differentiation during the analysis, and since the signal from an individual component is not subject to matrix effects.

While the methodology for analyzing the spectra of minerals is relatively straightforward, the methodology involved in obtaining spectra suitable for analysis is not. The absorption spectra of solids are well known to exhibit a marked particle size dependency. For a mineral dispersed in a KBr pellet, the nature of the particle size is given by [1]

$$A = (m/KSpd)\log((1-K)+K\theta) \tag{1}$$

The absorption (A) measured for a given mass (m) of mineral having density (p) can be seen to depend in a complex fashion on the particle diameter (d), the fraction (K) of the pellet cross sectional area (S) covered by the mineral, and the transmittance (θ) of the mineral particle. K, which is a measure of the dispersion of the mineral in the pellet, is also a function of the particle diameter, increasing with smaller particle size. The particle diameter also occurs in the transmittance term ($\theta = \exp(-kd)$ where k is the mineral absorptivity), since it is the pathlength of the absorbing particle. The absorption can be seen (Fig. 1) to increase as the particle size decreases until a maximum value is reached. This maximum occurs at the size for which K equals unity, i.e. when the particle size is sufficiently small that the total cross section of the pellet is covered by the mineral. In this case, no radiation passes through the pellet without encountering a mineral particle in its path. The absorbance becomes independent of particle size, and depends directly on the mass and absorptivity of the mineral:

$$A = mk/Sp \tag{2}$$

It should be noted that the dispersion criteria (K = 1) can be accomplished at larger particle sizes by increasing the mass of sample used in preparing the pellet. This is however a self-defeating proposition since the resulting absorbances are then generally too high for use in quantitative work and since other undesirable phenomena (scattering, Christiansen Effect) also result from the larger particle size. While it is conceivable to make quantitative measurements on minerals at any point on the graph providing that the particle size of sample and reference are matched, the particle size of components in a unknown mixture is

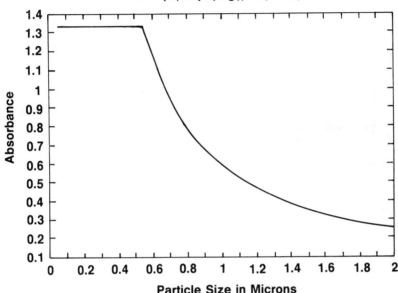

FIG. 1. The calculated intensity of the 1085 cm^{-1} absorption for quartz as a function of particle size. Calculated for 0.1 mg of quartz uniformly dispersed in a 13 mm diameter KBr disk. Variables are defined in the text.

generally not a known or controllable parameter. It is thus necessary to make measurements in the particle size range for which the absorbance is maximized and equation (2) is valid, and this in turn will generally require reduction of particle size to the micron and sub-micron level. We note that since a mineral mixture which is to be analyzed may contain components of various hardnesses, the particle size reduction technique used in the pellet preparation must reduce the harder minerals without destroying the morphology of the softer minerals.

The affects of particle size on infrared absorbance are illustrated in Fig. 2 which shows the spectra of 0.1 mg of quartz obtained using various pellet preparation methodologies. From studies such as these on the affects of initial sample mesh size, grinding times and dilutions, etc., we developed a method for obtaining spectra typified by the one shown at the top of Fig. 2; spectra which we feel represent the maximum absorbance per unit weight of sample. The method we have developed involves

- Pregrinding the mineral to 325 mesh; (usually in a pestle and mortar)

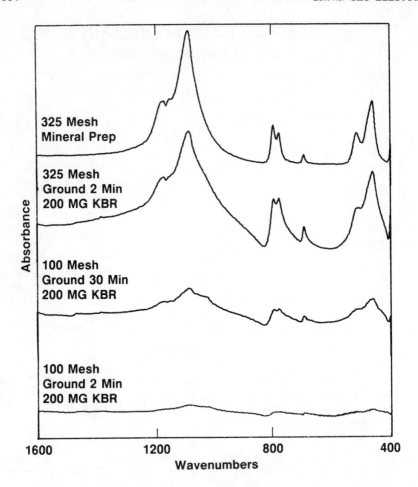

FIG. 2. Spectra of 0.1 mg of quartz as a function of preparation method. The bottom three spectra represent quartz of a
given mesh size ground in the presence of 200 mg KBr for the
indicated times. The top spectrum was obtained using the
preparation method described in text.

· Weighing 0.2 to 0.3 milligrams of the ground mineral (6-place
   balance) and 25 milligram of KBr (5-place balance) into a
   stainless steel grinding capsule;

· Grinding the mixture for 5 minutes on a grinder mill;

· Adding an additional 175 milligrams of KBr;

· Dispersing the mixture for 5 additional minutes on the grinder
   mill;

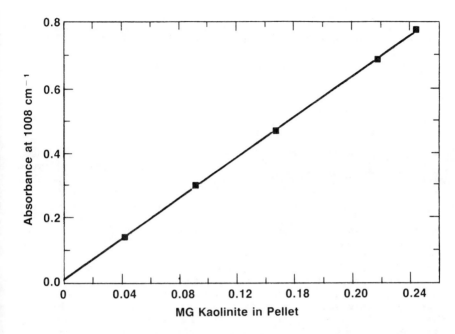

FIG. 3. Beer's Law Plot, absorbance versus concentration, for
Kaolinite in a KBr pellet prepared by the method described in
the text.

· Transferring the mixture to a pellet die, evacuating, and
pressing the pellet;

· Weighing the pellet and correcting for any lost material.

In general we find that reducing the particule size (to 325 mesh)
prior to pellet preparation is necessary to maximize the absorbance
for the harder minerals. The pellet preparation involves separate
grinding and dispersion steps since we have found that the minerals
are not effectively ground in the presence of the 1000 fold excess
(0.2mg/200mg) of KBr. The small amount of KBr used in the grinding
step serves two functions, to prevent loss of material during the
manipulation of the sample and to protect the softer minerals which
might otherwise be destroyed. We have found this method to yield
spectra of minerals suitable for quantitative analysis. The
linearity of the spectra as a function of sample concentration is
illustrated in Fig. 3 which shows the absorption of the 1008 cm^{-1}
band for kaolinite clay as a function of sample concentration.

In the analysis of actual mineral mixtures we make use of a modification of the spectral fitting program (CURVEFIT) originally
introduced by Antoon and Koenig [2]. The spectral fitting program
treats each spectrum as a vector whose dimension is equal to the

number of data points. The ordinates of the vector are the frequencies, and abscissas are the corresponding intensities. The dot
product of two such spectral vectors is then a sum of the products
of the intensities at each frequency. The vector for the mixture's
spectrum (S) is expressed as a linear combination of the vectors
for the pure mineral reference spectra (R_i);

$$S = \sum_i a_i R_i \tag{3}$$

where the coefficients (a_i) are the masses of the individual
minerals which are to be determined. By taking dot products among
the sample and reference spectra, a series of simultaneous
equations is generated which can then be solved for the individual
component concentrations. In our initial attempts to apply this
type of fitting to mineral problems, we found that the baseline
differences among the samples and references caused by variations
in scattering resulted in the fits being more sensitive to baseline
matches than absorbance matches. By including a term in the fit
which is linearly dependent on frequency, i.e. a straight line with
slope and offset, these scattering differences can be minimized so
as to allow for the fitting of mineral absorbances rather than
baseline. Similarly, we have found it desirable to allow for
multiple frequency regions in the fitting so as to be able to avoid
potential interferences (e.g. water, organics), to eliminate automatically negative results from the fit, to synthesize automatically the final linear combination spectrum which the program
determines to be the best fit, and to normalize the results based
on the sample weight rather than 100%. These changes have been
incorporated into a mineral analysis program we call MINFIT.

Figure 4 demonstrates the application of the mineral analysis
methodology to a synthetic mixture of minerals prepared for us by
colleagues who were conducting a round robin study of various
mineral analysis methods. The linear combination spectrum provides
an excellent fit to the actual sample spectrum, and the results
obtained are in good agreement with the composition of the synthetic mixture. Note that the set of reference spectra initially
supplied to the fitting program included six feldspars, of which
three were rejected during the fit on the basis of giving negative
results. The three feldspars which are found by the fit are
representative of the actual feldspar used in the mixture. Note
that the minor error in the clay determination (illite and
chlorite) results principly from the fact that the specific
chlorite used in preparing the mixture was not available in our
spectral reference library.

Figure 5 shows the application of the mineral analysis methodology
to an actual mineral sample, Green River Oil Shale. In this case,
of course, the sample composition is unknown and the predicted
composition can not be directly checked. The difference spectra
between the actual and synthesized spectrum indicates a reasonable
degree of agreement between the actual and predicted compositions.
Another check on the analysis is provided by elemental data. By

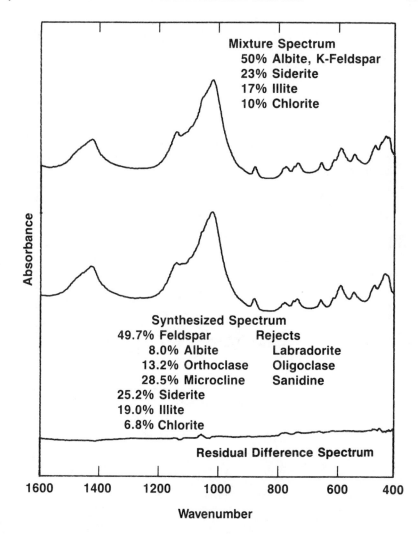

FIG. 4. FT-IR Mineral Analysis of a prepared mixture.

using the elemental analyses of the reference minerals and their predicted concentrations, we can generate an overall elemental composition of the shale to check against its actual elemental composition (Table 1). The predicted elemental data is in relatively good agreement with that determined directly, and the total amount of minerals accounted for is consistent with the organic content of the shale. Note that we can also use the IR data to predict the distribution of a particular element among various mineral phases. Figure 6 and Table 2 show similar data for a Morroco shale.

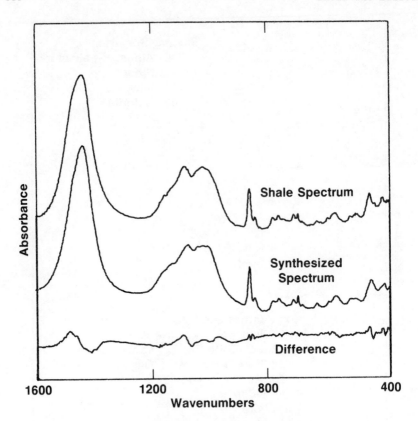

GREEN RIVER SHALE

FIG. 5. FT-IR Mineral Analysis of Green River Shale. The
synthesized spectrum represents the summation of mineral
reference spectra corresponding to the composition given in
Table 1.

We have now successfully applied this methodology in a variety of
mineral applications. We find that our ability to quantitatively
determine minerals is not determined by spectroscopic limits, but
rather by our ability to identify the components in a mixture so as
to provide the fitting program with a representative reference set,
and the availability of these pure mineral phases to use as
references.

TABLE 1.   FT-IR Mineral Analysis of Green River Shale

Mineral Component	Wt %	Mineral Component	Wt %
Calcite	20.8	Analcite	4.7
Dolomite	16.8	Siderite	3.9
Aragonite	13.6	Sanidine	3.2
Quartz	9.9	Illite	1.3
Microcline	7.9	Kaolinite	0.9
Albite	7.8		

Elemental	By FT-IR	By Analysis
Na	1.14	1.76
Mg	2.50	2.55
Al	2.84	2.71
Si	12.03	10.96
K	1.46	0.76
Ca	17.50	16.63
Fe	1.60	1.60
Carbonates as CO_2	24.72	23.60

TABLE 2.   FT-IR Mineral Analysis of Morroco Shale

Mineral Component	Wt %	Mineral Component	Wt %
Na-Montmorillonite	16.2	Sanidine	6.6
Calcite	14.2	Kaolinite	5.6
Dolomite	12.8	Siderite	5.4
Quartz	10.0	Montmorillonite	3.8
Illite	8.4	Gypsum	0.8
		Total	83.8

Elemental	By FT-IR	By Analysis
Na	0.25	0.12
Mg	2.44	2.18
Al	4.86	4.58
Si	14.59	14.24
K	1.51	0.83
Ca	9.00	10.92
Fe	2.94	2.32
Carbonates as CO_2	14.56	13.95

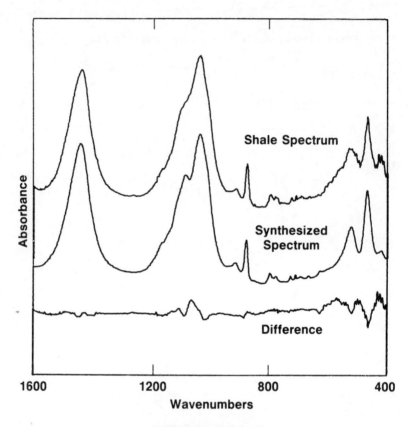

MOROCCO SHALE

FIG. 6.   FT-IR Mineral Analysis of Morroco Shale.   The synthesized spectrum represents the summation of mineral reference spectra corresponding to the composition given in Table 2.

INTERACTING SYSTEMS: ORGANIC BLENDS

For organic mixtures where intercomponent interactions are present, the absorption of an individual component is not independent of the concentration of the other components and the spectral fitting technique discussed above is not longer applicable.   In this case, calibrations must be developed based on the spectra of mixtures. Such calibrations can be obtained through the application of a matrix form [3] of Beer's law

$$A = K * C \qquad\qquad\qquad (4)$$

where A is a matrix of absorption data, K is a matrix of absorptivity data and C is a matrix of concentration data. If f, s, and c are the numbers of frequencies, spectra and components used in the analysis, the dimensions of the three matrices are (A) fXs, (K) fXc and (C) cXs. The inversion of the matrix equation to obtain the absorptivity (K) matrix provides the least square solution to the calibration for an overdetermined (more mixtures and/or frequencies than components) analysis

$$AC^t(CC^t)^{-1} = K \tag{5}$$

Further manipulation of the equation can provide a form which can be directly applied to absorbances of unknowns to obtain concentrations.

$$(K^tK)^{-1}K^tA = C \tag{6}$$

This K matrix mathematics [3] is the basis of Digilab's QUANTATE® program [4] for multicomponent analysis, which is the analysis software used in the example we will now discuss.

Although the matrix mathematics software for establishing the calibration is available, it is left to the user to determine methods for obtaining spectra suitable to analyze and to choose the absorptions which are to be included in the analysis. In the example to be discussed, neither of these problems is straight forward. The problem of interest involved FT-IR multicomponent analysis for chemical plant quality control applications. In particular, for a product which consisted of a blend of four chemicals (henceforth referred to as components A-D) plus a diluent oil, a calibration was needed which would allow for the determination of whether or not the blend composition was within specifications. It was thus not necessary to measure all components over the range from 0% to 100%, but rather to determine accurately whether each component fell within a narrow range about its target concentration. To complicate matters, the viscosity of the product blend is too high to allow for loading fixed pathlength cells with neat product. Since all the components in the blend are potentially variable no internal thickness calibration is possible. While the product is soluble in a variety of organic solvents, most typical infrared solvents (CCl_4, $CDCl_3$, etc.) would obscure regions where measurements are likely to be made. Since the aliphatic regions of the spectra were not likely to provide useful quantitative data, cyclohexane was chosen as a solvent at levels (50% wt/vol) just sufficient to lower the viscosity enough for the loading of the cells. Higher dilutions were undesirable since longer pathlengths would be required to obtain measurable product absorptions, increasing the potential for solvent interferences.

Clearly, before one attempts to set up a multicomponent analysis based on mixtures, it is useful to determine that intercomponent interactions do in fact exist. Figure 7 compares the carbonyl region of the spectra for two test blends. Component A which is responsible for the carbonyl absorptions is present at the same

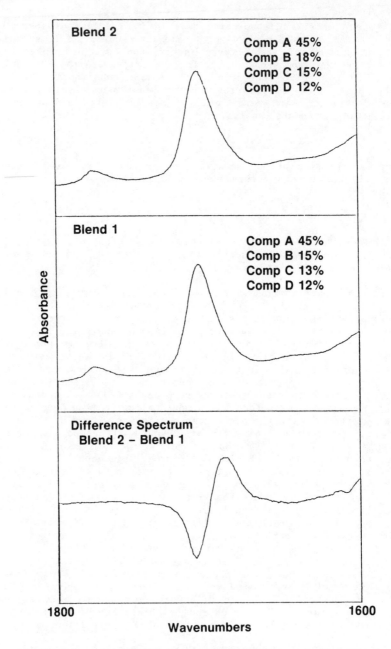

FIG. 7. An example of intercomponent interactions. The difference spectrum (expanded 25 times) demonstrates the effects on the Component A carbonyl absorption due to changes in the concentrations of Components B and C.

level in both blends, whereas components B and C are present at higher levels in the second blend. Subtraction of the two spectra clearly demonstrates that the frequency of the carbonyl absorption for component A is sensitive to the level of the other components. Other intercomponent interactions can be determined by similar application of difference spectra, justifying the use of the mixture based calibration.

Picking the frequencies at which to measure absorbances for use in the calibration is by no means straightforward. This is particularly true of baseline frequencies because of the zero-intercept requirement imposed by the K matrix method (i.e., the K matrix equation (4) does not allow for any background absorbance, absorbance going to zero when concentration is zero). Fortunately, another program available from Digilab, ACTIVITY® [4], can be employed in choosing the analysis frequencies. ACTIVITY® generates a spectrum corresponding to the standard deviation in absorbance among a series of spectra. By applying the program to spectra of blends for which only one component (and diluent) is varied, the deviation in absorbance corresponding to the variation in that component can be obtained. Since this deviation will include intercomponent affects, it provides a more reliable means of choosing analysis frequencies than do the spectra of the neat components. Points of maximum deviation provide maximum sensitivity to the component variation, whereas points of minimum deviation are ideal candidates for baseline endpoints. Figure 8 compares the spectra of a target product blend, the standard deviation spectrum generated for component A, and the spectrum of component A. Since component A comprises nearly 50% of the total product, its absorptions are relatively easy to pick out in the product spectrum, and the standard deviation spectrum is very similar to that of the component itself. Note, however, that there are some differences between the spectrum of the component and the standard deviation spectrum. In particular, the middle of the three methyl bands observed for the component ($\sim$1377 cm^{-1}) is dominated in the spectra of the blends by methyl absorptions due to other components and is not observed in the standard deviation spectrum. If we look at similar data for component B which comprises only about 15% of the product (Figure 9), we can no longer readily identify the component absorptions in the target product spectrum. The standard deviation spectrum still resembles the spectrum of the component with some exceptions. The methyl vibrations ($\sim$1380 cm^{-1}) of component B represent such a small contribution to the total methyl absorption that they are only weakly observed in the standard deviation spectrum. The interaction of component B with the carbonyl in component A is clearly shown, the dispersive-shape difference spectrum previously shown now being observed as two positive bands in the standard deviation spectrum. More importantly, the intensity of the 1230 cm^{-1} band for component B is relatively larger in the standard deviation spectrum than the component spectrum, indicating that this band, which is subject to less interference from absorptions due to other components, is in fact more sensitive to the component B concentra-

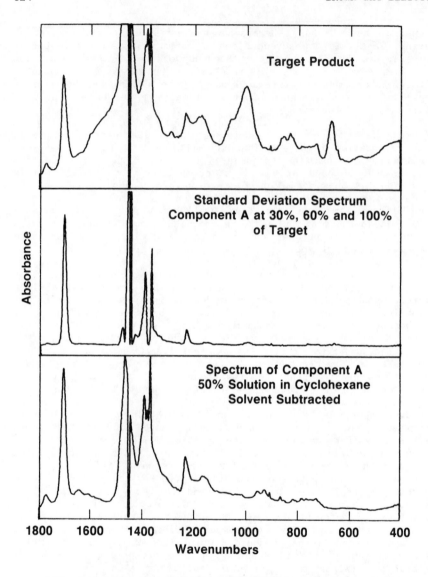

FIG. 8. Application of Standard Deviation Spectrum for determining analysis frequencies. Application to Component A which
comprises approximately 49% of target product.

tion than are some of the intrinsically stronger component B
absorptions.

Based on these types of standard deviation spectra it is possible
to determine suitable peak and baseline frequencies for use in the
analysis. These data are then supplied to the QUANTATE® program in

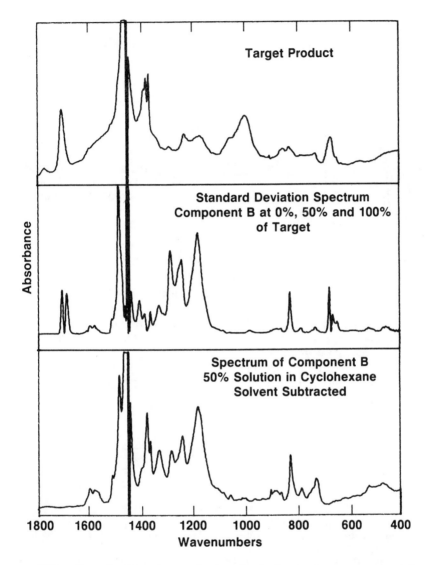

FIG. 9. Application of Standard Deviation Spectrum for determining analysis frequencies. Application to Component B which comprises approximately 16% of target product.

the form of a reduction script which defines which spectral points are to be used in the matrix calculations. The reduction script we used for the calibration of this product is given in Fig. 10 along with a graphical representation of the absorption measurements it defines. Points represent peak height measurements relative to a baseline (B). Area measurements and/or grids (multiple peak

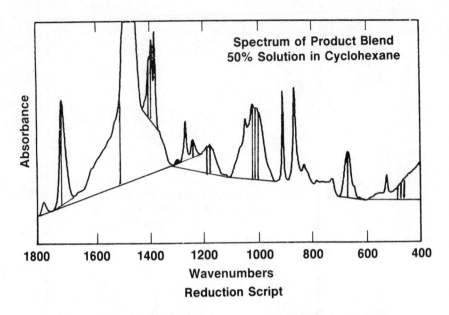

Reduction Script

Point 1181 B(1300 1100)   Grid 450 430 5 B(600)
Area 1300 1275 B(1305 1270)   Point 1706 B(1740 1660)
Point 669 B(705 600)   Point 1231 B(1270 1200)
Grid 1010 990 5 B(1100 925)   Point 1389 B(1405 1350)
Point 1495 B(1800 1300)   Point 1377 B(1405 1350)
Point 1172 B(1300 1100)

FIG. 10. A graphical representation of the measurements involved in the FT-IR multicomponent analysis of the product blends. The reduction script defines the points (peak heights), areas (integrated areas) and grids (multiple peak heights within a range) defined relative to the baselines (B).

heights over a defined range) are used in some instances to increase the sensitivity to certain lower concentration components.

The concentrations of the reference samples used in developing a multicomponent calibration are of equal importance to the choice of analysis frequencies. Since we were only interested in measurements over a relatively narrow range of concentrations in the quality control application, our reference samples need not include the entire range of possible concentrations. However, since we are interested in accuracy at the extremes of the specification range, our reference concentrations must span a larger range than the allowable specification range. In general, if a component has a target concentration t, and an allowable concentration range from t-d to t+d, we choose to include concentrations over the range t-1.5d to t+1.5d in the calibration. Generally, we choose five

TABLE 3. FT-IR Multicomponent Analysis for Quality Control of
Chemical Blends

Component	A	B	C	D	Diluent
Target Concentration	49%	16%	15%	12%	8%
Specification Range					
Minimum	45%	15%	14%	11%	
Maximum	53%	17%	16%	13%	
Calibration Range					
Minimum	43%	14%	13%	10%	
Maximum	55%	18%	17%	14%	
Calibration Data, Average Absolute Error for the Measurement of 30 Calibration Test Blends					
	0.5%	0.3%	0.1%	0.2%	1.0%

possible concentrations for each component in the reference blends,
the two extremes, the minimum and maximum specifications and the
target. Since all possible 5^4 combinations of the concentrations
of the four components cannot be handled within the constraints of
the software, nor do we have time to make up that number of calibration solutions, we randomly choose combinations subject only to
the constraints that the relative proportions of any two components
are not constant throughout the series; that at least two components (excluding diluent) are at different levels in each blend,
and that all five levels are represented for each component within
the calibration blends approximately the same number of times.
Thirty blends are produced, of which nineteen are used in
establishing the calibration and eleven as calibration checks.

Applying the reduction script described previously to the spectra
of a series of blends prepared in this fashion, we obtain the
calibration described in Table 3. The calibration provides for the
measurement of the four components to an average error of between
0.1 to 0.5 weight percent (1-2% relative). Even for the diluent
oil, which is low in concentration and weakly absorbing, the
average error in the measurement is only 1 weight percent
(absolute). This type of calibration is clearly suitable for our
quality control applications since the measurement errors are
significantly less than the allowable product compositional variations.

CONCLUSIONS

It is clear that FT-IR multicomponent analysis can be successfully applied to a variety of complex materials, both inorganic and organic. The methodology chosen for a particular analysis will depend on the nature of the materials to be analyzed. Other mathematical techniques such as factor analysis and cross-correlation make possible additional types of quantitative applications. It should be noted, however, that the mathematical tools for multicomponent analysis represent only part of the solution to mixture analysis; the users ability to define the system to be analyzed and to obtain spectra suitable for analysis often being the limiting factor in real world applications.

ACKNOWLEDGMENT

The authors would like to acknowledge the valuable assistance of John Baltrus, Karen Graf, Donald Bachert, and Janet Quodomine.

REFERENCES

1.   H. W. van der Marel and H. Beutelspacher, Atlas of Infrared Spectroscopy of Clay Minerals and their Admixtures, Elsevier, New York, 1976, p. 59, and references therein.
2.   M. K. Antoon, J. H. Koenig and J. L. Koenig, Appl. Spectrosc., 31, 518 (1977).
3.   C. W. Brown, P. F. Lynch, R. J. Obremski and D. S. Lavery, Anal. Chem., 54, 1472 (1982).
4.   The QUANTATE® software package is a product of Bio-Rad Laboratories, Inc., Digilab Division.
5.   The ACTIVITY® program, which is part of the QUANTPAK® software package, is a product of Bio-Rad Laboratories, Inc., Digilab Division.

VIBRATIONAL SPECTROSCOPIC STUDIES OF THIN FILMS

John F. Rabolt

IBM Research Laboratory
San Jose, California 95193

INTRODUCTION

The importance of organic thin films in the fabrication of microelectronic devices is undisputed, however, the development of
nondestructive techniques for the characterization of such thin
films has proceeded slowly. The main barrier to progress arose
because of the exceedingly small amount of material present in
submicron films [1-6] requiring highly sensitive state-of-the-art
techniques [7-10]. Recently with the advances in IR and Raman
instrumentation coupled with the development of novel analytical
techniques [3-9], there has been an explosion of activity in the
study of thin films. As expected, IR spectroscopy led the way with
reports of studies on submicron films by ATR and specular reflectance measurements [5-9]. These novel techniques allowed the
investigation of both thin organic films as well as Langmuir-
Blodgett (L-B) multilayers [8-10].

Progress in obtaining Raman measurements from thin films has proceeded much slower since, in general, the Raman scattering cross
section of a molecule is 10-12 orders of magnitude lower than its
infrared absorption cross section for the same vibrational normal
mode. Thus, Raman studies of submicron films of L-B multilayer
assemblies have been unsuccessful in the past. Two possible ways
to overcome this problem and increase the spontaneous Raman
scattering from a material are to increase the scattering cross
section [the mechanism responsible for resonance Raman scattering
(RRS)] and/or increase the scattering volume of material. Recent
surface-enhanced Raman scattering studies [11-14] from small
molecules adsorbed on metal surfaces report enhancements of 10^6 in
band intensities of the surface species and in all probability
contain at least some contribution from RRS to the enhancement
mechanism. Although this technique has generated considerable
excitement, its general applicability is somewhat compromised by
the fact that experiments must be conducted in either an electro-
chemical cell or in high vacuum.

Recently, a novel approach [15,16] utilizing integrated optical
techniques succeeded in obtaining the Raman spectrum of submicron
films by actually coupling (via prism) laser light into the film

Waveguide Experiments

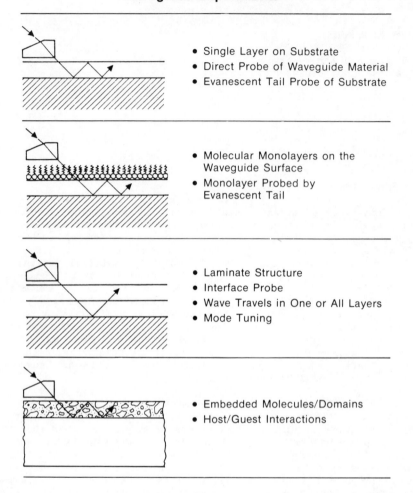

- Single Layer on Substrate
- Direct Probe of Waveguide Material
- Evanescent Tail Probe of Substrate

- Molecular Monolayers on the Waveguide Surface
- Monolayer Probed by Evanescent Tail

- Laminate Structure
- Interface Probe
- Wave Travels in One or All Layers
- Mode Tuning

- Embedded Molecules/Domains
- Host/Guest Interactions

FIG. 1. Schematic illustration of the use of integrated optics to obtain Raman scattering from a thin film, a L-B monolayer, a polymer laminate and an embedded molecule or domain.

whose spectrum was desired (see Fig. 1). As the light traversed the film from one end to the other via total internal reflection, the number of molecules from which light was inelastically scattered increased significantly. In addition, the initial confinement of a 150 micron diameter laser beam into a submicron thick film caused a dramatic increase ($\sim10^3$) in the optical field intensity within the film. Since Raman intensity is proportional to both the scattering volume and laser intensity, these two factors combine to allow the recording of spectra of ultrathin films. Using variations of this technique, it was shown [17,18] that the RRS of a single L-B monolayer of dye could be observed (Fig. 1b) and the perturbations of its vibrational bands detected when in contact with polymer or glass surfaces. Because these techniques are new, the full range of flexibility has not yet been explored but initial results show considerable promise.

RESULTS AND DISCUSSION

Waveguide Raman Spectroscopy
Raman spectroscopy is an ideal tool for studying structure and morphology in polymeric materials because it is a nondestructive technique. Recently [15,18], Raman scattering has been coupled with integrated optics in order to investigate organic and polymer films in the submicron regime.

It is well known [19] in the field of integrated optics that a single or multilayered film deposited on a substrate can guide light by total internal reflection at the film interfaces. This is illustrated at the top of Fig. 1. For a guided mode to be supported in a thin film, the refractive index of the film must be larger than that of the substrate. When laser light, whose focused beam diameter is on the order of 200 microns, is coupled into such a film using a prism, the beam size within the film is constrained (in one dimension) to the thickness of the film. This results in a very high optical field intensity within the film thereby allowing the recording of its Raman spectrum.

When a film is below a certain thickness (0.6 - 0.8 microns at visible excitation wavelengths) it will no longer support a guided wave and a sample arrangement shown in the second panel of Fig. 1 must be used. By depositing a primary guiding layer of high index glass on the substrate a second thinner layer (polymer or Langmuir-Blodgett monolayer) can then be deposited and its Raman spectrum obtained using the evanescent field generated in the vicinity of its interface with the primary guiding layer. This evanescent field decays exponentially with distance into the thinner non-guiding layer but is sufficiently intense to excite Raman scattering from this layer. Using this sampling geometry a Raman spectrum of 0.08 micron polystyrene (PS) film and a single dye monolayer (0.003 micron) have been obtained [17].

In order to study polymer laminates, the composite waveguide structure illustrated in panel 3 of the accompanying figure has been used [20].

Calculations of the optical field intensity distribution within such a slab waveguide indicate that specific modes can be selected where a maximum optical field intensity exists either within one of the component layers, or at the interface between the two adjacent layers. Corresponding Raman measurements have been obtained from two component waveguides composed of polyvinylalcohol (PVA), poly-methylmethacrylate (PMMA) and polystyrene (PS). Results [20] suggest that in addition to studying polymer/polymer and polymer/substrate interfaces, this technique provides a nondestructive method for investigating coatings.

In the final panel of Fig. 1, the most recent application of WRS to the study [21] of guest molecules or domains embedded in a polymer matrix is shown. If the guest is a domain, certain precautions regarding size must be taken so as to prevent the scattering of light from interfering with the waveguiding characteristics of the film.

In the case of small anisotropic molecules it was found [21] that by adjusting the concentration of the guest molecules in the host polymeric waveguide, Raman spectra of quasi-isolated molecules (in the absence of crystal field effects) could be obtained. In addition, upon orientation of the polymeric waveguide during the casting process it was found that anisotropically shaped guest molecules were also oriented. These two observations together suggest that, in cases where single crystals cannot be grown, polarized Raman spectra of organic molecules can be obtained in oriented waveguide structures providing information about vibrational band assignments. The implications of such results are exciting and studies involving these composite waveguides are currently in progress in a number of government, academic and industrial laboratories.

Infrared Spectroscopy of Thin Films

Determination of Molecular Orientation. In the case of anisotropic thin films it is often desirable to determine the orientation of chemical groups relative to the surface of the substrate on which they are deposited [22]. This can be done by polarized infrared studies if a suitable method for obtaining IR spectra with the polarization of the incident electric field vector, both perpendicular and parallel to the substrate, can be found. Such a method composed of grazing incidence reflection (GIR) and standard transmission techniques is schematically illustrated in Fig. 2. Greenler [23] has shown that although the electric field component parallel to the reflective surface undergoes a phase shift of 180° independent of incident angle, the component perpendicular to the surface experiences a phase shift of ~180° only near grazing angles (88°) of incidence. At these angles the amplitudes of the electric field are additive and hence an enhancement of the IR spectrum of thin films deposited on metal surfaces is observed. In fact it has also been demonstrated [24] that this quantitative field enhancement of the perpendicular component is not only sensitive to the angle of incidence but also to the nature of the metal surface. However, when highly reflective metals, such as silver (Ag), gold (Au), etc., are used for reflection, high S/N polarized spectra can be routinely obtained using [24] an incident angle of 81° relative to the surface normal.

As is also illustrated in Fig. 2, combining GIR measurements with those obtained from standard transmission (in which the electric

Reflection
(at Grazing Incidence)

Transmission

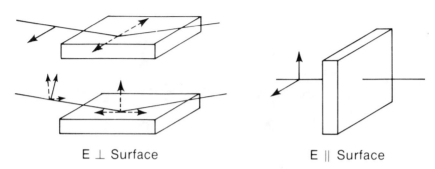

E ⊥ Surface   E ∥ Surface

FIG. 2. Direction of the polarized electric field vectors in grazing incidence reflection (left) and standard transmission experiments (right).

field vector is parallel to the surface) allows the determination of the orientation of molecular groups in anisotropic films.

These techniques have recently been applied to the study [9] of Langmuir-Blodgett multilayered films of cadmium arachidate, $(H_3C(CH_2)_{18}COO)_2Cd$. Not only was it determined from the IR spectra that the alkyl tails were oriented perpendicular to the surface to within $8 \pm 5°$, but it was also observed that crystal field interactions resulting from 2 molecules per unit cell caused the splitting of IR bands. After the orientation and band splitting was taken into account it became clear that the CdA monolayers crystallized in an orthorhombic crystalline habit with alkyl tails perpendicular to the surface as depicted in Fig. 3a.

Structure of L-B Films at Elevated Temperatures. Since the characterization of L-B films is at an early stage, the effect of temperature has been contemplated but no such studies have appeared. The ramifications of such investigations to determine the structural integrity of L-B films at elevated temperature as a function of time could be significant since very little is known about physical aging (how the structural integrity of a material changes with time, temperature, etc.) of a monolayer. The implication of such studies on the service of life of microelectronic devices [25] fabricated with L-B films could be substantial.

Shown in Fig. 4 is a schematic diagram of the device used in this laboratory to obtain GIR measurements at elevated temperatures [26]. The thin film heater is embedded in a brass backing plate which is held in contact with the monolayer sample supported on a 200 nm Ag film on glass. The thermistor allows regulation of the

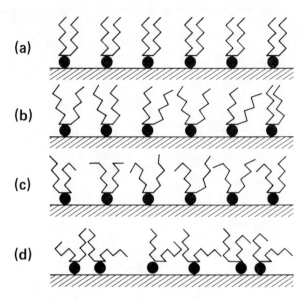

FIG. 3. Schematic diagram of the gradual melting process in
L-B monolayers of CdA. (a) ordered room temperature solid;
(b,c) gradual disordering of alkyl tails with ordered head
group lattice; (d) irreversible breakup of head group lattice.

temperature to within $\pm 1°C$. The entire sample assembly and plane
mirror is held in a stainless steel platform allowing temperatures
in excess of 200°C to be routinely achieved.

As seen in Fig. 5 the GIR spectra of cadmium arachidate monolayers
as a function of temperature has been obtained [10]. Thermal
analysis indicated that bulk crystals of this material melted at
110°C. As shown in the figure, the CH stretching vibrations
($3100-2800$ cm^{-1}) begin to change in relative intensity at 65°C,
more than 45° below the melting point. The symmetric CO_2
stretching vibration at 1432 cm^{-1}, on the other hand, remains
invariant until the temperature is in excess of 100°C. This latter
vibration [10] is characteristic of the head group represented by
the $Cd(COO)_2$ group and is indicative of the order maintained by the
group throughout the temperature range. At 125°C, a broadening of
this band and a dramatic change in intensity is observed due to the
breakup of the head group "lattice". Thus, the melting process
appears to be two stage and involves the disordering of the hydrocarbon tails long before the melting of the head group "lattice".
This was confirmed by a series of cyclic temperature experiments
which revealed that the disorder introduced at intermediate
temperatures was reversible upon return to ambient temperature.
However, once the sample temperature exceeded ~110°C, this was no
longer the case since interlayer diffusion accompanied the breakup
of the cadmium "lattice". In addition, the specific introduction
of disordered conformations into the alkyl tail are supported by a

Grazing Incidence Infrared Reflection Attachment

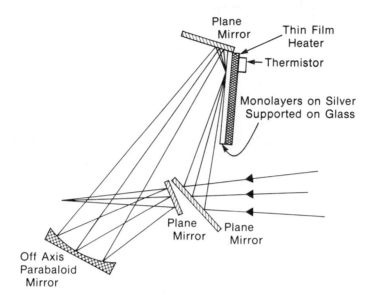

FIG. 4. Sample holder for grazing incidence reflection experiments at elevated temperatures.

gradual increase, with temperature, of the intensity of bands observed at 1340 cm-1 attributed [10] to tg(t-trans; g-gauche) sequences and at 1305 cm-1 assigned to the wagging vibration of CH_2 groups characteristic of gtg' conformational defects.

The molecular picture which emerges from the IR data is shown in Fig. 3. As the temperature of the L-B monolayers of CdA is raised above ambient a gradual reversible pretransitional increase in disorder is observed which is followed by a catastrophic breakup of the cadmium "lattice".

The implication of these measurements is that annealing of a L-B film does not improve its order as it does in semicrystalline polymer films. On the contrary, annealing an L-B monolayer introduces defects which tend to disorder and disorient the monolayer components.

CONCLUSIONS

Several techniques have been described which allow the recording of IR and Raman spectra of submicron films. Integrated optical techniques when used in conjunction with Raman spectroscopy have provided a means of investigating ultrathin films (<.08 micron), L-B monolayers, polymer-polymer interfaces and interactions between small molecules and the polymer matrix in which they were embedded.

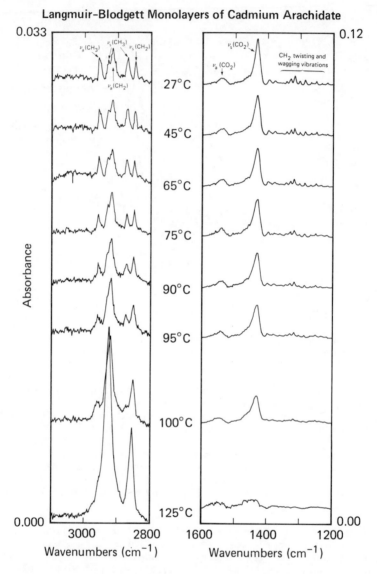

FIG. 5. Grazing incidence reflection IR spectra of CdA showing the structurally important bands as a function of temperature. Bands are labelled with their assignment.

Infrared spectroscopy is extremely useful for the study of anisotropic films since by using a combination of GIR and transmission
techniques, information about the specific orientation of chemical
groups can be obtained. This has been specifically illustrated in
the case of L-B monolayers of cadmium arachidate where specific

band assignments could be made after polarized IR spectra were obtained. Upon heating these films, GIR measurements indicated that prior to melting, pretransitional disorder in the alkyl tails is introduced as evidenced by the appearance of IR bands characteristic of gauche sequences. If the temperature exceeds 110°C there is an irreversible breakup of the ionic cadmium "lattice" followed by interlayer diffusion producing a randomly ordered film.

ACKNOWLEDGMENTS

The author would like to thank Drs. J. D. Swalen (IBM) and N. E. Schlotter (Bell Communications Research) for their contributions to many of the experiments described.

REFERENCES

1. S. A. Francis and A. H. Ellison, J. Opt. Soc. Am., 49, 131 (1959).
2. T. Takenaka, K. Nogami, H. Gotoh and R. Gotoh, J. Coll. Interf. Sci., 35, 395 (1971).
3. T. Takenaka, K. Nogami and H. Gotoh, J. Coll. Interf. Sci., 40, 409 (1971).
4. F. Kopp, W. P. Fringeli, K. Mühlethaler and H. H. Günthard, Biophys. Struct. Mech., 1, 75 (1975).
5. P. Chollet, J. Messier and C. Rosilio, J. Chem. Phys., 64, 1042 (1976).
6. T. Ohnishi, A. Ishitani, H. Ischida, N. Yamamoto and H. Tsubomura, J. Phys. Chem., 82, 1989 (1978).
7. P. A. Chollet, Thin Solid Films, 52, 343 (1978).
8. D. L. Allara and J. D. Swalen, J. Phys. Chem., 86, 2700 (1982).
9. J. F. Rabolt, F. C. Burns, N. E. Schlotter and J. D. Swalen, J. Chem. Phys., 78, 946 (1983).
10. C. Naselli, J. F. Rabolt and J. D. Swalen, J. Chem. Phys., (in press).
11. M. Fleishman, P. J. Hendra and A. J. McQuillian, Chem. Phys. Lett., 26, 163 (1974).
12. D. L. Jeanmarie and R. P. Van Duyne, J. Electroanal. Chem., 84, 1 (1977).
13. R. P. Van Duyne, J. Phys. (Colloq.) C5, 38, 239 (1977).
14. J. A. Creighton, M. G. Albrecht, R. E. Hester and J. A. D. Mathew, Chem. Phys. Lett., 55, 55 (1978).
15. J. F. Rabolt, R. Santo and J. D. Swalen, Appl. Spectrosc., 33, 549 (1979).
16. J. F. Rabolt, R. Santo and J. D. Swalen, Appl. Spectrosc., 34, 517 (1980).
17. J. F. Rabolt, N. E. Schlotter, J. D. Swalen and R. Santo, J. Polym. Sci.-Polym. Phys. Ed., 21, 1 (1983).
18. J. F. Rabolt, R. Santo, N. E. Schlotter and J. D. Swalen, IBM J. Res. and Dev., 26, 209 (1982).
19. D. Marcuse, Theory of Dielectric Optical Waveguides, (Academic Press, New York, 1974).
20. J. F. Rabolt, N. E. Schlotter and J. C. Swalen, J. Phys. Chem., 85, 4141 (1981).

21. N. E. Schlotter and J. F. Rabolt, Appl. Spectrosc., <u>38</u>, 208 (1984).
22. J. D. Swalen and J. F. Rabolt, in <u>Fourier Transform Infrared Spectroscopy</u>, Vol. 4, J. Ferraro and L. Basile, editors (Academic Press, 1985).
23. R. G. Greenler, J. Chem. Phys., <u>44</u>, 310 (1966).
24. J. F. Rabolt, M. Jurich and J. D. Swalen, Appl. Spectrosc. (in press).
25. G. G. Roberts, Sensors and Actuators, <u>4</u>, 131 (1984).
26. N. E. Schlotter and J. F. Rabolt, Appl. Spectrosc. (in press).

VIBRATIONAL SPECTROSCOPY FOR THE STUDY OF ORDER-DISORDER PHENOMENA
IN ORGANIC MATERIALS

Bernard J. Bulkin

Department of Chemistry
Polytechnic Institute of New York
Brooklyn, New York 11201

INTRODUCTION

Prior to the 1960's there had primarily been sporadic rather than
systematic investigations of phase transitions in organic materials
using vibrational spectroscopy. Most of what had been done related
to crystal field splittings which disappeared on melting or dis-
solution. Similar splittings in polymers were also associated with
crystallinity. Much of the early work on vibrational spectra of
organic crystals is summarized in a review article by Dows [1].

Even prior to the major advances in experimental vibrational spectroscopy, the development of the diamond cell led to a considerable
body of work on the transformation of organic crystals under
elevated pressure. This work certainly indicated the potential of
vibrational spectroscopy as a sensitive technique for studying
order-disorder phenomena in organics. The great advances -- laser
excited Raman spectroscopy and Fourier transform infrared spectroscopy -- which we associate with the 1960's, greatly broadened
the range of systems accessible to systematic investigation. In
many cases, these new techniques enabled us to attack old problems
with a response from vibrational spectroscopy which was not previously possible.

This paper will deal with aspects of these problems which have been
the central concern of my laboratory for the past seventeen years.
The problems include order-disorder phenomena in organic crystals
which form liquid crystals, and in polymers. Such materials have
phase transitions which are complex. They are often of greater
than first order, or are referred to as complex first order transitions [2]. Because they go beyond simple melting or crystal-
crystal phase transformations, molecular level data are necessary
to provide insight into what is actually happening in these
systems. Vibrational spectroscopy has managed to provide that
insight.

In this paper, I will not dwell on what it is we learn about liquid
crystals or polymers from their vibrational spectra during phase
transitions. Rather, the theme will be how the vibrational spectrum reports on these order-disorder phenomena in organic

139

materials, so that these examples may be seen as a guide to future research.

THE CRYSTAL-LIQUID CRYSTAL PHASE TRANSITION

In the 1960's, interest in the understanding of liquid crystalline phases had suddenly increased. These phases were well known in the early part of the twentieth century, and were the subject of numerous papers and books up through the early 1930's. At that time, however, interest declined rather rapidly.

It has been said that scientists got the idea, particularly from a large Faraday Society meeting in 1933, that liquid crystals were well understood and that there was nothing more to be done. I do not believe this to be the case. Rather, characterization techniques available at that time had been exploited to the maximum extent. In addition, since there were not any applications of liquid crystals in sight, there was no driving force for synthesis of new materials. They were regarded, by those who even knew of their existence, as a scientific curiosity.

Interest was revived when all of these aspects were cured at about the same time. Knowledge of liquid crystals was greatly increased by a review article by Brown and Shaw [3], and by a classic book by George Gray [4]. Several research groups, but most notably those at RCA led by Williams, Heilmeier and Goldmacher, and at the U.S. Army Night Vision Laboratory led by Ennulat and Elser, saw and began to develop new applications of liquid crystals to electro-optic devices. Characterization by vibrational spectroscopy, X-rays, nmr, and other techniques was simultaneously possible to a great degree of sophistication. Moreover, the theory of phase transitions had itself made great strides. Pierre G. DeGennes had been able to bring a systematic approach to superconductivity and now turned his attention to liquid crystals.

Thus by the early 1960's many forces came to make for a favorable climate for intense scientific and engineering study of liquid crystals. The results, both in universally accepted applications to displays and in scientific knowledge, were dramatic.

Our laboratory concentrated on just a few aspects of this problem. The unifying themes of this work are the role that vibrational spectroscopy can play in improving our understanding of liquid crystals, and the nature of order-disorder transitions in organic materials.

In 1968 Dr. Dolores Grunbaum and I began studying the infrared spectrum of 4,4' bis(methoxy)azoxy benzene, known almost universally as PAA for its non-systematic name of paraazoxy anisole, as well as homologs of this compound up to C_7.

We found that there were numerous bands in the infrared spectrum of crystalline PAA and the homologs which disappeared at the crystal-nematic liquid crystalline phase transition. In each case, the bands are relatively sharp, medium intensity bands separated from more intense bands by 15-100 cm^{-1}. Two examples are shown in Fig. 1.

What could be the origin of these "disappearing" bands? The molecule has either C_s or C_1 symmetry [5] so selection rules are probably not involved. Even the site symmetry or factor group symmetry, characterized subsequent to our initial work as C_{2h}^5 [6] could not account for additional bands in the crystal through selection rule arguments. The factor group could split the vibrational modes through coupling of the four molecules in the unit cell. This was discounted because the observation was always that the bands which disappeared were individual bands rather than multiplets which collapsed to a central frequency.

The explanation of these bands is nonetheless linked to strongly interacting molecules in the crystalline phase. It is that the disappearing bands are sum and difference modes between lattice vibrations and internal vibrations. Such modes are well known in the literature of ionic crystals.

Now if there are bands in the infrared spectrum which result from a coupling of the molecules to the lattice, it should be possible to use these to study the processes by which these molecules uncouple. This is exactly what happens during melting, and it is what happens in a more complex way when melting occurs to a liquid crystalline phase.

When the peak intensity of these bands is followed as function of temperature. just below the crystal-nematic (c-n) phase transition, one finds that the change is gradual rather than abrupt. There are pretransition effects occurring several degrees below the established transition temperature. The pre-transition effects are a manifestation of orientational and positional disorder in the crystalline phase. They occur because of the great geometric anisotropy of PAA and molecules like it. And these effects show up in the infrared spectrum through the coupling of lattice vibrations and internal vibrations.

To pursue this point, Dr. F. Prochaska and I examined the Raman spectrum of PAA between 10 and 150 cm^{-1} and W. B. Lok and I looked at the far infrared spectrum in the same region. The Raman spectrum of a single crystal of PAA is shown in Fig. 2. It shows a number of well defined maxima, and these can also be seen in powder

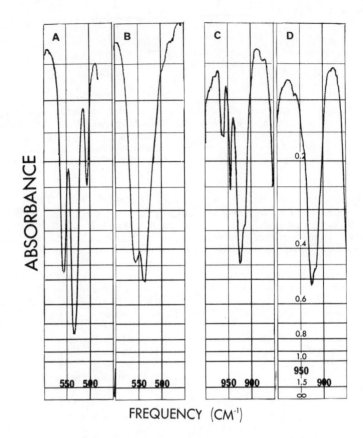

FIG. 1. Examples of bands in the infrared spectrum of alkoxyazoxy benzenes which disappear in the nematic phase. A and C
are crystalline phase spectra, B and D are nematics.

spectra, though with many of the shoulders and smaller peaks being
unresolved. I might note in passing that PAA was the first
compound whose Raman spectrum I ever ran in the low frequency
region. I do not think I ever found another one which gave as good
a spectrum!

If pre-transition effects connected with the lattice are visible
"indirectly" in the mid-infrared, we reasoned that they should be
seen directly in the low frequency Raman and far infrared data.
This is indeed the case. In the Raman spectrum, at temperatures
just below the c-n transition, we observed a series of changes.
These could only be explained as follows: A band, located at room
temperature near 80 cm^{-1}, moves across the spectrum towards zero
frequency as the phase transition is approached. This very
dramatic change appears in a similar way in the far infrared, but
it is much more difficult to make observations in that region. The

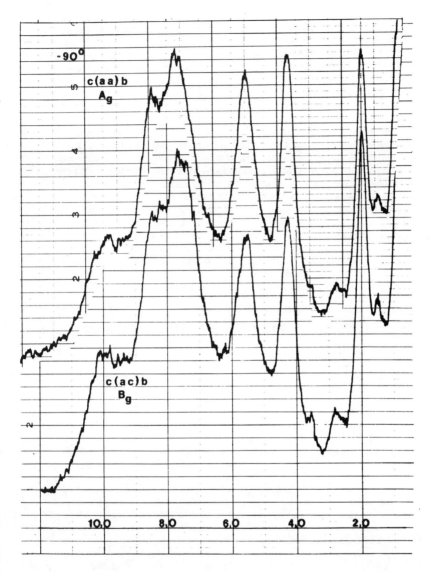

FIG. 2. Raman spectrum of PAA in the lattice vibration region. Spectrum shown is two analyzer positions for an oriented single crystal at -90°C.

samples must be very thin, and crystallites the size of the wavelengths tend to distort the bands. Moreover, the crystallization on a surface such as a quartz plate may lead to a different crystalline form, hence a different lattice vibrations spectrum from that found in the bulk sample studied in the Raman spectrum.

This is a particular problem in studying the vibrational spectra of materials such as PAA which readily assume a number of solid forms.

The observation in the Raman spectrum is what is known as a soft mode, a lattice vibration which becomes a free rotation or translation as the phase transition is approached. Such soft modes had been seen in the Raman spectra of ferroelectrics, but never in the spectra of organics.

Dr. Grunbaum and I attempted to understand this soft mode better by a calculation. In this calculation we set up rigid molecules of PAA in the lattice. The potential energy was constructed from atom-atom or group-atom potentials. This was then differentiated to obtain the force constants and harmonic frequencies for the lattice vibrations. Such calculations could be carried out routinely on crystals, and the interatomic and group-atom potentials available in the literature are a tool for vibrational spectroscopists which should be more widely used.

The results of the calculation, compared with experimental Raman and far infrared data, are shown in Table 1. The columns in the Table are organized into approximate directions of rotation and translation, keeping in mind that there is considerable mixing. The rows are representations of the C_{2h} unit cell, with A_g and B_g modes appearing in the Raman spectrum, and A_u and B_u in the infrared.

If one looks down a particular column, it is possible to see how a given lattice motion couples with the molecules in the unit cell. We can assume that for an isolated molecule in a uniform well there is one translational frequency which depends only on molecular weight, and three rotational frequencies depending on moment of inertia. As the four molecules in the unit cell couple, the potential becomes anisotropic and splitting occurs. The stronger the interaction, the greater the splitting, which is manifested by the spread in frequencies down a particular column.

As the calculation clearly shows, the highest and lowest frequency lattice modes are arising from the same coordinate, strongly coupled. Near 80 cm^{-1}, we have modes which occur at nearly the same frequency for all four representations. These are the origins of the soft modes.

Thus the lattice vibration region of the spectrum can give us information on order-disorder phenomena as well, and the information is interpretable based on calculations.

We can now summarize the picture which vibrational spectroscopy gives of the order-disorder phenomena occurring at the c-n transitions of PAA. In the crystalline phase, beginning a few degrees below the c-n transition, molecules become partially decoupled from the unit cell structure. Orientational or translational disorder, or both, are present. This occurs because the intermolecular forces in the unit cell are very anisotropic. Some of the lattice

TABLE 1. Observed and Calculated Lattice Vibration Frequencies
of PAA

Symmetry Species	Translatory			Rotatory		
Ag						
observed	30	52	70	16	74	91
calculated	28	55	62	20	69	94
Bg						
observed	30,37	52	70	16	74,90	95
calculated	37	60	69	25	82	101
Au						
observed	50	70	--	135,150	84	50?
calculated	51	69	0	130	89	55
Bu						
observed	--	--	70?	115	84?	50?
calculated	0	0	67	119	78	52

modes manifest a strong intermolecular interaction, while for
others it is very weak. It is this anisotropy which is the key to
liquid crystalline phase formation, and which makes melting a very
different process for nematogenic molecules.

Vibrational spectroscopy can illuminate this order-disorder phenomenon in quite a different way. The alkyl and alkylcyano biphenyls,
first synthesized by George Gray [7] and coworkers, are liquid
crystalline materials which are chemically very stable, and can
yield nematic and smectic phases at room temperature and below. As
a result they are important for electro-optic displays. For some
time, we have been studying one member of this series, 4,cyano,4'-
octyloxy biphenyl, known as 8OCB. This compound has a crystal-
smectic (c-sm) phase transition at 54.5°, sm-n at 67°, and n-1 at
80°. From the viewpoint of the Raman spectroscopist, this molecule
is useful because the nitrile stretching vibration, near 2220 cm^{-1},
is well isolated from other bands in the spectrum. This is relatively rare in mesogenic molecules, which tend to be of low symmetry, complex, and with numerous overlapping bands in the vibrational spectrum.

Dr. T. Kennelly and I first looked at the Raman spectrum of 8OCB,
and were somewhat surprised to see that the CN stretching frequency
was a doublet in the crystalline phase. (Figure 3). The relative

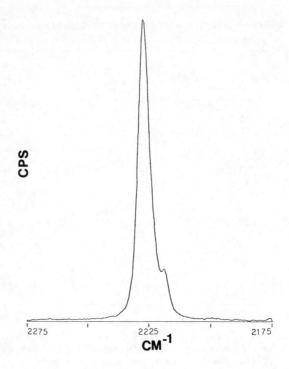

FIG. 3. Raman spectrum of crystalline 8OCB in the CN
stretching region. Spectrum is that of the stable crystalline
phase.

intensity of the two bands is about 5:1, but this varies somewhat
with samples selected at random. As the temperature was raised
close to the c-sm transition, the relative intensity of the weaker,
low frequency component, increased (Fig. 4). This is the opposite
effect from that seen in PAA, i.e., the weaker band increases in
intensity as the transition is approached. So we need to examine
in what new way the vibrational spectrum is affected by the
transition.

We now understand this weaker CN component to arise from crystal
imperfections along the direction of smectic phase formation. The
energy surface of this crystal has rather low barriers, so that
such imperfections occur readily. As we will see, further Raman
studies show that the situation is even more complex than this two
band picture would indicate.

Rather than repeat the same sort of study which we had carried out
on the alkoxyazoxy benzenes, and which had already been confirmed
for other systems (for a summary see Ref. [8]), a different
approach was taken with 8OCB. It is well known that mesogens often

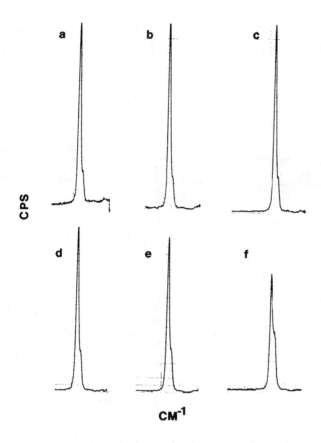

FIG. 4. Change in the intensity of the weaker band in the CN
stretching region of the Raman spectrum of 80CB near the c-sm
phase transition. Spectra a and b are well below the
transition. Spectrum c is about 2° below the transition, and
spectra d-f are taken at 0.3° progressively higher increments
of temperature near the transition, but still in the crystalline phase.

show crystalline polymorphism. This is not surprising considering
the ease with which disorder along certain directions occurs. With
80CB, it was found that the nematic liquid crystal could be
quenched into ice water to form a different crystalline phase.
This phase has a different appearance optically, i.e., it is glass-
like rather than polycrystalline. Figure 5 shows the Raman spectrum of this metastable crystalline phase in the CN stretching
region. Now a single sharp band is seen. There are also a number
of other differences, small but reproducible, in other regions of
the spectrum. Note also that both of the liquid crystalline phases
and the isotropic liquid also show single CN stretching bands.

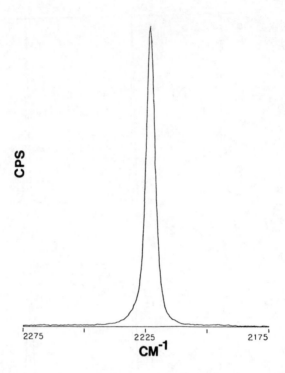

FIG. 5. Raman spectrum of the metastable crystalline form of
8OCB in the CN stretching region, showing the single peak.

We will return to a discussion of the uses of Raman spectroscopy to
understand order in these various phases in a subsequent section.
However, to continue with the discussion of the crystal-liquid
crystal phase transition, we describe now a set of experiments
which involve annealing the disordered crystal, that formed by
quenching, back to the stable crystalline phase. This annealing is
thus the reverse process from our previous heating of the stable
crystal to produce a disordered crystal.

To follow the annealing process, Dr. J. Sloan and I used Rapid
Scanning Raman Spectroscopy (RSRS). This approach is one which I
feel is so promising that it dominates all of our activity today.
The technique, originally developed in Delhaye's laboratory in
Lille [9], allows us to monitor the annealing process in a quasi-
continuous manner.

RSRS works in a mechanical fashion to scan a region of the spectrum. In our realization of the technique, done in cooperation
with the manufacturer, Jobin-Yvon, a stepping motor is set up so
that it only needs to rotate in a single direction in order to
repeatedly scan the spectrum. Almost any variable which is controllable in a regular Raman experiment can be controlled during an

RSRS scan. The entire experiment is run by a Data General Nova 2 computer, which also collects the data for later data reduction. Details are available elsewhere. [10]

The strategy for using RSRS is as follows: In the middle frequency region, we observe functional group modes, such as the CN stretching vibration of 80CB. RSRS allows us to monitor the kinetics and mechanism from the viewpoint of a particular functional group.

Now we know from many vibrational spectroscopic studies that functional groups are not very sensitive to long range order. Once one gets past a few nearest neighbors, it is unlikely that there will be much more change in vibrational frequencies. But the situation for lattice vibrations is quite different. Such modes only begin to appear when a few unit cells have been built up, and should change until the crystallite sizes have become greater than the wavelengths of light being used in the experiment. Thus, RSRS provides a possible way of distinguishing crystallization's early stages from the growth stages.

In addition to the distinction between functional group modes and lattice modes, we expected and found that RSRS would provide a means for looking at mechanism of annealing as well as kinetics. By this we mean that the crystallization probably is not a single step process, but a multi step one, and we expect to be able to catch these steps in RSRS.

Figure 6 shows the Raman spectrum of 80CB in the CN stretching region as a function of time, obtained by RSRS. The spectroscopic region studied is 2205 to 2245 cm^{-1}, and each scan takes 12 seconds. The spectrum is sampled at 0.8 cm^{-1} intervals, although the results are displayed as continuous curves. We begin, at the right hand side of the figure, with the single band characteristic of the metastable crystalline phase, and end with the two bands of the stable phase. But this happens in several stages.

First, there is a long induction period. This is truncated in the figure. This period is temperature dependent and can be used to derive an activation energy for the onset of nucleation and growth.

After some time, the intensity begins to increase, and increases over the course of 30-40 sec to a maximum, still maintaining a single band. The intensity then decreases rapidly, and during this decrease the spectrum itself changes. As seen in an expansion of this region, in Fig. 7, as many as five distinct maxima can be seen. The wavenumbers are summarized in Table 2.

By the end of this decrease in intensity, the spectrum is that of the stable crystal. However, the overall intensity begins to increase and does so for some time after the main transition is completed.

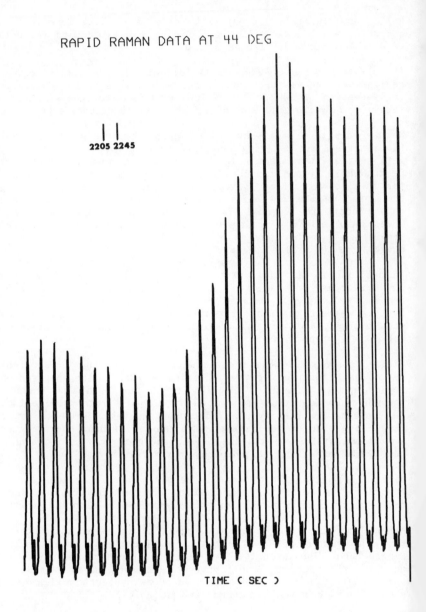

FIG. 6. Series of rapid Raman (RSRS) data for 8OCB in the CN stretching region, as molecules undergo transition from metastable to stable crystalline phase.

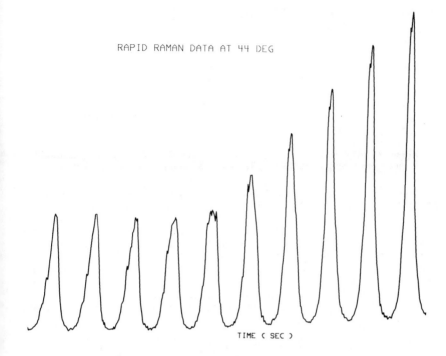

RAPID RAMAN DATA AT 44 DEG

TIME (SEC)

FIG. 7. Expansion of data from Fig. 6 in region where the main transition is occurring.

TABLE 2. Raman Spectroscopic Maxima in the 2200-2250 cm^{-1} Range of Crystalline 80CB

Metastable	Intermediate	Stable
	2217	2217
	2220	
	2222	
2225	2225	
	2228	2228

We thus observe the following sequence:

1. An induction period in which the spectrum is unchanged.
2. A period of intensity increase in which the spectrum is other-
    wise unchanged from that of the metastable form.
3. A rapid period of intensity decrease coupled with spectro-
    scopic changes from the spectrum of the metastable to that of
    the stable form, with several intermediate spectra seen.
4. A slow period of intensity increase in the spectrum of the
    stable form.

Similar results have been obtained at 42, 43, 44, 45, 46, and 47°C,
with the rate constants being directly dependent on temperature.

The data shown in Figures 6 and 7 give us the beginnings of a new
insight into the metastable crystalline phase of 80CB.

We begin with a discussion of the results in period 3. This is the
main phase transition, and we postulate that initial nucleation is
over before period 3 begins. The spectroscopic changes in this
region are then associated with growth of nuclei and phase propaga-
tion. For such a process Avrami kinetics should be obtained. This
is demonstrated for the intensity at 2225 cm^{-1} in period 3 by the
plots in Fig. 8. The series of parallel plots yield an Avrami
parameter n = 2.27 ± .07.

It is by now generally agreed that simply determining n from such
plots cannot give much insight into a nucleation mechanism or
dimensionality of growth as was once supposed. Many combinations
of nucleation/dimensionality of growth yield n values near 2. It
is nonetheless comforting that one of these is heterogeneous
nucleation with two dimensional growth, which is quite reasonable
for a crystalline phase of mesogenic material.

An activation energy for this process can be determined by well
established techniques. The value found from these data is 46.5
kal/mole. While this may seem high, it is similar to values found
some time ago by Bulkin et al. for the pretransition process.

It is significant that the number of distinct frequency maxima in
the CN stretching region, and the positions of the bands, changes
during the transitions. From a spectroscopic point of view there
can be several causes of this change. The number of bands are
primarily determined by the unit cell symmetry, or the site symmetry in a crystal. Even a reorientation of the molecules on their
sites could in principle change this symmetry, but for the CN
stretch to be affected this would need to be a significant reorientation about one of the axes approximately normal to the long
axis of the molecule.

The number of bands may also reflect a number of distinct sites.
This differs from the previous explanation as follows: In the
previous paragraph, we said that a unit cell may have several
molecules on sites of identical symmetry. Their CN groups may be

I (2225 CM-1) AVRAMI PLOT

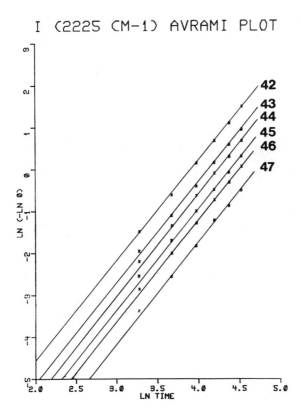

FIG. 8. Avrami plots obtained from data on main phase transition for 80CB at a series of temperatures.

coupled together to yield a number of modes. But as the phase transition progresses, molecules can move from one site to another. These distinct sites may have CN groups in slightly different mean fields, leading to different CN stretching frequencies. Of course, the complex spectrum observed may result from both of these mechanisms.

Given the formation of the metastable phase from a quenched nematic, and the difficulty of rotating molecules normal to the long axis, it is reasonable to conclude that crystallization occurs by progressive translations of molecules in a plane containing the director. These translations either bring the molecules through successive local potential energy minima until they are part of a growing phase, or there are several routes molecules follow to the growing phase, leading to the complex of bands observed along the way. In either case, the key idea is that the progress of the transition viewed from the vantage point of an oscillating CN group is: 1. Begins in an isotropic environment where it is uncoupled

to other CN groups in the crystal. 2. Undergoes successive translations, through several distinct sites where there may be interactions between oscillators. 3. Ends up in a crystalline form
which probably involves two distinct sites, one of which is more
populated than the other (see Ref. [10] for more details on this
assignment).

Turn now to a consideration of the intensity change observed in
period 4, the long time at the end of the transition. This slow
tail following the main transition does not follow Avrami kinetics.
During this period no frequency changes are seen, only intensity
building. We believe that this is an observation of the phenomenon
known as crystal perfection. A crystal, once formed with molecules
on sites, undergoes two processes. The molecules reorient slightly
to "perfect" the crystal, actually to minimize the overall lattice
energy. Crystallites may also realign slightly so as to build
overall dimensions of crystalline regions. It is interesting that
Raman spectroscopic intensities are sensitive to such a process.

Finally, we turn to period 2, in which, after a long induction,
intensity increases by about 10% during a 30-40 second period. If
one examines this as a function of temperature, the increase in
intensity always occurs in the same time period relative to the
main transition, i.e., we never observe the beginning of the main
transition, then an increase, nor do we see the increase followed
by a plateau. Always the main transition is coupled to a preceding
step which manifests itself as the intensity increase. Again the
spectrum during this period remains that of the metastable phase.

It is not possible to conclusively explain this spectroscopic event
on the molecular level without more investigation. However, we
believe that it is associated with the formation of nuclei in the
sample. During the period, the concentration of nuclei increases
and the effects of this on the crystal begin to propagate even
before the metastable form collapses to the stable crystal. We
have conducted experiments to attempt reversal of the phase transition or freezing out by quenching in period 2, but this appears to
be impossible. Such experiments always yield the final crystalline
phase. Thus once nuclei have formed it is not easy to stop their
growth.

USE OF RAMAN BAND SHAPES FOR STUDYING ORDER

There has long been interest in retrieving information about intermolecular interactions from a study of the shapes of bands in
vibrational spectra. In solution, Raman bands have half widths of
about 10 cm^{-1}, far greater than the natural linewidth, and these
widths are associated with the relaxation processes. The problem
of such a study is thus a classic inverse problem, that of calculating the relaxation processes given a particular transition and
band contour.

Pioneering work in this area was carried out by Gordon, who developed the formalism for use of correlation functions to study reorientational relaxation from infrared and Raman band shapes. If A(t) is the time dependent variable associated with the spectroscopic transition, the correlation function C(t) is defined as C(t) <A(0)* · A(t)>. Gordon showed how one could calculate a correlation function from an appropriately normalized set of vibrational spectroscopic intensities to obtain a correlation function of the molecular reorientation, itself normalized so that C(t) = 1 at t = 0. The molecular reorientations for small molecules (say 5 atoms or fewer) were shown by many authors applying Gordon's formalism to be an efficient relaxation mechanism responsible for most of the band broadening corresponding to times up to 4-5 picoseconds.

For heavier molecules (say molecular weights of greater than 70) reorientation on the 0-4 psec time scale becomes less likely, or occurs only through smaller angles. In this situation other routes to the relaxation arise. In particular, a group of mechanisms collectively called vibrational relaxation begin to occur. Nafie and Peticolas [11] showed how one can separate vibrational and reorientational relaxation from Raman polarization measurements.

As molecules become still larger, it would be expected that reorientation would not take place on the picosecond time scale at all. This might be expected to narrow the bands as the result of a relaxation mechanism disappearing. In fact, this is not observed. For a wide variety of molecules, ranging from small molecules to polymers, in a variety of condensed phases at room temperature (e.g., neat liquids, solutions, liquid crystals, solids) the half widths remain at 10 cm^{-1} to within a factor of 2-3. Indeed, this small range of half widths is what makes infrared spectroscopy so useful as a qualitative analysis tool.

What must be occurring is the rise of other relaxation mechanisms as reorientation becomes less efficient. This section deals with these vibrational relaxation mechanisms in the large molecule case.

Vibrational relaxation mechanisms have been classified in various ways. Probably the most useful of these is the division into T_1 and T_2 mechanisms. The former are energy relaxation, while the latter are dephasing. Each of these may occur by several different routes. Indeed, it is possible that not all of these routes are yet known. Energy relaxation, for example, may occur to like molecules (resonance energy transfer), to solvent, or internally. Dephasing may take place by several mechanisms as well.

Dr. Kenneth Brezinsky and I began to work in this area with investigations of the correlation functions of 8OCB obtained in a variety of solvents and phases, focussing on the well isolated CN stretching mode. Later, Dr. Ellen Miseo and I extended this work to other bands and to an examination of the different mechanisms involved.

In these studies, the calculation procedure is as follows: The vibrational spectrum of a band centered at v_0 with full width at half height $v_{\frac{1}{2}}$, is measured from $v_0 - 5v_{\frac{1}{2}}$ to $v_0 + 5v_{\frac{1}{2}}$. However, if there is interference from overlapping bands, either side can be measured from the center and the results reflected to create the full data array.

To interpolate additional points in the correlation function, we fill the array of data points with zeroes to double its length.

Two arrays are measured in the Raman spectrum, I_{vv} and I_{vh}. The vibrational relaxation portion is calculated as

$$I_{vib} = I_{iso} = I_{vv} - \frac{4}{3} I_{vh}$$

The correlation function for vibrational relaxation is then

$$C(t)_{vib} = \int_{-\infty}^{\infty} I_{vib}(\omega)e^{-i\omega t} \, d\omega$$

To check for reorientation, we define

$$C(t)_{vh} = \int_{-\infty}^{\infty} I_{vh}(\omega)e^{-i\omega t} \, d\omega$$

and $C_{rot}(t) = C_{vh}/C_{vib}$.

From the band intensity data, we compute a second moment M_2

$$M_2 = \int_{-\infty}^{\infty} I(v)(v-v_o)^2 dv / \int_{-\infty}^{\infty} I(v)dv$$

From the correlation function, a correlation time is computed

$$\tau = \int C(t)dt$$

The correlation time for C_{vib} is τ_v. τ_v and M_2 allow calculation of a collision time τ_c

$$\tau_c = 1/\langle\omega^2\rangle\tau_v$$

and $\langle\omega^2\rangle = 4\pi^2 c^2 M_2$. Combining these, the Kubo function for pure dephasing processes is calculated.

We have examined [12] the correlation functions obtained from the Raman spectrum of 80CB in the CN stretching vibration region (2225 cm^{-1}) in crystalline, liquid crystalline, isotropic liquid, and solution phases. It was shown that:

a. the correlation functions were pure vibrational relaxa-
    tion, with no reorientational component, by the criterion
    of Nafie and Peticolas [11].
b. when the correlation function for the isotropic neat
    liquid was plotted as log C(t) vs. time, it was quite
    non-linear.
c. in dilute solution in CCl_4 or benzene, the form of log
    C(t) changed and it became linear to 4 psec.
d. by contrast, in $CHCl_3$ or CH_3SCN, dilution did not affect
    C(t).

It is possible to build an interpretation of these results in terms
of the theory of T_2 band shapes developed by Kubo [13] and extended
by Bratos [14], which several workers, originally Rothschild [15]
have applied to vibrational spectra. For our purposes, the most
important equation describes the correlation function in terms of
two quantities which can be derived from the measured data. These
are $\langle\omega^2\rangle$, calculated from the second moment of the band, and τ_c, a
collision time, which is related to the correlation time for vibra-
tional relaxation by $\tau_c = 1/\langle\omega^2\rangle\tau_{vib}$. The so-called Kubo equation
is then

$$C(t) = \exp[-\langle\omega^2\rangle(\tau_c^2(\exp(-t/\tau_c)-1) + t\tau_c)]$$

Most workers have looked at two limits of this equation, $\langle\omega^2\rangle^{\frac{1}{2}}\tau_c \ll$
1, known as fast modulation, and $\langle\omega^2\rangle^{\frac{1}{2}}\tau_c \gg 1$, the slow modulation
limit. These are convenient because in the fast and slow modula-
tion limits the band shapes become Lorentzian and Gaussian, respectively. These cases are easy to recognize from the correlation
functions by plotting log C(t) vs. time, yielding a straight line
for a Lorentzian band and a parabola for a Gaussian.

With this background, we interpreted the results described above as
others have done [15], by saying that in the neat liquid there are
many structures in the clusters around the CN group. This was
thought by us to be the slow modulation condition of dephasing, in
which oscillators must relax in a semi-rigid lattice on the picosecond time scale.

In dilute benzene or carbon tetrachloride solution, particularly at
times beyond 0.7 psec, the clusters of 80CB molecules were thought
to be broken up. This leads to progressively larger numbers of the
CN groups being surrounded by solvent molecules. These do not
interact strongly with the CN group. Motional averaging occurs on
the psec time scale, leading to fast modulation.

This approach is confirmed in the more polar solvents, $CHCl_3$ and
CH_3SCN where the local structures of the 80CB molecule are replaced
by those arising from the strongly interacting solvent.

Thus we believed that our initial results could be interpreted
completely in terms of dephasing mechanisms. Subsequently, Miseo's
work showed that when we attempt to quantify the results in terms

of the Kubo equation the qualitative interpretation given above does not hold up. It is necessary to invoke T_1 mechanisms as well.

This is a result of examining a number of additional bands as well as comparing the previous results with the Kubo equation. In particular, Miseo obtained:

1. Results for two additional bands in isotropic liquid and solution phases. The 1600 cm^{-1} band is a C-C stretching mode of the biphenyl moiety, primarily of the bond joining the two rings. The 1175 cm^{-1} band is a C-H out of plane bending vibration. Both are approximate descriptions of these modes.

2. For each case studied in this and in previous work, the second moments and correlation times have been computed.

3. Infrared spectra of the CN stretching mode for several cases, and correlation functions computed from them.

Figure 9 shows the correlation functions obtained from three bands (2225, 1600, and 1175 cm^{-1}) in the spectrum of 80CB in the neat isotropic phase at 85°C. The relaxation is rapid and non-exponential. All three modes show identical relaxation behavior within experimental error. Using the method developed by Nafie and Peticolas, a correlation function C_{vib} may be computed from the polarized Raman spectra. This computation shows that the correlation functions of Fig. 9 are completely vibrational relaxation correlation functions, with no reorientational component.

Figure 10 shows the correlation function for the 1600 cm^{-1} band of 80CB in $CHCl_3$ at several concentrations. The isotropic phase correlation function is included for visual reference. As concentration decreases, the relaxation slows and appears to become more nearly exponential at longer times. This is in contrast to the behavior observed for the 2225 cm^{-1} band in $CHCl_3$ solution. Figure 11 indicates this most clearly by comparing the correlation functions for the three bands at 0.37 mole fraction in $CHCl_3$ solution.

This difference in behavior is also seen in methylthiocyanate. As in $CHCl_3$, one sees a slower, more nearly exponential decay at 1600 and 1175 cm^{-1} than at 2225 cm^{-1} in CH_3SCN. In CCl_4, however, the concentration behavior of the correlation functions for the three bands is virtually identical. In each case, lowering the concentration slows the relaxation markedly at times beyond 1.5 psec.

Most authors have interpreted results such as those obtained here in terms of dephasing being the dominant or at least the predominant relaxation mechanism. This interpretation has usually been built around the goodness of fit of the results to the Kubo equation. In certain cases our results do fit well with the Kubo equation, but in others there is a significant deviation at longer times. For all cases studied by us, the deviation is always that the experimentally observed correlation function relaxes more rapidly than the Kubo equation would predict.

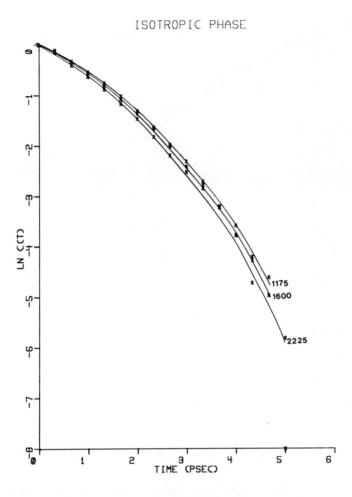

FIG. 9. Correlation functions for three bands of 8OCB obtained in the isotropic phase.

We interpret this additional relaxation as being due to energy
relaxation, without making any assumption at this point as to
whether the energy relaxation mechanism is inter- or intramolecular.

LIPID-WATER GELS

A transitional case of order-disorder transitions in organic
materials between small molecules and polymers is lipids. Indeed,
many workers have gone from the study of lipids to polymers such as
polyethylene because of the relationship in the spectroscopic
problems involved.

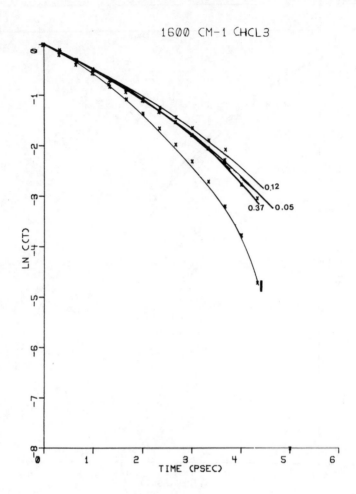

FIG. 10. Correlation function for the 1600 cm-[1] band of 80CB
at a series of concentrations in $CHCl_3$. The isotropic phase
function is included for visual reference.

Lipid-water gels are important to a number of central issues involving membranes for biological systems, and transport across such
membranes. So much work has been done in this field in the past
decade that it is hard to realize how little was firmly established
about these systems and their phase transitions as recently as
1970. Raman spectroscopy, and more recently infrared spectroscopy,
have played a very useful role in this process. They are likely to
continue to do so as more complex systems are studied.

Our work on lipid-water gels began early in this process, and ended
before the main force of vibrational spectroscopy was brought to
bear. In the early 1970's, Dr. Krishnamachari and I were able to

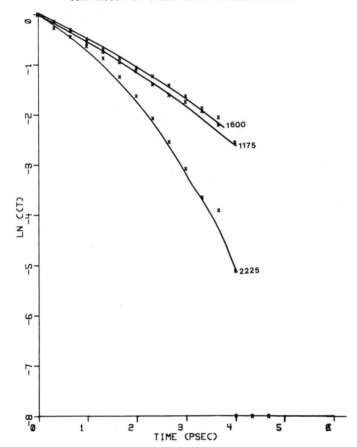

FIG. 11. Comparison of the correlation functions for 8OCB in CHCl$_3$ for three different bands, showing how those for 1175 and 1600 become nearly linear on the logarithmic scale, while the 2225 cm^{-1} band does not change.

show that several bands in the infrared and Raman spectra of various phospholipids were sensitive to thermal phase transitions observed in the scanning calorimeter [16]. With time, interpretation of these effects on both CH and CC bond motions emerged, and it was clear that the vibrational spectrum was following a melting of the hydrocarbon chains of the lipid, going from a nearly all trans chain to one with a high population of gauche conformers.

This melting manifests itself in both expected and, to us, somewhat surprising ways in the spectra. For example, the CC stretching vibrations on the 1000-1150 cm^{-1} region change from a split peak to

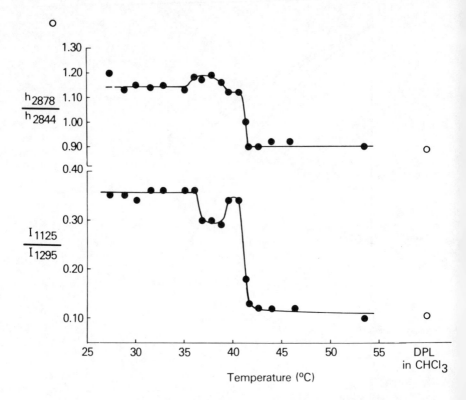

FIG. 12. Raman intensities for diplamitoyl lecithin in the CH and CC stretching regions as a function of temperature. Sample was a gel containing 20% water.

a broad central peak, as one would expect in chain melting. There are also changes in half width of the methylene deformation vibrations in the 1440-1500 cm-^{1} region of the infrared spectrum.

Somewhat surprising is the considerable change observed in the CH stretching region during the phase transition. As vibrational spectroscopists, we were brought up on the logic that carbon-carbon bonds are a poor coupler of information about carbon-hydrogen bond vibrations. This led us to expect that changes in dihedral angle about the carbon-carbon bonds would not be seen very much in the C-H stretching region. This turned out not to be the case.

Figure 12 shows an example of the sensitivity of the C-H stretching region to temperature for diplamitoyl phosphatidyl choline. Not only is the main melting transition seen, but a lower temperature transition, known for some years as the pretransition, is also observable. This transition is now understood to involve a rotation of apposed chains in the lipid bilayer, as well as a rippling

of the bilayer. Vibrational spectroscopy is one of the few techniques which can provide molecular level information on such a
phase transition.

ANNEALING OF POLYMERS

Polymers show very similar phenomena in the infrared and Raman
spectrum to liquid crystals, as order-disorder transitions occur.
In this section I will describe some recent work by Dr. M. McKelvy
and Mr. F. DeBlase, which we are carrying out. Professor M. Lewin
of the Israel Fiber Institute is also a collaborator on these
studies.

The order-disorder transition to be discussed here is the crystallization of an amorphous solid polymer. Such phases can be produced by heating a polymer to the melt, and then quenching the
sample in ice water. If such a sample is heated at appropriate
temperatures, depending on the particular polymer, crystallization
takes place. This crystallization manifests itself in many changes
in properties, including density, mechanical properties, thermal
properties, etc. It is possible to follow the progress of the
development of the crystalline phase by a variety of techniques,
but few of them give information at the molecular level. To date
most kinetic studies of crystallization of polymers have used bulk
measurements such as dilatometry. Our studies have focussed on
polyesters. In particular I shall describe here some work on
poly(ethylene terephthalate), known as PET or 2GT, using Raman
spectroscopy as the probe. Several other papers by Koenig and
coworkers have discussed complementary work using infrared spectroscopy.

Figure 13 shows the Raman spectrum of PET, a bulk sample before and
after annealing. This indicates what are by now well known changes
which occur in the spectrum as crystallinity develops [17-20]: The
carbonyl stretching vibration narrows considerably, with half band
widths decreasing from a maximum of about 30 cm^{-1} in the most
amorphous samples to about 10 cm^{-1} in highly crystalline samples.
A new band grows in at 1092 cm^{-1}. It has also been observed that
there are intensity changes in the band at 273 cm^{-1}.

These sorts of changes are of a type with which we are already
familiar in vibrational spectroscopic studies of order-disorder
transitions. The half band width change for the carbonyl
stretching vibration is usually understood in terms of a restric-
tion on the number of distinct environments of the carbonyl groups.
This occurs in two ways. The carbonyl group conformation becomes
more restricted as crystallization occurs, with the groups lying
predominately in the plane of the phenyl ring, in a trans conformation to each other (on opposite sides of the ring), as contrasted with a number of different angles in the amorphous
material. In addition, the environment surrounding carbonyl groups
becomes more uniform as the chains pack into an ordered array of
unit cells. It is also possible that some of the narrowing of band
occurs as a result of selection rules. The maximum frequencies for

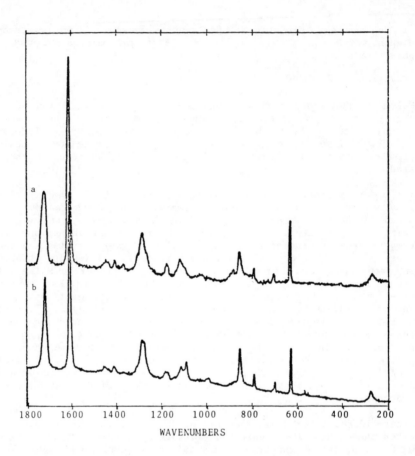

FIG. 13. Raman spectrum of a. amorphous and b. highly annealed, crystalline PET in the 200-1800 cm- region.

the infrared and Raman active carbonyl stretching vibrations of crystalline PET have been reported to differ by 2-4 cm^{-1}. In the amorphous sample, there is no symmetry around the ring and all carbonyl stretching vibrations are active in both spectra. In the crystalline spectrum, the center of symmetry reduces the Raman and infrared activity to the symmetric and asymmetric stretching vibrations, which are however, weakly coupled across the ring.

The growth of the band at 1092 cm^{-1} has been less easy to understand, despite many calculations. However, it is possible to conclude that it is connected to the conformational change which occurs in the glycol portion of the polymer repeat unit. This section undergoes a change from a mixture of <u>gauche</u> and <u>trans</u> in the amorphous material, to <u>trans</u> in the crystalline. The 1092 cm^{-1} band is associated with the <u>trans</u> linkage.

The crystallization of polymers differs from that of small molecules in many ways. Most important is that one never achieves high degrees of crystallinity in polymers such as PET, although one can in other special cases, such as polyethylene. Thus, a highly annealed sample of PET may still contain large quantities of gauche conformer (say 40-60%). This needs to always be kept in mind when examining the vibrational spectra of such systems.

Another important feature of the discussion already given for PET needs to be emphasized as well. The crystallization process involves internal changes in the chains as well as interchain alignments. Over many years vibrational spctroscopy has often been used to examine crystallinity in polymers, and certain bands become known as "crystallinity bands". In fact, some of these bands, probably most of them, are not associated with long range order at all, but with the conformational change internal to the chain, which is a part of the crystallization process.

When one studies fibers, this situation becomes even more complex. Now there are three sorts of changes which occur in the vibrational spectrum: crystallinity, conformational change, and chain orientation. This is because it is possible to produce fibers with a significant degree of oriented amorphous material. The separation of these different effects is only now beginning to be possible.

Dr. McKelvy and I have studied the crystallization process of PET using RSRS. Figure 14 shows the evolution of the band at 1092 cm^{-1} for a bulk sample. From data such as these, it is possible to obtain kinetic information. Once again, as in the case of 80CB, we find that the crystallization process is a multi-step one. Stage 1, not shown in Figure 14, is a long induction period. Temperature dependence studies of this period can be used to derive an activation energy for nucleation.

Following the induction period, there are three successive stages (2-4) which are seen most clearly in reduced data such as the Avrami plot of Fig. 15. In stage 2, the band grows in rapidly. Then in stage 3, the process comes to a nearly complete stop. After some time, this barrier is overcome, and the process continues. We have discussed the possible explanation of this complex crystallization process in another publication [21], and intend here only to point out that once again, for polymers, vibrational spectroscopy is able to provide rather detailed information on the transition.

Kinetic data such as these allow us to examine the behavior of different functional groups during the annealing process. Thus we have contrasted the kinetics as viewed from the glycol linkage with those seen from the perspective of the carbonyl group. I believe that there is no other technique which currently allows such information to be obtained. The data are collected with a time resolution of just a few seconds, in phases varying from glassy to highly crystalline, and with good structural resolution. Moreover,

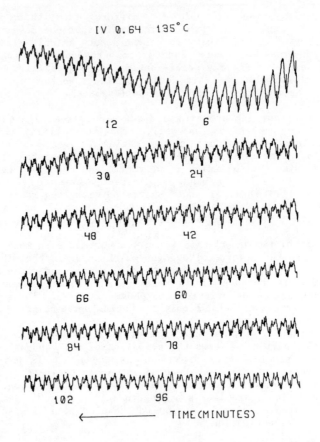

FIG. 14. RSRS data for PET annealed at 135°C, showing evolution of the 1092 cm-[1] as a function of time (45 sec per scan). Sample has an intrinsic viscosity of 0.64.

it should be possible to extend such measurements to crystallization from melts or solution.

But it is possible to go beyond the functional groups in using the Raman spectrum as a probe of annealing. By extending measurements to the low wavenumber region (below 150 cm-[1]) it should also be possible to study lattice vibrations. Surprisingly, there has been almost no work in the past on the low frequency spectra of PET.

Figure 16 shows the Raman spectrum of an amorphous and a crystalline sample contrasted. It is clear that annealing produces a major change in this region. New bands at 129 and 73 cm-[1] grow in. Moreover, there are probably other bands, unresolved from the background, which grow in as well. It is likely that these modes

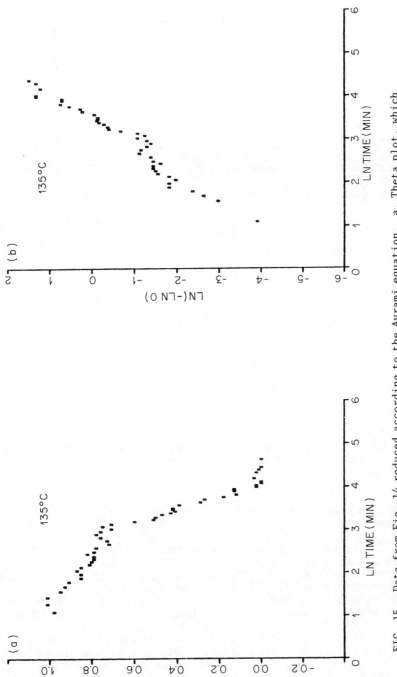

FIG. 15. Data from Fig. 14 reduced according to the Avrami equation. a. Theta plot, which normalizes the intensity change, and b. the so-called Avrami plot.

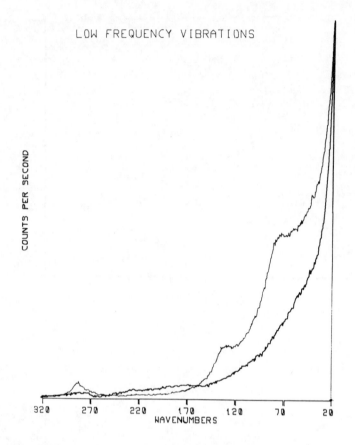

FIG. 16. Raman spectrum of PET in the low frequency region.
The lower trace is an amorphous sample, the upper trace is a
highly annealed sample.

arise from both internal torsional modes of the <u>trans</u> chains, and
lattice vibrations.

It is also useful to contrast the spectra of fibers with those of
bulk samples in the region. Figure 17 shows the low frequency
Raman spectrum of a fiber of PET which is spun with a take-up speed
of about 3300 m/min. Under these conditions, one produces fibers
which are still highly amorphous, but show significant degrees of
orientation, as measrued by birefringence. These fibers are known
as partially oriented yarns or POY. Figure 17 gives an indication
of the effect of this orientation on the Raman spectrum. The
spectrum of the POY shows a greater scattering in this region than
does the bulk sample. Figure 18 makes this more visible by
plotting the difference between the two spectra. The spectra have
been normalized to make the scattering equal at 20 cm-[1], so that

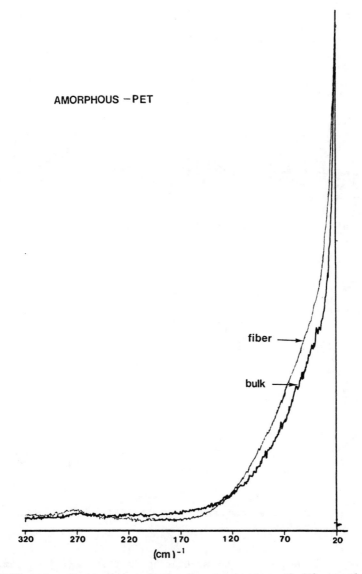

AMORPHOUS −PET

fiber ⟶

bulk ⟶

320   270   220   170   120   70   20

(cm)⁻¹

FIG. 17. Comparison of the Raman spectra of fiber and bulk amorphous PET samples.

the "band" seen in the difference spectrum is dropping off steeply, perhaps artificially so, on the low frequency side. However, it must, in any case, be a broad scattering which is unsymmetrical to the high frequency side. This broad scattering disappears when these fibers are annealed with their ends free, a process which also decreases the birefringence markedly, indicating a loss of

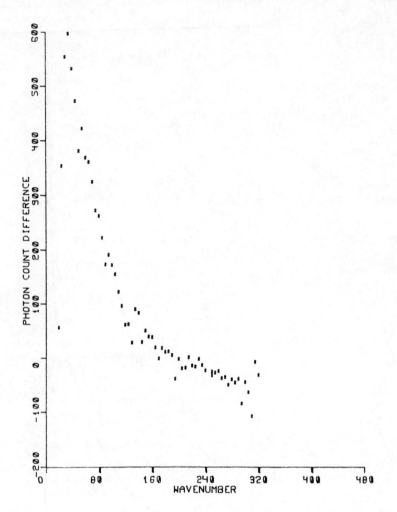

FIG. 18. Difference spectrum obtained by subtracting bulk from fiber data.

amorphous orientation. The scattering in this region is reminis-
cent of that seen in the low frequency region of many materials in
their liquid crystalline phases.

CONCLUSION

In the development of vibrational spectroscopy, there has been a
constant interleaving of advancing basic understanding of the
vibrational spectrum using more and more sophisticated techniques
on small molecules, and application of this understanding to
chemical problems. Each experimental or theoretical advance has

expanded the horizons for the range of applications which are accessible.

In this paper, I have tried to illustrate this through the example of order-disorder phenomena in complex organic materials. Thirty years ago, many of these spectra would not have even been possible to acquire, let along analyze. Yet now, it is possible to acquire high quality data and to derive detailed, often quantitative information from those data which contribute substantially to the problem areas involved.

The analytical techniques available to the chemist have continued to expand. Vibrational spectroscopy, one of the oldest of the new techniques, still has much to offer. This paper has tried to emphasize the combination of studies which cross phase boundaries, in some cases involving great differences in molecular mobility, over a range of temperatures. In most cases, the data are acquired on the time scale of seconds to a few minutes. Both magnetic resonance and X-ray techniques, which provide more direct interpretation and often superior structural resolution, have difficulty in cases such as these.

I look forward to being a part of the evolution of vibrational spectroscopy in the coming decades, and to a continuing vital role for the Coblentz Society in this process.

ACKNOWLEDGMENTS

The work described in this paper has been supported by the American Cancer Society and by the Polymers Program of the National Science Foundation, Grant DMR-8304220. I am grateful to my many collaborators whose contributions are mentioned.

REFERENCES

1.   D. A. Dows, R. Fox, M. Labes, and A. Weissberger, Chemistry and Physics of the Organic Solid State, Interscience, New York, 1963.
2.   J. E. Mayer and S. F. Streeter, J. Chem. Phys., 7, 1019 (1939).
3.   G. H. Brown, W. G. Shaw, Chem. Rev., 57, 1049 (1957).
4.   G. W. Gray, "Molecular Structure and the Properties of Liquid Crystals", Academic Press, New York, 1962.
5.   B. J. Bulkin, F. T. Prochaska, and D. Beveridge, J. Chem. Phys., 55, 5828 (1971).
6.   W. R. Krigbaum. Y. Chatani, and P. G. Barber, Acta Cryst. B26, 97 (1970).
7.   G. W. Gray, K. J. Harrison, and J. A. Nash, Liquid Crystals and Ordered Fluids, 2, 617 (1974).
8.   B. J. Bulkin, Vibrational Spectroscopy of Liquid Crystals, in R. Hester and R. Clark, eds., Adv. Infr. and Raman Spectr., Heyden, 1980 pp. 151-176.
9.   J. M. Beny, B. Sombret, and F. Wallart, J. Mol. Str., 45, 349 (1978).

10. J. Sloan, Ph.D. Thesis, Polytechnic Institute of New York, 1982.
11. L. A. Nafie and W. L. Peticolas, J. Chem. Phys., 57, 3145 (1972).
12. B. J. Bulkin and K. Brezinsky, J. Chem. Phys., 69, 15 (1978).
13. R. Kubo, in Fluctuation, Relaxation, and Resonance in Magnetic Systems, D. TerHaar, ed., Plenum, New York, 1962.
14. S. Bratos and E. Marechal, Phys. Rev., A4, 1078 (1971).
15. W. G. Rothschild, J. Chem. Phys., 57, 991 (1972).
16. B. J. Bulkin and N. Krishnamachari, J. Amer. Chem. Soc., 94, 1109 (1972).
17. A. J. Melveger, J. Polym. Sci., A2, 317 (1972).
18. F. J. Boerio, S. K. Bahl, and G. E. McGraw, J. Polym. Sci. Polym. Phys. Ed., 14, 1029 (1976).
19. I. M. Ward, Chem. Ind. Long., 1102 (1957).
20. C. Y. Liang and S. Krimm, J. Molec. Spectr., 3, 554 (1959).
21. B. J. Bulkin, M. Lewin, and M. L. McKelvy, Spectrochim. Acta, (in press).

STRUCTURAL AND DYNAMICAL PROPERTIES OF MODEL AND INTACT MEMBRANE
ASSEMBLIES BY VIBRATIONAL SPECTROSCOPY

Ira W. Levin

Laboratory of Chemical Physics
National Institute of Arthritis, Diabetes, and
    Digestive and Kidney Diseases
National Institutes of Health
Bethesda, Maryland 20205

INTRODUCTION

Since the introduction nearly fifteen years ago of the basic fluid
mosaic model for the arrangement of functioning components in
biological membranes [1], numerous areas of biophysics and biochemistry have vigorously responded to the challenges presented in
unraveling the complex behavior exhibited by a wide variety of
lipid-protein aggregates. In particular, the versatility of vibrational spectroscopy in monitoring the conformational, dynamical and
packing properties of both model and intact membrane assemblies has
led to recent dramatic successes in understanding the interactions
of, for example, proteins, polypeptides, antibiotics and anes-
thetics with the lipid bilayer matrix [2-5]. The advantage of the
vibrational technique over many other physical methods lies in the
sensitivity of innumerable spectral features to bilayer reorganizations; thus, one has the ability to probe specific bilayer perturbations within a given structural region of the membrane. In this
discussion we apply vibrational spectroscopy toward clarifying the
rich polymorphic behavior of biomembranes by first examining the
Raman spectral responses to the thermotropic behavior of model
membrane systems comprised of aqueous dispersions of either symmetric or asymmetric chain phospholipid molecules. We will then
proceed to more complicated membrane systems and investigate, also
by Raman spectroscopic procedures, the contrasting bilayer effects
of polymyxin B sulfate, a peptide antibiotic, in reconstituted
bilayers composed of either neutral or acidic phospholipid
molecules. The diversity of interactions generally available to
bilayer components is emphasized in the thermal behavior of this
particular lipid-polypeptide recombinant in both the ordered gel
state and the higher temperature, fluid liquid crystalline phase.
As a final example, the C-H stretching mode intervals of the
infrared spectra of the natural membranes of coated vesicles,
uncoated vesicles and synaptic membranes isolated from bovine brain
tissue will be discussed in the context of the unique properties
and functions of the protein coated pit regions of cellular membranes.

MEMBRANE BILAYER ORGANIZATION

Because the distribution of bilayer lipids varies signficantly between membranes of different sources, generalizations regarding the lipid composition of biomembranes are difficult to frame. Phospholipids, however, tend to predominate and constitute approximately 40 to 90% of the total dry weight of a membrane [6]. In particular, the zwitterionic phosphatidylcholine species represents perhaps the most common lipid present in animal cell membranes. For the purposes of most of the present discussion, we will be primarily concerned with saturated, diacyl phosphatidylcholine bilayer dispersions. Since these molecules exhibit amphipathic properties, the presence of water spontaneously leads to thermodynamically stable lamellar structures, or liposomes, in which the polar headgroups face outward toward the aqueous medium, while the acyl chains form the hydrophobic bilayer core. In general, model systems composed of multilamellar dispersions prove to be extremely tractable for spectroscopically investigating bilayer behavior since the membranes can be easily perturbed either by thermal changes or by the introduction of additional bilayer components. These liposomes, which form concentric lamella, are several microns in diameter, with the individual lamella, or onionskins, separated by several layers of water. Depending upon lipid composition, pH and temperature, nonbilayer structures relevant to biological membranes may be generated for spectroscopic observation.

The physical properties of aqueous dispersions of synthetic phosphatidylcholine bilayers have been studied extensively by a wide variety of physical techniques. These multilamellar systems undergo a highly cooperative endothermic phase transition from the $L_{\beta'}$ gel phase to the L_{α} liquid crystalline state when the bilayers are heated above a characteristic temperature T_m. For dipalmitoylphosphatidylcholine (DPPC or diC(16)PC) multilayers, a phospholipid with chain lengths of sixteen carbon atoms, this first order phase transition occurs at ~41°C and is thermodynamically described by a marked increase in entropy; conformationally, the phase transition is accompanied by a sudden increase in intramolecular acyl chain disorder and a decrease in lateral, chain-chain interactions. These highly disordered chains exhibit a time averaged orientation perpendicular to the bilayer plane. In addition to the primary phase transition, the gel phase of DPPC displays an interesting polymorphism by undergoing at least two gel→gel order/disorder transitions. For dispersions incubated at 0°C for several days, a subtransition is observed at ~18°C, which represents a transition from a dehydrated lamellar crystalline L_C phase to the more hydrated lamellar, but metastable, $L_{\beta'}$ gel phase. The nearly all-trans, tilted phospholipid acyl chains probably pack in a highly ordered orthorhombic subcell lattice in the lower temperature crystalline phase. In passing to the higher temperature gel phase, the hydrocarbon chains remain in their tilted, extended conformations, but disorder rotationally, forming a distorted hexagonally packed lattice. The next low temperature phase change, or pretransition, occurs at ~35°C. In this $P_{\beta'}$ phase the one dimensional lamellar structure transforms into a two

dimensional lattice in which the bilayer ripples. In the pre-
transition phase the acyl chains pack in a hexagonal lattice.
Recent x-ray diffraction results suggest that in this phase the
chains may orient parallel to the bilayer stacking axis. The
enthalpy of the L_c to L_β', gel→gel subtransition of DPPC has been
reported as 3.2 Kcal/mole, while the enthalpies of the pretransition and main gel to liquid crystalline phase transition are
approximately 1.09 and 8.5 Kcal/mole, respectively, for multilamellar dispersions.

PHASE TRANSITION BEHAVIOR OF SYMMETRIC CHAIN PHOSPHOLIPID
DISPERSIONS: C-H STRETCHING MODE REGION SPECTRA

As described in considerable detail elsewhere [2], Raman spectra of
the aqueous phospholipid dispersions in the C-H stretching mode
region are particularly sensitive to the conformational and lattice
order of the lipid acyl chains. The most dramatic spectral changes
occur at the highly cooperative gel to liquid crystalline phase
transition, an endothermic event involving both intrachain
trans/gauche isomerization and a lateral expansion of the hydrocarbon chain lattice. These structural reorganizations are spectroscopically reflected by an increase in Raman intensity in the
2935 cm^{-1} region accompanied by a decrease in intensity of the 2884
cm^{-1} gel phase feature. We briefly note here that the 2850, 2884
and ∼2935 cm^{-1} features provide extremely useful spectral markers
for monitoring the dynamical, packing and conformational properties
of lipid bilayers. In particular, the three spectral transitions
are assigned, respectively, to the acyl chain methylene C-H symmetric stretching modes, the methylene C-H asymmetric stretching
modes and, in part, a Fermi resonance component of the acyl chain
terminal methyl C-H symmetric stretching mode. Figure 1 displays
the Raman spectra for diC(20)PC multilayers in the gel state at
51.4°C and in the liquid crystalline phase at 75.4°C. The apparent
intensity and frequency changes in the 2884 cm^{-1} feature on
increasing temperature arise predominantly from the disappearance
above T_m of an underlying broad background band. This background
has been assumed to arise from a Fermi resonance interaction
between the symmetric methylene stretching fundamentals at ∼2850
cm^{-1} and the continuum of binary combinations of the ∼1450 cm^{-1}
methylene scissoring modes of the nearly all-trans, extended acyl
chains [7,8]. Thus, in the liquid crystalline phase the 2884 cm^{-1}
linewidth increases several fold, while the Fermi resonance interaction involving the 2850 cm^{-1} lines is weakened because of the
vibrational decoupling introduced by the formation of gauche chain
conformers. Since the empirical peak height Raman intensity ratio
I_{2935}/I_{2880} is a measure of the interchain and intrachain
order/disorder processes of the bilayer lipid acyl chain, we will
use this derived spectral parameter to compare in Fig. 2 the series
of phospholipids from diC(14)PC to diC(22)PC. Although the C-H
stretching mode region is a congested spectral interval encompassing complex sets of vibrational transitions, the I_{2935}/I_{2880} and
I_{2850}/I_{2880} peak height intensity ratios provide extremely sensitive probes for monitoring conformational chain disorder simultaneously with intermolecular chain-chain disorder. Since different

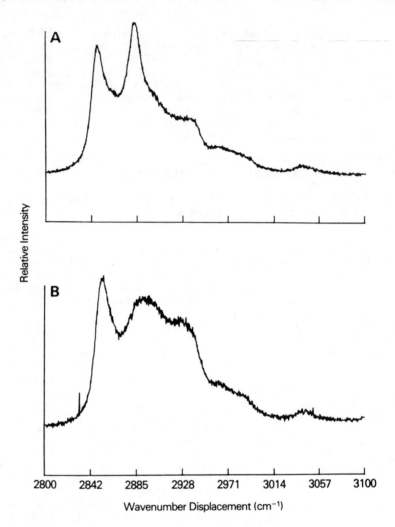

FIG. 1. Raman spectra of diC(20)PC bilayer dispersions in the 2900 cm^{-1} C-H stretching mode region for (A) the gel state at 51.4°C and (B) the liquid cyrstalline phase at 75.4°C.

selection rules are applicable to the infrared spectra of the extended gel phase acyl chains, it is more convenient to monitor either frequency shifts in the 2850 and ∿2920 cm^{-1} methylene C-H symmetric and asymmetric stretching mode regions, respectively, or a linewidth characteristic of the 2850 cm^{-1} feature of the infrared spectrum of the liposomal dispersion [9]. Although we are dealing with large, complex phospholipid molecules, their vibrational spectra are greatly simplified when viewed in the context of the spectra of even or odd hydrocarbon chains.

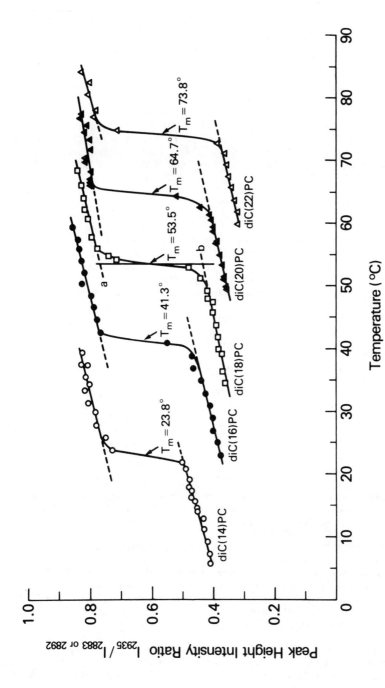

FIG. 2. Temperature profiles for saturated, symmetric chain phosphatidylcholine dispersions using Raman spectral peak height intensity ratios derived from the C-H stretching mode region.

In this discussion we derive the main phase transition temperature T_m from a temperature profile constructed from the Raman intensity ratios. T_m is operationally defined as the midpoint temperature at which the intensity ratio equals 1/2 (a+b), where in Fig. 2, a and b are the Raman ratios at a given temperature in the extrapolated linear portions of the upper and lower ends of the sigmoidal transition curves. The T_m's may also be determined systematically by fitting the intensity ratios defining the temperature profile to an analytical expression reflecting a two state model for the gel to liquid crystalline phase transition [10]. The two state thermo-dynamical model, however, conveys no information on the molecular nature or origin of the phase transition. One can, however, extract from the analytical fits of the Raman data the van't Hoff enthalpies, quantities usually determined from calorimetric measurements. The values of the subsequently derived cooperative units undergoing the melting phenomenon have in this specific model a slightly different interpretation from that usually obtained from calorimetric data. This aspect arises since our model assumes a distribution of domain sizes that is independent of temperature. These considerations and derivations will be discussed elsewhere. As shown in Fig. 2, the transition temperatures are 73.8, 64.7, 53.5, 43.1 and 23.8°C, respectively, for the lipid dispersions of diC(22)PC, diC(20)PC, diC(18)PC, diC(16)PC, and diC(14)PC [11]. These values, summarized in Table 1, agree very well with the data determined by scanning calorimetry.

In general, the lower temperature pretransition is not revealed prominently in the temperature profiles constructed from the I_{2935}/I_{2884} peak height intensity ratios. Figure 3 displays the temperature curve for the diC(12)PC liposomes in which two transi-tions are, however, conspicuously observed at -4.3 and 3.3°C. The lower transition, which is extremely dependent on sample history, is sharper than the upper transition. Since the higher temperature order/disorder transition is essentially insensitive to the thermal history of the sample, analogous to the behavior exhibited by the primary transition of all the longer chain diacyl systems, the -4.3 and 3.3°C order/disorder transitions are associated with the pre-transition and main transition, respectively, of the diC(12)PC multilayers [11]. This system represents an interesting example for which the vibrational data corrected and clarified the con-clusions derived from calorimetric determinations [12].

Figure 4 displays the temperature profile for the diC(10)PC disper-sions in which a phase transition is observed at -8.5°C. This is ostensibly a surprising result since the changes in enthalpy, entropy and molar volume are estimated to be near zero on the basis of the gel to liquid crystalline thermodynamic data of the longer chain homologues. We have associated the observed order/disorder change for the diC(10)PC system with a rearrangement of a rela-tively disordered lipid bilayer ($I_{2935}/I_{2880} \cong 0.62$ in the gel phase compared to $I_{2935}/I_{2880} \cong 0.4-0.5$ for longer chain species) to the micellar phase. This interpretation is further supported by Fig. 4 in which the diC(10)PC profile is compared to that for 1-C(16)PC (1-palmitoyllyso PC; that is, a single chain molecule).

TABLE 1. Summary of the Order/Disorder Transition Data of
Phospholipid Dispersions Derived from Raman Spectral Parameters

Dispersions	T_m [a] °C	ΔI_R at T_m	T_m [b] °C	ΔS [b] (eu/mol)
diC(10)PC	-8.5	0.232		
diC(12)PC	-4.3	0.127	-1.8	6.3
	3.3	0.129		
diC(14)PC	23.8	0.252	24	18.2
diC(16)PC	41.3	0.299	41	27.4
diC(18)PC	53.5	0.340	55	32.3
diC(20)PC	64.7	0.368		
diC(22)PC	73.8	0.395	75	42.5

[a]Transition temperatures determined by peak-height intensity ratios
derived from the C-H stretching mode region.
[b]Thermodynamic data taken from Ref. [15].

Lyso PC's are well known to form micelles in excess water [13].
The comparison of the diC(10)PC and 1-C(16)PC above T_m yields
nearly identical intensity ratio values.

A comparison of the temperature profiles for the diC(14)PC to
diC(22)PC multilamellar dispersions at temperatures 10°C higher
than the observed T_m's indicates that the longer chain systems show
a trend to greater disorder in the liquid crystalline state [11];
that is, I_{2935}/I_{2884} is greater for the longer chain molecule. In
contrast, at 10°C below T_m in the gel state, the longer chain
systems tend to pack in a more ordered lattice [11], as suggested
by the smaller I_{2935}/I_{2884} intensity ratio. The intensity ratios
for the diC(12)PC liposomes, however, do not follow this general
pattern. The large value for I_{2935}/I_{2884} in the disordered state
above T_m for diC(12)PC system suggests a mixture of perhaps single
shell vesicle structures or smaller multilamellar structures with
the more usual larger liposomes.

The temperature profiles discussed above for the diacyl systems
were determined from synthetic, saturated chain DL-phospholipids.
We have recently examined the series of diC(14)-L-α-PC,
diC(16)-L-α-PC and diC(18)-L-α-PC liposomes on a Raman spectrograph

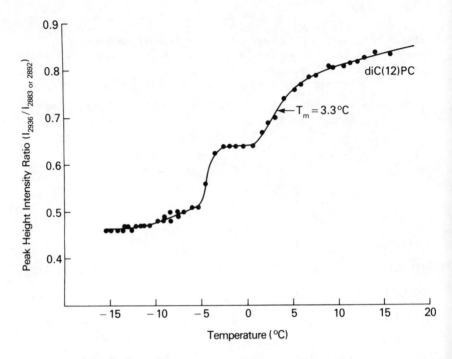

FIG. 3. Temperature profile for diC(12)PC. Note the two observed order/disorder transitions.

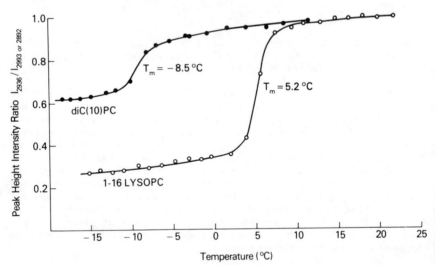

FIG. 4. Comparison of the temperature profiles for diC(10)PC and 1-C(16) lyso PC. The single chain 1-C(16) lyso PC lipid forms an interdigitated gel state bilayer at low temperature and a micellar phase on melting.

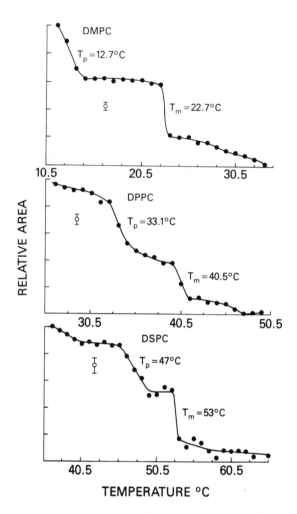

FIG. 5. Temperature profiles for diC(14)-L-α-PC, diC(16)-L-α-PC and diC(18)-L-α-PC dispersions derived from the intergrated intensities of the C-H stretching mode region.

utilizing diode detection techniques. Instead of plotting peak height intensity ratios, we construct temperature profiles from the integrated intensities of the entire C-H stretching mode region. Figure 5 displays these curves. As the temperature increases and the lattice expands, the band intensity decreases. The interesting point is that the greatest intensity decrease occurs at the pre-transition, which suggests that environmental changes occurring within the bilayer during the onset of the various dynamical processes involving the ripple bilayer structure induce significant perturbations to the bond polarizability changes. Although the subtransition is not always evident in the temperature profiles constructed from peak height intensity ratios, temperature curves based upon integrated intensities reflect a number of lattice rearrangements, including the subtransition, for samples undergoing prolonged incubation times [14]. For these dispersions the low temperature integrated intensity data indicate subtle acyl chain rearrangements [14].

Spectral studies on the temperature behavior of the order/disorder transitions of these model membranes are clearly important for assessing the thermal information commonly obtained by scanning calorimetry. That is, the spectral interpreations provide a detailed molecular basis for discussing the thermal events. Thus, the main gel to liquid crystalline phase transition, for example, can be partitioned into effects arising from the lipid polar headgroups, the interface region and the hydrophobic acyl chain core of the bilayer. In a crude sense, one is "imaging" the effects of bilayer perturbations through an analysis of the spectral frequencies, intensities and linewidths associated with specific atomic groups of the membrane constituents.

The temperature profiles derived from the Raman spectral intensity data can be more directly related to the thermodynamic properties reflecting the phase transition behavior of the multilamellar dispersions. For example, the difference ΔI_R in the Raman intensity ratio across the main transition region at T_m behaves in a curvilinear manner [11], rather than in a linear fashion, as a function of acyl chain length. This effect is analogous to the behavior of the thermodynamic parameters ΔS, ΔH and ΔV for the phase transitions, which also follow a curvilinear relationship with chain length [15]. An interesting correlation is presented in Fig. 6 in which a linear relationship is observed for plots of the spectroscopically determined ΔI_R against the calorimetrically deduced entropy change ΔS. From this plot one defines an "effective" difference in Raman intensity ratios ΔI_R^{eff} at T_m. For the present series of diacyl liposomes, the empirical ΔI_R^{eff} is proportional to the transition entropy through the expression $\Delta I_R^{eff} = (6 \times 10^{-3}) \Delta S$.

RAMAN SPECTRA OF MIXED CHAIN BILAYER DISPERSIONS

The previous series of diacyl phosphatidylcholine dispersions reflects a membrane structure in which essentially two monolayers are juxtaposed to form the bilayer. Presumably, a small degree of coupling exists acrosss the bilayer center through interactions at the acyl chain termini. We will now contrast the diC(ii)PC series with multilamellar dispersions derived from saturated, asymmetric (or mixed) chain phosphatidylcholines. Specifically, we examine the Raman spectral changes as a function of temperature, in the hydrocarbon C-C and C-H stretching mode regions, for the series of molecules in which the sn-1 acyl chain is fixed in length at 18 carbon atoms and the sn-2 chain increases in steps of two methylene units from 10 to 18 carbon atoms. We will anticipate a structural model for the highly asymmetric phosphatidylcholines that suggests

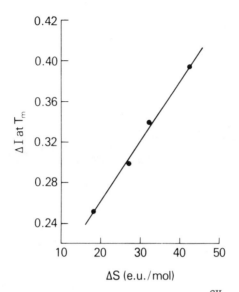

FIG. 6. Linear correlation between ΔI_R^{CH}, the change in I_{2935}/I_{2880} at T_m, and the transition entropy ΔS for the diC(14)PC, diC(16)PC, diC(18)PC and diC(22)PC multilayers.

an interdigitation of the acyl chains in the gel phase bilayer [16].

Several C-C skeletal stretching modes, found in the 1050 to 1150 cm^{-1} region, have proved useful in determining the degree of intra-chain disorder in the bilayer. The prominent features, shown in Fig. 7, used for this purpose are the 1065 and 1130 cm^{-1} out-of-phase and in-phase C-C stretching modes for the all-trans chain states, respectively. A weaker feature at ∿1106 cm^{-1} is assigned to the k=1 member of the C-C stretching mode progression for an all-trans chain whose phase differences between adjacent oscilla-tors are other than 0 or π. When either the temperature of the bilayer increases or other appropriate perturbations arise, a feature at ∿1088 cm^{-1}, arising from C-C stretching modes for gauche conformers, either appears or increases in intensity while the other three vibrational features decrease in intensity. Tempera-ture profiles determined from either the I_{1088}/I_{1065} or I_{1088}/I_{1130} peak height intensity ratios are sensitive indices for monitoring specifically the intramolecular order/disorder characteristics introduced into the hydrophobic region of the bilayer matrix.

From the temperature profiles shown in Fig. 8, the relative gauche/trans populations along the acyl chains in the mixed chain lipid dispersions in the gel state may be compared. For temperatures 10°C below their respective T_m's, the greatest relative gauche/-trans population occurs for the C(18):C(14)PC species; the degree of intrachain disorder follows the sequence C(18):C(14)PC >

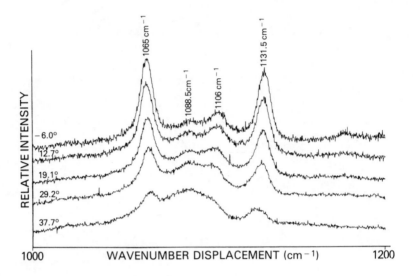

FIG. 7. Raman spectra of the 1100 cm^{-1} C-C stretching mode region for C(18):C(14)PC dispersions at temperatures spanning the gel and liquid crystalline phases. T_m = 30.0°C.

C(18):C(16)PC > C(18):C(12)PC > C(18):C(10)PC = C(18):C(18)PC. We understand this behavior by first considering that as the two chains become inequivalent in length, the terminal segment of the longer sn-1 chain undergoes trans/gauche isomerization to fill the void space under the contiguous, shorter sn-2 chain and near the bilayer center. It was proposed from calorimetric data that when the sn-1 chain exceeds the sn-2 chain length by four or more methylene units, the asymmetric lipids pack in an interdigitated arrangement across the bilayer center in the gel state. That is, an interdigitated structure compensates for chain inequivalence since the stability of the bilayer would be increased. For the series C(18):C(18)PC, C(18):C(16)PC and C(18):C(14)PC, we proposed that the bilayer transforms from one where the two leaflets are discrete [C(18):C(18)PC and C(18):C(16)PC] to one where the leaflets just begin to interdigitate [C(18):C(14)PC]. [16] In the latter C(18):C(14)PC species, the close proximity of the four sterically large terminal methyl groups would strongly disorder the acyl chains near the bilayer center. As the series proceeds to the C(18):C(12)PC and C(18):C(10)PC highly asymmetric chain systems, we now expect an increase in intrachain order as the bilayer inter-digitated packing arrangement induces the extended, ordered chain configuration. Thus, the intra- and interchain disorder would follow the sequence C(18):C(14)PC > C(18):C(12)PC > C(18):C(10)PC, which, in general follows the Raman data [16].

Figure 9 presents the temperature profiles using the I_{2935}/I_{2880} intensity ratios as indices. We note that at a common re-duced temperature the intra- and interchain disorder for the

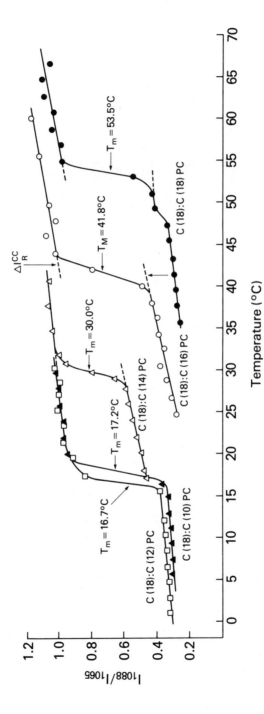

FIG. 8. Temperature profiles for the mixed (asymmetric) chain stearoylphosphatidylcholine dispersions derived from the I_{1088}/I_{1065} peak height intensity ratios.

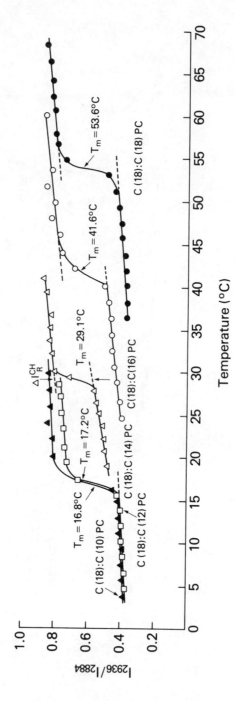

FIG. 9. Temperature profiles for the mixed chain stearoylphosphatidylcholine dispersions using
C-H stretching mode parameters as indices.

C(18):C(10)PC bilayer is equal to that for the symmetric chain C(18):C(18)PC bilayer. If the asymmetric C(18):C(10)PC bilayers were not interdigitated, we would expect a high degree of conformational disorder as the sn-1 chain would fold to fill the void volume created by a shortened sn-2 chain.

The gel state packing trends for the asymmetric series is summarized in the C-H stretching mode region spectra displayed in Fig. 10. By focusing specifically on the I_{2850}/I_{2880} ratios, which primarily reflect intermolecular interactions within the bilayer matrix, we note first a disordering of the acyl chain lattice as the series proceeds through the C(18):C(18)PC, C(18):C(16)PC, C(18):C(14)PC species and then a reordering of the bilayer lattice as the series progresses to the C(18):C(12)PC and C(18):C(10)PC liposomes. (An increase in disorder is reflected by an increase in the I_{2850}/I_{2880} intensity ratio.) Since ΔI_R^{CH}, defined above, is proportional to the transition entropy, a plot of ΔI_R^{CH} for the mixed chain dispersion, as a function of acyl chain length, should run parallel to the calorimetrically determined entropy changes for acyl chain melting. The behavior of ΔI_R^{CH} for these systems is shown in Fig. 11.

A question regarding the details of the chain packing arises for the interdigitated, highly asymmetric lipid species. That is, (1) does the shorter chain of one lipid in a leaflet pack end to end with the longer chain of another lipid in the opposing monolayer, or (2) do the two shorter chains pack end to end? The first arrangement follows the usual bilayer arrangement with two chains per headgroup at the lipid-water interface region, while the second lattice packs with three chains per headgroup, an uncommon structure. A recent x-ray determination for the C(18):C(10)PC bilayers with high water content is consistent with the second, more novel packing arrangement [17]. The present Raman data, however, do not distinguish between the two interdigitated models for the acyl chains. Perhaps both models are applicable for various interdigitated lipid bilayers.

In summary at this point, we propose two types of chain packing configurations for the lipid bilayer. In the first type the lipid molecules within the two opposing leaflets pack independently, as in the C(18):C(18)PC and C(18):C(16)PC dispersions. For the lipid systems with inequivalent chains, the molecules whose sn-2 chains range from C(14) to C(10), an interdigitated chain arrangement couples the bilayer leaflets. This type of membrane packing may be relevant in intact bilayer systems since many natural membrane lipids, such as glycosphingolipids, sphingomyelin, cerebrosides and sulfatides, reflect large asymmetries in the lengths of their hydrocarbon chain moieties which could serve to couple the two bilayer leaflets. Interdigitated bilayers may provide a means for the integral lipid components to convey information across the membrane bilayer through conformational changes. Since the effects of structural changes within the polar headgroup are transmitted into the acyl chain region [18], interdigitated bilayers, or interdigitated domains in intact membranes, could, in principle, allow a

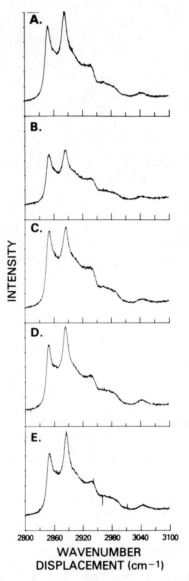

FIG. 10. Comparison of the C-H stretching region Raman
spectra for the mixed chain species at ∼10°C below their
respective T_m values. (A) C(18):C(18)PC; (B) C(18):C(16)PC;
(C) C(18):C(14)PC; (D) C(18):C(10)PC; (E) C(18):C(12)PC.

mechanism by which extrinsic membrane molecules at one face of the
membrane may influence bilayer behavior in the polar region of the
opposing lipid leaflet.

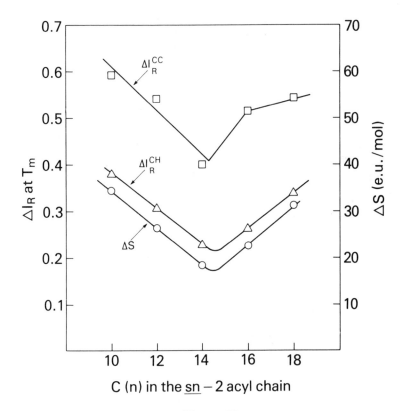

FIG. 11. Behavior of ΔI_R^{CH}, ΔI_R^{CC}, and ΔS as a function of the length of the sn-2 chain. (ΔI_R^i represents the change in amplitude of either the CH or CC peak height intensity ratios at T_m.)

POLYPEPTIDE-PHOSPHOLIPID INTERACTIONS AND BILAYER REORGANIZATIONS

By increasing the complexity of bilayer behavior through the addition of a membrane perturbant, we can design model systems for characterizing the predominant effects governing lipid-protein interactions. Our aim is to separate and individually elucidate the effects arising from either an extrinsic (outside) or intrinsic (inside) membrane component. Bilayer recombinants containing the antibiotic polymyxin B, a polycationic cyclic peptide with a branched acyl side chain, enable one to examine in detail the interactions of a tractable amphipathic membrane constituent with both charged and neutral phospholipid bilayers. Specifically, an understanding of the associations of polymyxin B with bilayers separately composed of either the acidic phospholipid, dimyristoyl phosphatidic acid (DMPA) or the neutral species, dimyristoyl phosphatidylcholine (DMPC), enhance our understanding of the

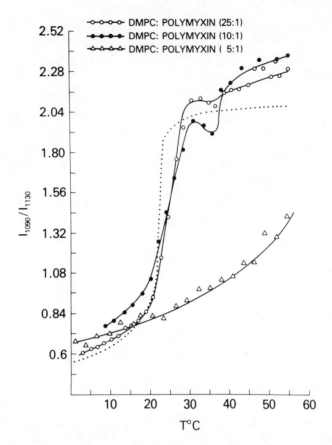

FIG. 12. Temperature profiles for DMPC/polymyxin B recombinants using the I_{1090}/I_{1130} peak height intensity ratios as indices.

structural prerequisites required for inserting membrane constituents and forming intramembrane complexes.

Although it had been reported that polymyxin B does not bind to zwitterionic species, the temperature profiles in Fig. 12 reflecting intrachain disorder imply significant bilayer perturbations within the DMPC assembly. That is, as the concentrations of polymyxin B are increased, the gel to liquid crystalline phase transition is elevated and broadened in addition to the appearance of a second, higher temperature order/disorder transition [19]. The second transition at 38°C is associated with the further fluidization of a class of immobilized lipids [20]. The temperature profile suggests a slight reordering property may arise from the insertion of the antibiotic's hydrophobic eight carbon chain tail into the bilayer. When this hydrophobic chain is removed, as in a colistin/DMPC reconstituted system, the second transition is eliminated. After the melting of the immobilized lipids in the polymyxin B containing disperions, the disorder of the liquid crystalline bilayer is greater than that of pure DMPC dispersions. By assuming that the entire melting curve encompassing both order/-disorder transitions reflects the fluidization of the lipid matrix, we estimate that about three lipid molecules are immobilized by each antibiotic molecule. For increased concentrations of polymyxin B (5:1 DMPC/polymyxin B mole ratio), the liquid crystalline phase is significantly ordered as a consequence of the extreme broadening of the phase transition. Temperature profiles based upon the lipid C-H stretching mode region support the conclusions derived from the C-C stretching mode data [19]. The second order/-disorder transition, however, is not observed in the profiles reflecting predominantly lateral interactions.

In contrast to the behavior of polymyxin B in zwitterionic species, Fig. 13 presents the temperature profiles for bilayers of DMPA/-polymyxin B in a 10:1 mole ratio. Several order/disorder transitions characteristics are quite evident. Figure 14 displays the C-H stretching mode region spectra corresponding to the order/-disorder transitions observed in the temperature profiles. Compared to pure DMPA at low temperatures, the polymyxin B liposomes exhibit considerable disorder. The lowest transition at 24°C is associated either with the melting of domains of lipids which are only hydrophobically bound to polymyxin B or, alternatively, a simple acyl chain reorganization from a more ordered to a less ordered lattice subcell. This transition may be analogous to the subtransition noted in the pure lipid bilayers. Since x-ray data indicate that polymyxin B induces acyl chain interdigitation in negatively charged bilayers [21], bilayer interdigitation may also be important in the DMPA assembly. The second order/disorder transition at 37°C is associated with the melting of antibiotic/ lipid domains in which polymyxin B is complexed to DMPA through both electrostatic and hydrophobic interactions. The third transition at 46°C represents the melting of bulk lipid, while the fourth transition at 56°C is attributed to a bilayer to micellar phase transition. The C-H stretching mode region spectra in Fig. 14 display the characteristic and diagnostic patterns for the bilayer liquid crystalline and micellar states, respectively. The intense 1002 cm^{-1} trigonal ring distortion of the phenyl group of the phenylalanine residue within the positively charged headgroup of the antibiotic was used to distinguish between the order/disorder transitions involving the gel state hydrophobically bound antibiotic/lipid complexes and the complexes bound through both hydrophobic and electrostatic interactions [19].

NATURAL MEMBRANES: LIPID RESPONSE TO THE CLATHRIN COAT PROTEIN

It is clear from the numerous examples in which vibrational techniques have been applied to diverse, complicated model membrane systems that new insights and perceptions have been developed in our comprehension of membrane behavior. We now discuss the utilization of infrared spectroscopy in understanding the interactions

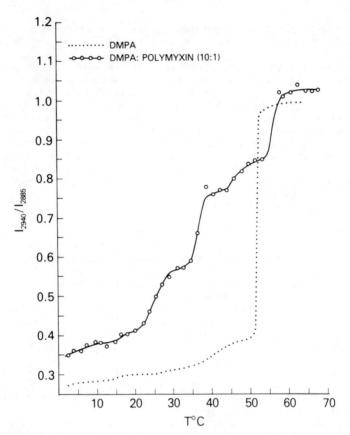

FIG. 13. Temperature profiles for a dispersion of a 10:1 DMPA/polymyxin B mole ratio derived from C-H stretching mode region parameters. (The temperature profile was determined from spectra for which the polymyxin B contributions were subtracted.) The dashed line represents the profile for a pure DMPA system.

occurring within intact, natural membrane assemblies. In particular, we will be concerned with the membrane lipid response to the clathrin coat protein in vesicles isolated from bovine brain tissue.

Clathrin, the major structural protein associated with both the coated pit regions of plasma membranes and coated vesicles, has been implicated in a number of fundamental intracellular functions in eukaryotic cells. The involvement of coated pits and coated vesicles in the endocytosis of biological macromolecules by specific receptor systems is an area of considerable attention and activity. The basic unit of the membrane protein coat is the triskelion (M_r = 650,000), a trimer of clathrin subunits (M_r =

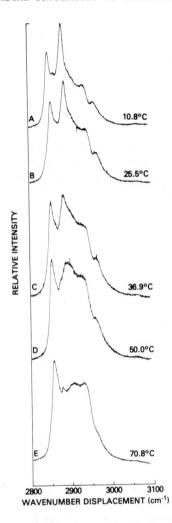

FIG. 14. Raman spectra in the C-H stretching mode region for a 10:1 DMPA/polymyxin B mole ratio dispersion at temperature spanning the gel, liquid crystalline bilayer and micellar states. (The minor spectra contributions from polymyxin B have not been subtracted in the presentation.)

180,000) in combination with three associated small proteins (M_r = 30,000 - 36,000) [22]. The triskelions first assemble in a planar array of hexagons on the membrane surface; their subsequent growth forms the closed polyhedral lattice structure associated with the coated vesicle [23]. In an attempt to define the mechanisms involved in the transition from uncoated membranes to clathrin coated pits and then to coated vesicles, we investigated by infra-red spectroscopy the lipid perturbations arising from the inter-

actions of the clathrin coat with the bilayers of the intact
membrane [24,25]. Specifically, the 2850 cm^{-1} feature, assigned to
the acyl chain methylene symmetric C-H stretching modes for the
coated vesicles, uncoated vesicles and intact synaptic membranes is
sensitive to the number of <u>gauche</u> conformers along the acyl chain.
By monitoring these different membrane systems as a function of
temperature, we concluded that clathrin increases significantly the
number of <u>gauche</u> chain conformers in the bilayer matrix of the
coated vesicle systems [25]. For example, the increase in lipid
disorder at 21°C, accompanying the observed 0.44 cm^{-1} frequency
increase between coated vesicles and uncoated vesicles, is approximately equivalent to the acyl chain rotameric disorder incurred by
heating liquid crystalline DMPC by 10°C [25,9].

Figure 15 presents the spectra for the coated and uncoated vesicles
and the synaptic membranes at 21°C. Buffer and water spectra were
subtracted. The peak frequencies, determined by gaussian fits of
141 data points for each spectrum, show that the stretching modes
associated with the coated vesicles are shifted to higher frequencies by 0.44 cm^{-1} compared to that of the uncoated vesicle
system, which, in turn, is 0.2 cm^{-1} higher than the modes for
synaptic membranes. At 38°C the coated vesicles show a 1.00 cm^{-1}
shift to higher frequency compared to the uncoated system. Table 2
summarizes the vibrational data for the three membrane systems.
Since biochemical evidence suggests that the lipid environments of
the coated pit and coated vesicle are not significantly different
from uncoated membrane domains [25,26], the bilayer disorder
induced by the clathrin coat may be a major factor required for
membrane invagination and subsequent coated vesicle formation. It
is also interesting to note that Raman spectral studies of model
membranes involving clathrin interactions with DPPC disperions
demonstrate dramatic perturbations to the bilayer lipid matrix
[27].

CONCLUDING REMARKS

In this necessarily brief excursion into an area of overlapping
biophysical and biochemical interests, we emphasize the cogent role
that vibrational spectroscopy has assumed as perspectives of
membrane behavior and function are constantly being reassessed and
modified. Although our present discussion followed a sequence in
which we first increased the complexity of several model systems
and then examined isolated, natural biological membranes, we
recognize the importance of encouraging the simultaneous study of
both types of systems. As, for example, we find spectroscopic
evidence in model bilayers for the existence of differing lipid
morphologies, it becomes pertinent to consider the relevance of
these structures in intact membranes. Areas of future investigation for the vibrational spectroscopist lie in detailed studies of
both the nonbilayer and interdigitated lipid phases, since these
lipid organizations will probably become increasingly evident as
more exacting studies on the domain structure of natural membranes
continue. Since many of the structurally related studies of para-
mount interest to the membrane scientist are amenable to the

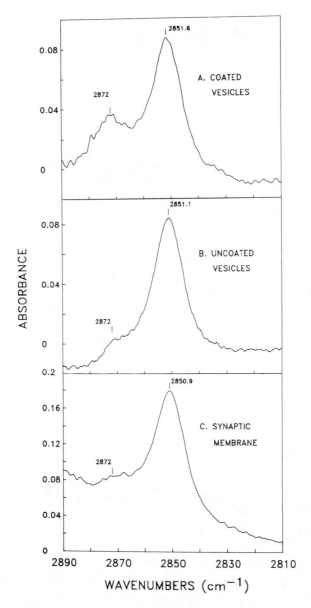

FIG. 15. The infrared absorbance spectra at 21°C of the lipid acyl chain methylene symmetric modes for (A) coated vesicles, (B) uncoated vesicles and (C) synaptic membranes. (Peak frequencies were determined from a gaussian fit of 141 data points between 2857 and 2843 cm-1. The band center of the gaussian curve is fit to an uncertainty of less than ±0.02 cm-1.)

TABLE 2. Summary of the Vibrational Frequencies of the Methylene Symmetric Stretching Modes for Coated Vesicle, Uncoated Vesicle and Synaptic Membrane Assemblies

Membrane Assembly	$21°C^a$	$38°C^b$	$50°C^b$
Coated vesicles	2851.47±0.16	2853.51±0.05	2854.11±0.05
Uncoated vesicles	2851.03±0.15	$2852.51±0.03^c$	$2852.87±0.04^c$
Synaptic membrane	2851.10±0.23	2852.02±0.02	2852.44±0.02

[a] Reported uncertainties are at the 95% confidence level for three separate preparations.

[b] Reported uncertainties are standard deviations of the gaussian fit (141 data points) to one highly purified sample preparation for each membrane class.

[c] The observed frequency increases for uncoated vesicles compared to synaptic membranes possibly reflect the decreased phospholipid/-cholesterol mole ratio in the uncoated vesicle assembly.

experimental approaches and thinking of the vibrational spectroscopist, collaborative studies between the spectroscopist and biologically oriented investigator provide yet another extraordinarily fertile area for applying and extending the techniques encompassed within the vibrational discipline.

ACKNOWLEDGMENT

I wish to thank the following colleagures who participated in the research studies described here; Dr. C. H. Huang, Dr. R. G. Adams, Dr. W. H. Kirchhoff, Dr. E. Mushayakarara, Dr. J. S. Vincent, Dr. C. J. Steer, and Dr. T. J. O'Leary.

REFERENCES

1. S. J. Singer and G. L. Nicolsen, Science, 175, 720 (1972).
2. I. W. Levin, in Advances in Infrared and Raman Spectroscopy, Vol. 11, (R. J. H. Clark and R. E. Hester, Eds.), John Wiley and Sons, New York, 1984, pp. 1-48.
3. R. C. Lord and R. Mendelsohn, in Membrane Spectroscopy (E. Grell, Ed.), Springer-Verlag, New York, 1981, pp. 377-426.
4. W. P. Fringeli and Hs. H. Gunthard, in Membrane Spectroscopy (E. Grell, Ed.), Springer-Verlag, New York, 1981, pp. 270-332.
5. D. F. H. Wallach, S. P. Verma and J. Fookson, Biochim. Biophys. Acta, 559, 153 (1979).
6. D. E. Metzler, Biochemistry, Academic Press, 1977, pp. 252-300.

7. R. G. Snyder, S. L. Hsu and S. Krimm, Spectrochim. Acta, 34A, 395 (1978).
8. R. G. Snyder and J. R. Scherer, J. Chem. Phys., 71, 3221 (1979).
9. R. Mendelsohn, R. Dluhy, T. Teraschi, D. G. Cameron and H. H. Mantsch, Biochemistry, 20, 6699 (1981).
10. W. W. Kirchhoff and I. W. Levin, unpublished results.
11. C. H. Huang, J. R. Lapides and I. W. Levin, J. Amer. Chem. Soc., 104, 5926 (1982).
12. S. Mabrey, J. M. Sturtevant, Proc. Natl. Acad. Sci. U.S.A., 73, 3826 (1976).
13. F. Reiss-Husson, J. Mol. Biol. 25, 363 (1967).
14. R. G. Adams and I. W. Levin, unpublished data.
15. J. T. Mason and C. Huang, Lipids, 16, 604 (1981).
16. C. Huang, J. T. Mason and I. W. Levin, Biochemistry, 22, 2775 (1983).
17. T. J. McIntosh, S. A. Simon, J. C. Ellington and N. A. Porter, Biophys. J., 45, 41a (1984).
18. S. F. Bush, R. G. Adams and I. W. Levin, Biochemistry, 19, 4429 (1980).
19. E. Mushayakarara and I. W. Levin, Biochim. Biophys. Acta. 769, 589 (1984).
20. I. W. Levin, F. Lavialle and C. Mollay, Biophys. J., 37, 339 (1982).
21. J. L. Ranck and J. F. Tocanne, FEBS Lett., 143, 171 (1982).
22. E. Ungewickell and D. Branton, Nature (London), 289, 420 (1981).
23. J. Heuser, J. Cell Biol., 84, 560 (1980).
24. J. S. Vincent, C. J. Steer and I. W. Levin, Biochemistry, 23, 625 (1984).
25. C. J. Steer, J. S. Vincent and I. W. Levin, J. Biol. Chem., 259, 8052 (1984).
26. T. Simon, D. Winek, B. Brandon, S. Fleischer and B. Fleischer, J. Cell Biol., 95, 249a (1982).
27. I. W. Levin and C. J. Steer, unpublished data.

BIOLOGICAL FT-IR: INDUSTRIAL AND ACADEMIC APPLICATIONS

R. J. Jakobsen, F. M. Wasacz, and K. B. Smith

National Center for Biomedical Infrared Spectroscopy
Battelle's Columbus Laboratories
Columbus, Ohio 43201-2693

INTRODUCTION

For twenty years biologists have shunned the use of infrared spectroscopy as either an analytical or a structural tool. There were
basically two reasons for this behavior:

(1) Twenty years ago, the infrared spectra of proteins were run as
     solids and not in the physiologically real state -- that of an
     aqueous solution.

(2) Only the strongest bands of the proteins (i.e., the Amide I
     and II bands) were used for spectral interpretation and with-
     out deconvolution techniques, these broad bands did not give
     enough definitive structural information to aid the biologist.

In the last five years, it has been amply demonstrated that not
only can aqueous solutions be routinely run using the current
generation of FT-IR instrumentation, but instrumental sensitivity
is such that small differences in solutes can be determined in
aqueous solutions. This sensitivity in the use of aqueous solutions has opened the door to vitally needed studies which are
essential to the capability to interpret the spectra of biological
molecules such as proteins. These solution studies consist of
varying parameters such as pH, ionic strength, concentration, etc.,
and determining the spectral changes due to the changes in parameters. By using parameter variations where the corresponding
structural changes have been documented (in the literature) by
other techniques, it is possible to obtain spectral-structure
correlations for proteins. The problem is that this type of
approach to the interpretation of the infrared spectra of proteins
is relatively new and such information is only beginning to be
available.

This paper describes part of our effort to provide such spectral-
structure correlations for proteins. Our basic approach to spectral interpretation involves two steps. The first is to obtain
transmission solution spectra of both single proteins and of
mixtures of proteins where parameters of the solution are systematically varied. The second is to obtain ATR spectra of flowing

solutions of both single proteins and mixtures of proteins with the
same type of parameter variation as used for the transmission
spectra. The transmission spectra permit spectra-structure correlations to be made for the proteins in solutions and the ATR spectra permit structural information to be obtained on the protein
film adsorbed on the ATR crystal. Both types of information are
extremely useful for interpreting the protein spectra.

Our efforts at obtaining spectra-structure correlation from transmission spectra of proteins have been given in several talks and
publications and will not be repeated here. This study is concerned with adsorbed protein films, and the study not only yields
spectra-structure correlations, but also illustrates applications
of FT-IR in the biological sciences. These applications are both
industrial, such as biotechnology and biocompatibility studies, and
academic, such as in studies of blood coagulation and studies of
protein structure.

ADSORBED PROTEIN DATA

Experimental
The basic experiment is to flow a protein solution through a dual
channel ATR flow cell. This flow cell, which has been described in
other publications, consists of two rectangular flow channels about
1×10 mm in width and 100 mm in length. The ATR crystal forms one
wall of the flow channel and the flow inlets are as parallel to the
length of the crystal as is possible, thus effecting nearly laminar
flow. The flow can be directed to either of the flow channels,
thus permitting a direct comparison of results under the same
experimental and instrumental conditions. As the protein solution
is pumped through the ATR flow cell, proteins adsorb onto the
surface of the ATR crystal. Spectra are continuously taken during
the flow period so the changes in protein adsorption with time can
be monitored. Pure solvent can then be pushed through the cell and
the desorption of any loosely bound proteins can be followed.

Albumin
An albumin solution in saline at pH 3 was pumped through one
channel of the ATR flow cell while an albumin solution at a pH of 9
was flowed through the other channel. Spectra of the final
adsorbed albumin film at each pH are shown in Fig. 1. From this
figure, it can be observed that at a pH of 9 the ratio of the 1310
cm^{-1} band to that of the 1245 cm^{-1} band is markedly higher than the
ratio observed at a pH of 3. The change with pH is very similar to
the change observed with pH in the transmission spectra of the
dissolved proteins. This last statement has interesting connota-
tions, i.e., these adsorbed protein films directly reflect the
nature of the protein in solution. On the one hand, this is not
unexpected since it is well known that solvents greatly influence
protein structure. However, on the other hand, this is a little
surprising since adsorbed films are often more governed by the
nature of the adsorption surface and the nature of the adsorbate-
surface interaction.

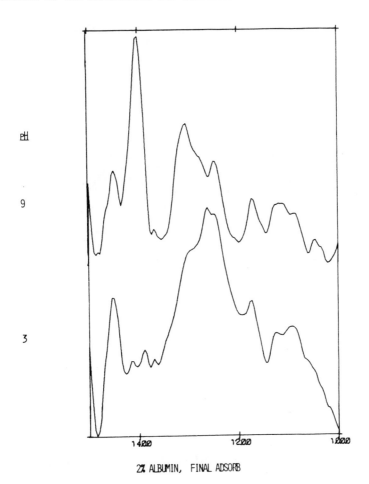

FIG. 1.   Spectra of the final adsorbed albumin film at a pH of 9 and 3.

Further evidence for the influence of the solvent comes from the spectra of Figs. 2 and 3. In Fig. 2 (bottom) the spectrum of the film adsorbed from a solution of pH 9 is again shown and compared to the spectrum of the desorption with saline at a pH near 7.4 (top, Fig. 2). Although the adsorbed film is already attached, its structure changes when subjected to the influence of a different pH (initially pH 9, then going towards pH 7.4). The observed changes are that the 1310/1245 cm^{-1} band ratio decreases as would be expected from a lowering of pH (see spectra of Fig. 1). In Fig. 3, the spectrum of the film adsorbed at a pH of 3 is shown (bottom) and this spectrum is compared to the spectrum resulting from desorption with saline at a pH near 7.4 (top, Fig. 3). Here, although there are interferences (1260 cm^{-1}) due to the silicone

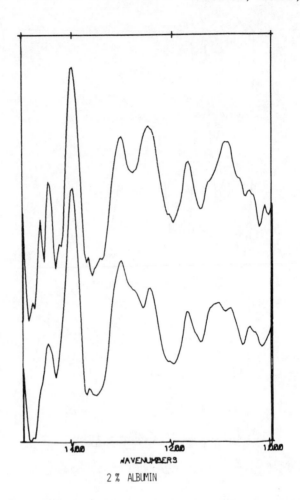

FIG. 2. Spectra of the final desorbed albumin film at pH 7.4
(top), and the final adsorbed film at pH 9.0 (bottom).

gasket, it still can be seen that as the pH is raised, there is a
definite increase in the 1310/1245 cm^{-1} band ratio. Thus, the
initial results indicate that the adsorbed protein film directly
reflects the nature of the solvent in contact with the film and
appears to not be affected by the nature of the surface or the type
of adsorbate-surface interaction. Figure 4 shows a comparison of
the spectra obtained from (top Fig. 4) the albumin molecules in
solution at pH 9 and (bottom) the adsorbed protein film where the
proteins are adsorbed from an albumin solution also at a pH of 9.
These spectra are virtually identical and further support the
hypothesis that adsorbed proteins indirectly reflect the nature of
the proteins in solution. However, Fig. 5 shows a comparison of
the spectra obtained from (top, Fig. 5) albumin dissolved in a

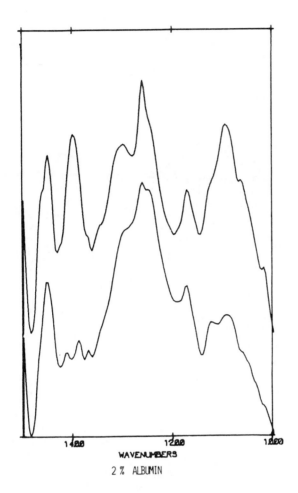

FIG. 3. Spectra of the final desorbed albumin film at pH 7.4
(top), and the final adsorbed film at pH 3.0 (bottom).

solution at a pH of 3 and from (bottom, Fig. 5) albumin adsorbed
onto the surfaces of the ATR crystal from a solution also at a pH
of 3. Here the spectra are similar in terms of the 1310/1245 cm^{-1}
band ratios indicating similarities in the secondary structure or
conformation of the albumin molecules. But there are also
differences between the two spectra. No 1205 or 1270 cm^{-1} bands
are seen in the spectra of the adsorbed films. From a previous
solution pH study, these are the bands that indicate unfolding of
albumin subunits in the elongation of the albumin molecule. We are
now in the process of getting spectra of adsorbed albumin at other
pH's and thus at the present time, all we can say is that for
adsorbed albumin (from a solution at pH 3) either the subunit
unfolding does not take place, or it is already complete by a pH

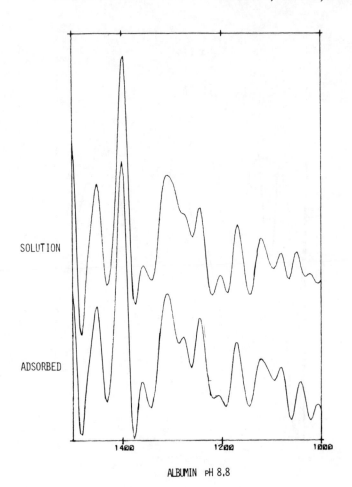

FIG. 4. Comparison of the spectra obtained from albumin
molecules in solution at pH 8.8 (top), and the adsorbed protein film where the proteins are adsorbed from an albumin
solution at pH 8.8 (bottom).

of 3. This, however, does indicate that adsorbed proteins can be
affected by factors other than the nature of the solvent.

Non-Aqueous Solvents

In order to learn more about how solvents affect the nature and
structure of adsorbed films and to attempt to determine the role
water plays in the structure of adsorbed proteins, we investigated
the adsorption behavior of gamma globulin in a solution of a non-
aqueous solvent, ethylene glycol. Gamma globulin is only slightly
soluble in ethylene glycol (the resulting solution is less than 0.1
weight percent) and at this concentration the only infrared signal

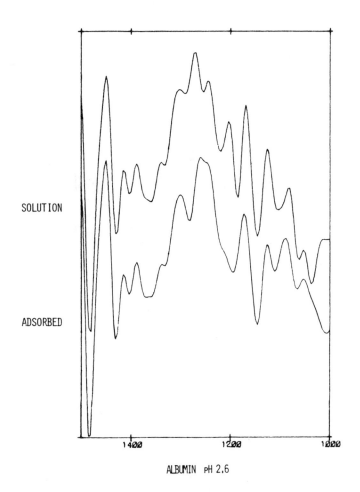

ALBUMIN pH 2.6

FIG. 5. Comparison of the spectra obtained from albumin dissolved in solution at pH 2.6 (top), and from albumin adsorbed onto the surfaces of the ATR crystal from a solution at pH 2.6 (bottom).

comes from either the solvent or adsorbed protein molecules and none of the signal comes from the protein molecules in solution. Figure 6 shows spectra of the Amide I and Amide II bands at various protein solution flow times. After 10 minutes of the protein solution flow, the spectrum of the adsorbed film (bottom, Fig. 6) looks much like the transmission spectrum of gamma globulin in saline solution. However, by about 60 minutes of flow time (middle, Fig. 6), changes are beginning to be observed which are mainly the appearance of a band near 1630 cm^{-1}. This 1630 cm^{-1} band continues to grow in intensity as compared to the 1645 cm^{-1} band (which indicates β-sheet structure in gamma globulin) as shown

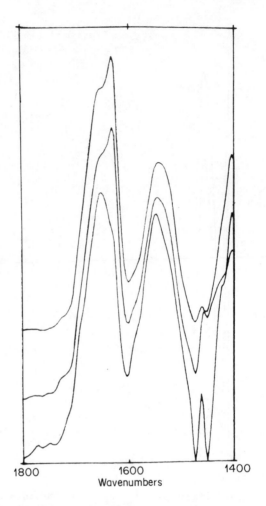

FIG. 6. Spectra of the Amide I and Amide II bands of gamma globulin in ethylene glycol at various protein solution flow times: 120 min. (top), 61 min. (middle), and 10 min. (bottom).

by the spectrum of 2 hours of flow time (top, Fig. 6). The appearance of a band near 1630 cm^{-1} has been observed in the spectra of other proteins when exposed to an air interface at room temperature for periods of time. This, coupled with the fact that ethylene glycol is known to be a solvent which partially unfolds proteins, indicates that the adsorbed gamma globulin film is unfolding with time under the influence of the ethylene glycol and the 1630 cm^{-1} band is the Amide I vibration of either the unfolded state or a new β-sheet structure formed from the unfolded structure. Thus, not only can solvents with known protein behavior be used to aid in

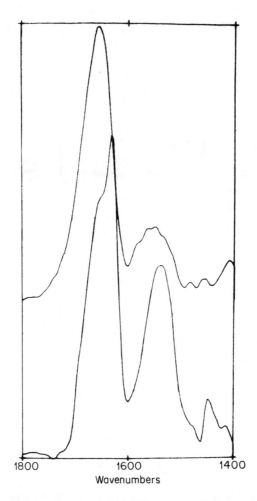

FIG. 7. Spectra of gamma globulin in ethylene glycol: desorption with ethylene glycol followed by a desorption with saline
(top), and desorption with ethylene glycol (bottom).

making spectra-structure correlations, but the adsorbed protein
film can unfold even without exposure to an air interface.

Gamma globulin was adsorbed onto the surface of an ATR crystal from
a flowing solution of ethylene glycol, and was then followed by
flowing pure ethylene glycol through the ATR cell in an attempt to
desorb any loosely bound gamma globulin molecules. Within experimental error, no protein was desorbed from the ATR crystal surface
and, as can be seen at the bottom of Fig. 7, there is little
difference between the spectrum of gamma globulin adsorbed from
ethylene glycol (top, Fig. 6) and the spectrum of gamma globulin

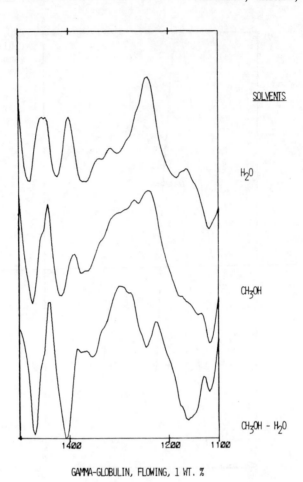

GAMMA-GLOBULIN, FLOWING, 1 WT. %

FIG. 8. Spectra of gamma globulin in ethylene glycol: adsorbed from a saline solution (top), when methanol is flowed
past the film adsorbed from saline (middle), and subtraction
of the two top spectra (bottom).

desorbed with ethylene glycol (bottom, Fig. 7). However, if the
desorption with ethylene glycol is then followed by a desorption
with saline, the spectrum shown at the top of Fig. 7 is obtained.
Here the major frequency of the Amide I band has returned to 1645
cm^{-1}, there is little or no 1630 cm^{-1} Amide I present, and the
resulting spectrum is similar to that seen (bottom, Fig. 6) early
in the adsorption process and before extensive unfolding had taken
place. This indicates that not only is the unfolding reversible,
but again the structure of the adsorbed protein film is largely
governed by the nature of the solvent surrounding the protein
molecules.

The spectrum at the top of Fig. 8 is the spectrum of a protein film
of gamma globulin adsorbed from a saline solution. This spectrum
is almost identical to the transmission spectrum of gamma globulin
desorbed in saline. Some solvents are known to promote α-helix
formation in proteins and methanol is one of these solvents. The
spectrum resulting when methanol is flowed past the gamma globulin
film adsorbed from saline is shown in the middle of Fig. 8. This
spectrum (using methanol) is quite different than the spectrum of
the adsorbed gamma globulin film from saline (top, Fig. 8) and the
difference is that there is increased adsorption in the 1270-1320
cm^{-1} region when using methanol. The subtraction of the two top
spectra of Fig. 8 (methanol minus saline) yields the spectrum shown
at the bottom of Fig. 8. This subtracted spectrum shows strong
adsorption near 1300 cm^{-1}, a frequency that we have been proposing
as the frequency of the Amide III vibration of proteins with
α-helix conformation or secondary structure. Since this 1300 cm^{-1}
band in gamma globulin results from the use of methanol, an α-helix
inducing solvent, the appearance of such a band lends credence to
our assignment.

Effects of Surfaces

Most of the previous studies have shown that the structure of the
adsorbed protein film is governed by the nature of the solvent,
i.e., by the solution structure of the dissolved protein. Is this
always the case? The answer to this question is no -- as in most
cases there are exceptions. Figure 9 shows spectra of adsorbed
hemoglobin films at various adsorption times. The top spectrum
shows the adsorbed film spectrum after three minutes of protein
solution flow, and the bottom spectrum after a flow time of 52
minutes. Initially there is a high 1310/1245 cm^{-1} band intensity
ratio which decreases as adsorption time lengthens. Figure 10
repeats the spectrum of the final adsorbed film (bottom) and
compares it to the transmission spectrum of hemoglobin in solution
(top, Fig. 10). By comparing the top spectra of Figs. 9 and 10 it
can be seen that hemoglobin initially adsorbs in the same conformation as the solution conformation, but with time of surface
contact, the hemoglobin changes to the structure shown at the
bottom of Fig. 9.

Other examples of where the nature of the surface influence protein
adsorption are shown in Figs. 11 and 12. In these experiments one
of the ATR crystals used in the dual channel is first coated with a
thin layer of an alkylated cellulose (such polymers are reported to
have good blood biocompatibility). In this way we can directly
compare the adsorption of proteins on an alkylated cellulose
surface (one channel) as compared to the germanium surface of the
ATR crystal in the other channel. Figure 11 shows spectra of gamma
globulin adsorbed on each of the two surfaces. The spectra of
gamma globulin adsorbed on a germanium surface (bottom, Fig. 11)
have their main Amide III frequency at about 1250 cm^{-1} with a
shoulder near 1270 cm^{-1}. Correspondingly, the proteins adsorbed on
the alkylated cellulose surface (top, Fig. 11) give the main Amide
III frequency as 1270 cm^{-1} with asymmetry or a shoulder near 1250
cm^{-1}. This likely indicates a reversal in the β-sheet/random

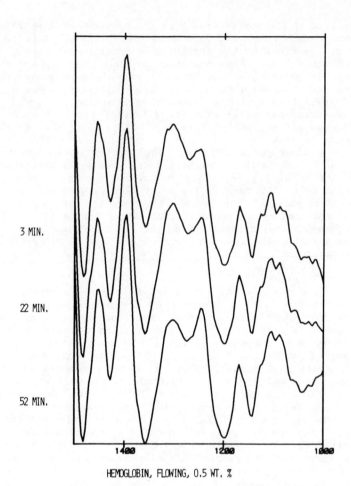

HEMOGLOBIN, FLOWING, 0.5 WT. %

FIG. 9. Spectra of adsorbed hemoglobin films at various adsorption times: 3 min. (top), 22 min. (middle), and 52 min. (bottom).

conformation ratio between the two surfaces, but at the very least shows protein structural differences due to two different adsorbent surfaces.

Such surface differences are even more dramatically illustrated in Fig. 12 where the same two surfaces are used, but the protein solution now contains a mixture of albumin and gamma globulin. The middle spectrum of this figure shows the protein mixture adsorption on the alkylated cellulose surface while the bottom spectrum shows the adsorption from this mixture onto a germanium surface. The bottom spectrum (Ge) looks predominantly like gamma globulin (see bottom spectrum of Fig. 11) with small amounts of albumin being

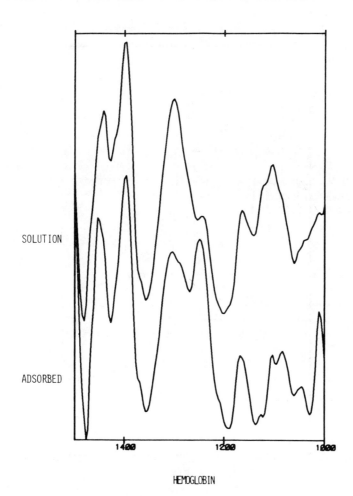

FIG. 10.   Comparison of the transmission spectrum of hemoglobin in solution (top), and the final adsorbed film
(bottom).

present (as indicated by the band near 1310 cm-^{1}). Yet the spectrum of adsorption onto an alkylated cellulose surface shows a much
higher percentage of albumin than gamma globulin (compare 1310/1240
cm-^{1} ratios in the two spectra). The subtraction result between
these two spectra is shown at the top of Fig. 12 and shows that a
1310 cm-^{1} band (albumin) is stronger in the alkylated cellulose
adsorption spectrum than in the germanium spectrum. Thus, not only
does the nature of the surface affect the structure of individual
adsorbed proteins, but it affects the competition between proteins
in mixtures for adsorption sites. Both the change in protein
structure upon adsorption and the competition between proteins can

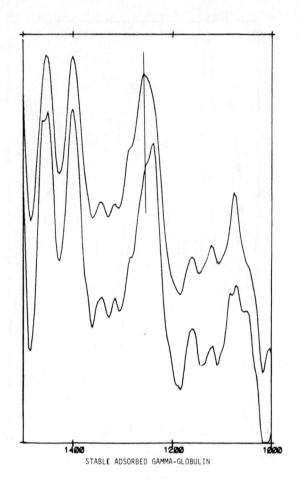

STABLE ADSORBED GAMMA-GLOBULIN

FIG. 11. Spectra of gamma globulin adsorbed on an alkylated cellulose surface (top), and on a germanium surface (bottom).

be useful as an aid in spectra-structure correlations and in establishing group frequencies for proteins.

Thus, not only does parameter variation in solutions of proteins yield spectra-structure correlations for the dissolved proteins, but the same parameter variation used with adsorbed protein films not only permits more dilute solutions and different solvents to be used, but in addition to spectra-structure correlations due to parameter variations one can also obtain information on the effects of surfaces and on competition between proteins for adsorption sites.

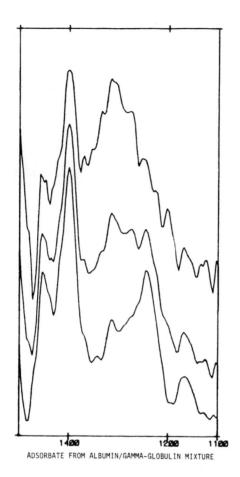

ADSORBATE FROM ALBUMIN/GAMMA-GLOBULIN MIXTURE

FIG. 12. Spectra of an albumin/gamma globulin mixture on germanium (bottom) and on alkylated cellulose (middle), and the spectrum resulting from the subtraction of the two spectra (top).

ACKNOWLEDGMENT

This work was partially supported by the National Institutes of Health under NHLBI Grant No. HL 24015 and DRR Grant No. RR01367.

CHEMISTRY AND STRUCTURE OF FUELS: REACTION MONITORING WITH DIFFUSE REFLECTANCE INFRARED SPECTROSCOPY

E. L. Fuller, Jr., N. R. Smyrl and R. L. Howell

Plant Laboratory, Department of Energy Y-12 Plant*
Martin Marietta Energy Systems, Inc.
P. O. Box Y
Oak Ridge, TN 37830

INTRODUCTION

Recent events have heightened our awareness of our precarious position with respect of our energy resources. Our continued maintenance of a high standard of living on this fragile planet depends upon our efficient utilization of the prerequisite natural supplies of energy. Detailed understanding of the chemistry and structure of these materials is tantamount to optimum utilization with a minimum insult to our environment. We also must obtain all of the information related to each and all of the chemical and physical steps involved in the winning of the stored energy in order to assure our judicious, conservative and economical reaping of this immeasurably valuable commodity. Recent works have shown that the Diffuse Reflectance Infrared Spectroscopy (DRIS) technique is extremely informative for evaluating the chemistry and structure of solid fossil fuels [1-3]. The DRIS technique has been shown to provide excellent means of studying the chemistry and structure of solids [4-6] with a minimum of sample preparation. Logically then it is to be expected that, since classical infrared analyses of coals and related materials has been so informative [7-10], the DRIS technique will be promising.

EXPERIMENTAL

The equipment and techniques used in this study are basically those described previously [1,2,11,12]. This work utilizes the versatility of the closed DRIS cell as a reaction chamber where the spectra are obtained for different degrees of reaction while the reaction proceeds. The examples reported here are for cases where conventional infrared techniques would not provide adequate data. DRIS is used to obtain good spectra of materials that are opaque to infrared radiation, free of the encumberance of a support of diluent (KCl, KBr, CsI, etc.) and/or distortions due to pressurization to form pellets or discs. The DRIS technique was used to characterize coal powder [1,3,10] and uranium dioxide powder [12] and reactions with gaseous reagents in each case. Both are visibly opaque and respond well to DRIS analysis, but are dissimilar in most all other respects. The uranium oxide was formed by reaction with water in the absence of other reactive species (oxygen, carbon dioxide, etc.) as a baseline study for the future studies of the synergy involved in the oxidation or uranium and related metals by the air of our atmosphere. Extreme care was taken to avoid exposure to any species other than water (with some exposure to argon in a "dry box" to effect transfer to the DRIS cell from the water reaction chamber. Transport to and position in the infrared spectrophotometer is done with the cell "valved off" from any reactants prior to the DRIS anaylses. The coal samples were powders much akin to those used for steam plant combustion [13] and somewhat finer than those used for fluidized bed combustion [2]. Diffuse reflectance from a sample depends markedly on the physical state of the sample [2,14,15] as well as its chemical composition. The spectral response is known to depend on the degree of deconsolidation (cominution) [2] and reconsolidation (compaction) [14] to some degree. The specific electromagnetic radiation interactions (diffuse reflection, multiple reflections, specular reflection, absorption, etc.) vary with respect to the infrared incident/exit geometry. Rigorous and exact analyses of these interactions is beyond the scope of this work [16-20] and is not needed for the analytical work described here. We monitor reactions on the same position on the same sample in order to reliably evaluate changes even though the absolute DRIS intensities may be perturbed by the aforementioned geometric factors [1,2]. In general the difference spectra are evaluated as unweighted differences between the original reference state spectrum and the spectrum of the same material after chemical or physical modification. The dedicated computer, so necessary for Fourier analyses and related calculations, is used to evaluate the difference spectra from the stored data with a high degree of precision [21].

All of the spectra reported in this study were obtained with our pyroelectric triglycine sulfate (TGS) bolometer to detect the reflected/ transmitted radiation. Adequate signals are sensed to allow us to acquire spectra with high signal to noise ratios. Repetative scanning was used for obtaining coadded interferograms (100 to 25000) in various instances, depending on the type of analysis in each case. We have not utilized the mercury cadmium telluride (MCT) photodetector even though DRIS analyses inherently deal with only a small fraction of the optical input power. We have found that the reference spectra (front surface mirror, powdered KCl, etc.) are not appreciably different for various temperatures (298 to 800K) if the TGS detector is used since the emitted radiation from the sample/reference is not great enough to "saturate" the detector (alter the electronic response to a given radiation power level increment). The greater sensitivity of the MCT detector leads to temperature dependent spectra, even for infrared inactive substrates, since the response is noted to occur into the nonlinear response regime of the MCT. The benefits of the MCT (fast response, greater sensitivity, lower noise level, etc.) can be realized for studies at ambient temperatures, especially in cases of low energy throughput. All of the spectra reported here were obtained with a Digilab Fourier Transform Interferometric Infrared Spectrometer (Model FTS 15) with full computer control with calculations made in double precision to assure maximum

resolution and minimum noise [27]. Unless otherwise stated, none
of the DRIS results quoted here have been subjected to any
mechanical and/or numerical "smoothing" for fear of introducing
bias or artifacts. This is not to cast aspersions on persons who
need to, or opt to, incorporate such procedures into their
operation either for cosmetic purposes or otherwise. This work is
designed to obtain the maximum amount of information and we do not
wish to "smooth" over any potentially informative trend in the
data. The reader will note a wide variation of noise level in the
graphic representations due virtually entirely to the number of
interferograms that are coadded in the signal averaging operation
prior to the Fourier transform to calculate the respective
spectrum. This operation decreases the noise/signal ratio to the
extent predicted by statistical analyses of the sources of noise.
We routinely use 10 to 25,000 scans in this operation with a trade
off so that we can follow the kinetics of rather fast reactions
with a relatively high noise/signal ratio and/or obtain accurate
data with a low noise/signal ratio for a reliable basis for
evaluating sensitive quantities such as derivatives, band shapes,
composite band summations, isotopic substitution frequency shifts,
etc. Our DRIS cell is designed to allow us to confidently obtain
both kinetic and static data by programmed parameter (temperature,
pressure, chemical composition, number of scans, etc.) around the
sample. Point of interest: if a fresh sample of coal is placed in
the purge environment of the spectrometer, a slow change occurs as
the moisture is lost (and oxidation occurs if oxygen is left in the
purge gas) over an extended period such as the 18+ hours required
to accrue 25,000 interferograms. We can precisely control the
environment around the sample in such a manner as to avoid chemical
and structural changes in the substrate while it is continuously
bathed by the incident light beam of the Fourier transform infrared
spectrometer.

RESULTS AND DISCUSSION

Recent developments in the field of infrared spectroscopy have
greatly enhanced the breadth and utility of the technique.
However, one can glean the maximum information from the spectra by
more or less following the train of thought that was involved in
the historic development of the field. The same computer system
that is used to efficiently acquire the spectra and to archivally
store the digital information is extremely useful in further
processing and correlations [2]. For exploratory studies one is
well advised to examine the graphical form of the spectrum in a
format similar to the chart records of the classical recording
spectrophotometers. In this manner, one can compare his results to
the libraries of spectra and/or tabulations of characteristic band
positions. Such exercises are helpful in gaining a general knowledge (fingerprinting) [28-31] of the chemical nature of the
material under study. Further analyses proceed to varying of trend
analyses and quantitation with correspondingly more information
obtained with respect to chemical and physical nature of the sample
and its reactions.

The following two examples are presented to show that the DRIS technique is uniquely capable of providing information related to the structure and reactions of both inorganic and organic feedstocks for energy production. The format of the presentation is somewhat unorthodox since it is inherently a combination of a research report, a report of new techniques, a report of alternate methods of data processing, an instructional report and a chronicle heralding the merits of the DRIS technique to this, the Thirteenth Anniversay of the Coblentz Society. In each case we have found it convenient to follow roughly the historic route of development of infrared spectroscopy. From the recognition of the phenomena to the identification of the responsible chemical and physical entities, to the evaluation of the amounts of material responsible for the absorption, to the evaluation of the structures, the kinetics and mechanisms and ultimately to the tailoring for efficient energy production. The analyst/researcher is well advised to relive this historic train of thought when approaching a new system or process. In this way the more complete perspective is acquired based on the total results of

(1) Qualitative analyses to assure we know what chemical species is involved;

(2) Quantitative analyses to do our very best to know how much of each entity is present;

(3) Structural evaluation to go beyond the "fingerprint" activity to evaluate the chemical and physical structure of equilibrium as well as intermediate states that may go unrecognized by cursory examinations; and

(4) Proceeding to evaluate the kinetics and mechanisms of commercially, politically, and potentially important processes. Improvement of any process requires intimate understanding of the nuances and interplay of the chemical and physical states if we are to advance beyond the crude and limited empirical optimization processes.

URANIUM/OXYGEN/HYDROGEN/WATER SYSTEM

Water vapor reacts readily with cominuted uranium powder to form amorphous hydrated oxide/hydroxide powder as shown in the photomicrograph of Fig. 1. The DRIS data of Fig. 2 were obtained in the closed sample cell under successively increased static vacuum states (atmospheric pressure to 10E-3 Pa) for the duration of each of the data acquisition periods.

Qualitative Analyses
Cursory examination of the data of Fig. 2 provides us with some knowledge of the chemical species present. Three general features of note are

(1) A broad, skewed (asymmetric) band at 3600 to 2400 wavenumbers that is noted in many systems where O-H groups are relatively tightly bound by hydrogen bonds;

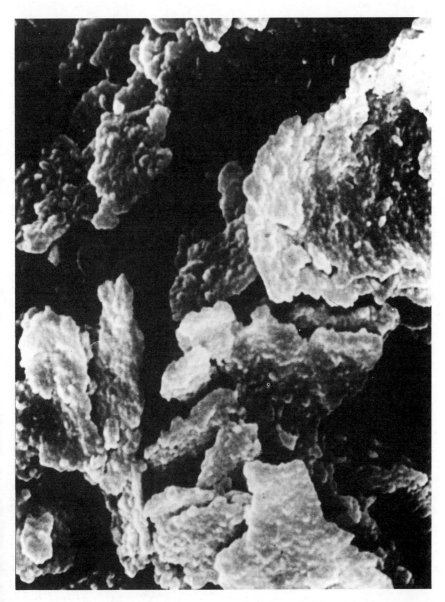

FIG. 1. Electron Photomicrograph of Uranium Oxide. These thin shards spalled from the metal substrate in the form of 1 to 4 laminae into the liquid water reaction medium at 100°C. Approximately 10 micrometers full scale.

DRIFT SPECTRA: DRYING OF UO$_2$ (C)

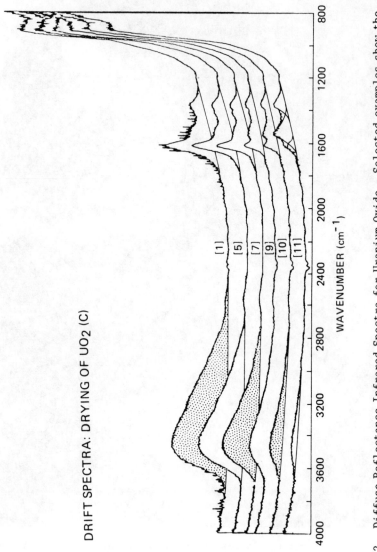

WAVENUMBER (cm^{-1})

FIG. 2. Diffuse Reflectance Infrared Spectra for Uranium Oxide. Selected examples show the change in spectral features associated with drying. Qualitative identification is very good but quantitative analyses are thwarted by the generally curved and somewhat oscillatory background due to light scattering and interference effects, respectively.

(2) An asymmetric feature at 1700 to 1300 wavenumbers that seems
     to have ca.3 components with the 1620 wavenumber feature due
     to the H-O-H bend of bound molecular water molecules and the
     1360 and 1450 bands associated with the solid substrate; and

(3) a shoulder feature in the region of 900 wavenumbers.

This qualitative analysis is valuable to show us that the majority
of the bound water is there in a hydrating form since the very
existence of the 1620 wavenumber band (and its corresponding decrease) is inconsistent with the existence of large amounts of
dissociated water in the form of a hydroxide. Extensive quantitative analyses are thwarted by the general curved nature of the
background curves and the periodic undulations due to the optical
interference phenomena within the solid particles.

Quantitative Analyses
Since all of the spectra of Fig. 2 were taken for the same portion
of the same sample, lying in the same position undisturbed for the
course of this study, we can rest assured that the scattering
geometry and the chemical components have not been altered (except
for the volatile components that we purposely removed with the
vacuum). We can then offer an alternate mode of data presentation
where the spectrum of the vacuum state is subtracted from each of
the preceeding spectra as shown in Fig. 3. Mathematically this
mode of presentation is equivalent to presenting each of the noted
single beam reflectance spectra, $R(n,i)$, ratioed to that of the
vacuum state, $R(14,i)$. The data of Fig. 3 is in the form
$-\log[R(n,i)/R(14,i)]$ for each of the n pressure states [13] and
each of the i wavenumbers (1600 at the noted nominal 2 (two) wavenumber resolution). The merits of the computer aided analyses are
well appreciated with respect for the tedium of the calculation
whether performed as described above with the single beam data or
from the double beam data of Fig. 2. The end result is identical
since the mathematical values for the inert reference (KBr, KCl,
Front surface mirror, etc.) are cancelled: $-\log[R(n,i)/R(ref,i) +
\log[R(14,i)/R(ref,i)] = -\log[R(n,i)/R(14,i)]$. In this case we have
acquired spectrum number 14 after prolonged evacuation to assure
that no more spectral changes were occurring due to volatile loss.
In this mode of presentation we note the analogy to the more
classical notation of transmission experiments where absorbance, A,
is expressed in terms of the respective transmission, T: $A =
-\log[T(n,i)/T(ref,i)]$. Functionally, the two values are equivalent
in that the amount of light falling on the detector is evaluated
and subjected to the same mathematical treatment. The only
difference is in the physical path of the analytical light beam for
the transmission and reflection experiments.

The spectra of Fig. 3 clearly show the changes wrought in exposure
of this sample to the vacuum for the first time since it was formed
at saturation from the uranium and water feedstocks. These data
represent the "difference spectra" and must be recognized as the
ratiometric comparison of the spectral features of the given state
to the vacuum/40 C state of prolonged exposure noted for n = 14 as

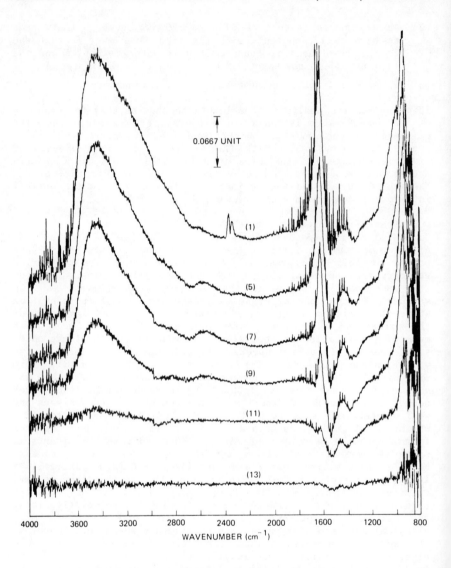

COMPARISON DRIFT SPECTRA: DESORPTION FROM UO₂ (C)

FIG. 3. Difference Spectra for the Drying of Uranium Oxide.
The nonlinear effects due to nonchemical optical effects is
not observed in this mode of presentation. The changes due to
the loss of hydrogen bonded water of hydration and the transi-
tory nature of a hydroxide specie is quite evident.

described above. Alternate interpretation of these spectra in any
sort of "absolute" sense is to be avoided at all cost since the
effects and spectral features are meaningful only in the sense that
they refer to modification or changes with respect to the alternate
reference state, 14, used for the calculations mentioned above. We
note that the water vapor bands persist 4000 to 3600 and 1900 to
1400 wavenumbers and serve to monitor the concentration in the head
space above the sample as does the 2350 band for gaseous carbon
dioxide. We note here that these bands can be used in conjunction
with the pressure gauge to monitor the activity of these components
individually for a much superior evaluation of the kinetics and
mechanisms of sorption and reaction processes. The reader is
cautioned to recognize that the intensities of these bands are
referenced to the aforementioned nonstandard state (pressure,
temperature, time, geometric configuration, etc.) and quantitative
analyses require cognizance of the difference between this
"relative" state and the "absolute" state for the best precision
and accuracy.

Structure Evaluation
Closer scrutiny of the band shape of the bound O-H vibrational band
show features characteristic of somewhat distinct modes at 3530,
3400, 3190, 2890, and 2510 wavenumbers corresponding to sequen-
tially increasing strength of the hydrogen bonding [32,34]. The
number of distinct orientations for hydrogen bond formation is
limited to the number of electron donor groups that can be oriented
in close juxtaposition to the hydrogen of the O-H group of
interest. A singlet involves only the participating hydrogen and
one single electron donor (water oxygen, oxide ion, or hydroxide
ion). Similarly, a doublet is comprised of two electron donors and
the covalent O-H entity, probably with the hydrogen positioned
coplanar with the three larger entities. More complex arrangements
are expected with lesser probability (less absorption) up to the
limit of the pentamer. This corresponds to hexagonal close packing
around the hydrogen: one proton donor, O-H, and five electron
donors. The energetics of the hydrogen bonds is known to give rise
to approximately proportional spectral energy (wavenumber) shift
[32-34] which is due in part to the greater charge interaction
between the electron donors and the proton. This electrostatic
interaction gives rise to the rectilinear plot of Fig. 4, where N
is the sequence number corresponding to the number of donor
electron pairs (partial residual excess charge) involved in each of
the sequential stretching vibrational frequencies noted above.
Algebraic minimum least squares analyses yields for the characteristic wavenumbers, $Y(N) = 3571.4 - 42.49N^{**}2$, with an index of
determination of 0.999989 and a standard error of determination
equal to 1.6 wavenumbers. The linear extrapolation to N equal zero
(no hydrogen bonding of the hypothetical hydroxyl) is markedly
different than the 3645 wavenumbers that is noted for the vibrational frequency of an isolated (non-hydrogen bonded) single gas
phase water molecule [32]. This would indicate that the bound
(adsorbed) species are all influenced by the attractive force field
of the surface as well as the hydrogen bonding discussed above.
The complex nature of the difference spectra in the 1700 to 1300

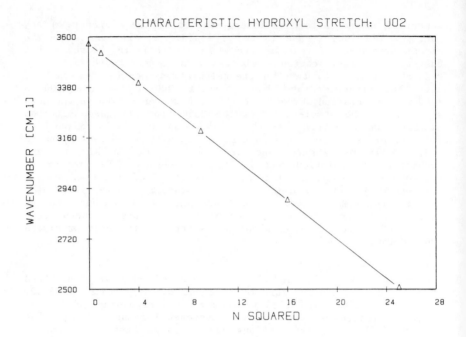

FIG. 4. Correlation of Hydroxyl Vibrational Energy to Strength of Hydrogen Bonding for Hydrated Uranium Oxide. The stronger hydrogen bonds formed by the interaction with the greater electron density tend to have characteristic lower vibrational frequencies (lower wavenumbers, reciprocal centimeters).

wavenumber region is due to dehydration, followed by a slower restructuring of the substrate to form the final U-O vibrating dipoles. The DRIS technique is uniquely suited to study this intermediate state. Isotopic effects are useful in further elucidating the complexities of the reactions between water and uranium [12]. These difference spectra also show the much greater discrete nature of the 870 wavenumber band (U-O) as we now observe only the changes that were previously distorted by the sharply sloping background.

This discourse is designed primarily to describe the DRIS technique and the reader is referred to other works where more details of the structure and chemistry of these systems are discussed [12].

Quantitative Analysis
It is quite obvious that accurate and precise quantitative analyses are more straight forward and less ambiguous when band area integration is performed on the difference spectra (Fig. 3) in reference to the data presented in Fig. 2. Details of the ana-

lytical methods and the interpretation of the results are given in Ref. [12] and are available to interested parties.

COALS AND RELATED MATERIALS

"Coal" is generally described in different terms by persons in different disciplines, depending on the method of observation and the end use for the specific processes. A field geologist may generally class coals as "black rocks" but soon must recognize coals as the product of "coalification" of plant tissues. In this context coals do not fit into a classical mineralogy even though the systematic study of the mineral matter in and around coalbeds is very information and an intimately intwined process that allows us to better understand the geological development of the given coal to its state of coalification. A coal miner/processor/vendor must view his product in terms of the economic import depending on the end use. He spends considerable effort and time to remove the "rocks" from his coal and does not appreciate such appelations for his "refined" product. A combustion engineer whose task is to supply heat, steam and/or electricity to a vigorous economy, is interested in the energy content which he can extract and wishes to minimize the burden and environmental impact of the inorganic mineral matter and noxious precursers (sulfur, nitrogen, etc.). In this context, it is worthy to recount that coal burning power plant release much more radioactivity in stack gases than nuclear fission reactors since the uranium and thorium levels are appreciable in most coal beds. Chemical firms (and the discipline now referred to as organic chemistry) have risen to prominence based on feedstocks derived from coal and, as our petroleum reserves wane, we will be forced to turn once again to this much greater and versatile reservoir of reduced carbon. Petroleum feedstocks have allowed the progress of synthetic formulations to spawn astute polymer scientist who now travel full circle and query as to the "polymeric" structure of coals themselves. Even our extremely important steel industry is based on the use of coal to provide the coke for the smelters. This industry has developed the knowledge of "coking" and related properties to an extent that much of the ASTM ranking is based on these requirements along with the BTU (heat content) and moisture (water content). Our mobile mechanized society requires vast amounts of liquid and gaseous fuels and current technology is such that considerable amounts of research are required to form adequate amounts of liquid/gaseous hydrocarbon fuels from our extensive coal reserves.

Even though direct utilization of coal is not the most prudent solution to our energy needs, interim solutions require that we "burn" this valuable resource by essentially instantaneous oxidation to water and carbon dioxide, fleetingly using the heat and electricity. In this process we are reversing the reduction process that involved years of plant metabolism followed by millions of years of anerobic digestion in the darkness of the waterlogged peat/lignite/coal bed. All of that ancient photochemical energy is virtually irretrievably lost to us in an instant within the fiery core of the furnace or hidden from view as a

turbulant explosion within the cylinder of an internal combustion
engine. Whenever we do have to consume this material it augers
well for us to best understand the chemistry and structure of this
solid material. Infrared spectroscopic techniques have been
extremely information for defining the chemistry and structure of
both inorganic [31] and organic [29,30] an classical techniques
have been very useful in defining the rudiments of the chemical
structure of coals and related materials [7,8,10]. Recent developments in instrumentation and techniques have developed the diffuse
reflectance infrared spectroscopy to a state that is superior based
on speed, convenience, versatility, quality of the spectra, and
simplicity of operation [1,2,4,16].

Qualitative Analyses
Powder beds generally serve as diffuse reflecting substrates as the
incident light is partially reflected from randomly oriented
surface (gas/solid interfaces). Some of the light is transmitted
through the uppermost particles and is subseuqently
reflected/transmitted at subsequent interfaces. A portion of the
incident light finds its way back through the upper surface of the
bed and is measured by the detector of the spectrometer. The
mathematical treatment of the data and the subsequent mode of
presentation of the spectra generally follows two schemes:

(1) the "absorbance" scheme as discussed above, and

(2) the "remission" scheme where the spectral intensity is
     reported [17,20] as the remission function, $f =$
     $\{1-[R(n,i)/R(ref,i)]\}$ **$2/\{2R(n,i)/R(ref,i)\}$.

Detailed numerical comparison have been made for the two methods
[5,17] to show the merits of the two techniques and that both are
quite useful mathematical aids for numerical and graphical
presentation of the DRIS data for coal powders as shown in Fig. 5.
A fourfold multiplicative scaling factor is used to "magnify" the
absorbance values to a scale comparable to that of the remission
(Kubelka-Munk) calculations for the same DRIS data. One is pressed
to say which method is "best" and/or of more use in identifying
species in the coal substance. In either case one can readily
recognize the "fingerprint" bands:

(1) the highly skewed (3600 to 2000 cm^{-1}) band for hydrogen bonded
     O-H stretching;

(2) the rather small band (3100 to 3000 cm^{-1}) for the aromatic and
     olefinic C-H stretch;

(3) inflections (1800 to 1650 cm^{-1}) due to residual C=O carbonyls
     on the side;

(4) the very strong band (1605 cm^{-1}) found for conjugated poly-
     nuclear aromatic compounds which overlaps the adjacent group
     of bands (1450 to 1000 cm^{-1}) that are generally characteristic
     of the bending modes of H-C-H and C-O-H (as well as stretching
     modes of C-O-C, O-C-C, O-C-O, etc.); and, ubiquitous for
     coals;

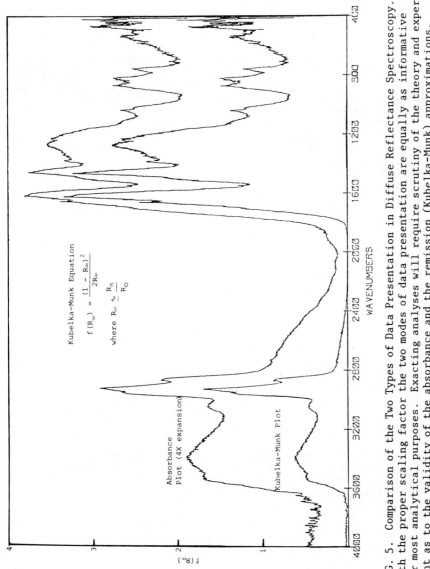

FIG. 5. Comparison of the Two Types of Data Presentation in Diffuse Reflectance Spectroscopy. With the proper scaling factor the two modes of data presentation are equally as informative for most analytical purposes. Exacting analyses will require scrutiny of the theory and experiment as to the validity of the absorbance and the remission (Kubelka-Munk) approximations.

(5) A very prominent triplet (900 to 700 cm^{-1}) arising from the
     out-of-plane bending mode of hydrogen moieties on an aromatic
     structure.

A wealth of information is available and one might be tempted to
favor the absorbance plot just because it seems to give the better
definition over the entire intensity realm, with greater delinia-
tion of the O-H bands in this case. We have opted to present all
of our DRIS data for this report in the absorbance format for
convenience and for the technical reasons tendered above. In this
format we can readily relate to the vast amount of "transmission"
spectra either visibly or by the use of a computerized library
routine. Figure 6 is the composite of the first five (5) molecular
species that were obtained as the "best match" by a computer search
algorithm comparison of the coal spectrum to the 60,000+ encoded
spectra of the Sadtler (TM) library. One is cautioned not to
assume that this is proposed as the chemical structure of the "coal
molecule" but must assuredly represent the types of structures that
are to be found in the melange of the naturally occurring material.
It is heartening that this structure possesses and embodies almost
all of the chemical and structural entities that have been found
and proposed by other researchers [35,36] based on various chemical
analyses.

A somewhat more detailed understanding of the structure and
chemistry coal is afforded by the versatility of the DRIS technique
when one compares the spectrum of the virgin coal with its full
complement of bed moisture (Fig. 7A) for a sample obtained from a
freshly opened face, sealed in an opaque container to prevent dehydration and/or oxidation by the air, and carefully maintained at
25(+/-3) degrees centigrade. Subsequently the sample was exposed
to vacuum in situ without moving the sample with respect to the
optics of the spectrometer to achieve the reference vacuum state
for spectrum 7B which was used to calculate the difference
spectrum, 7D. In this manner we can unequivocally differentiate
the spectral contribution of labile water O-H from that of the
bound O-H of hydration, alcoholic, phenolic, and/or acidic groups
in the coal structure. Some further detail are obvious in terms of
DRIS features:

(1) The vacuum labile water is characterized in terms of three
     components, i.e., the symmetric band at ca. 3600 cm^{-1} is very
     similar to that of liquid water, and the broad distribution of
     hydrogen bond energies corresponding to 3450, 3200 and ca 2600
     cm^{-1});

(2) There is an appreciable concentration of kaolinic mineral O-H
     as noted by the four sharp peaks at 3700 to 3600 cm^{-1} and
     other associated frequencies [1];

(3) The inflection points on Fig. 7B and 7C reveal distinct O-H
     entities within the organic structure giving rise to preferred
     absorptions at 3540, 3390, 3290, and ca 2550 cm^{-1};

(4) The hydrocarbon content of this material is stable with no

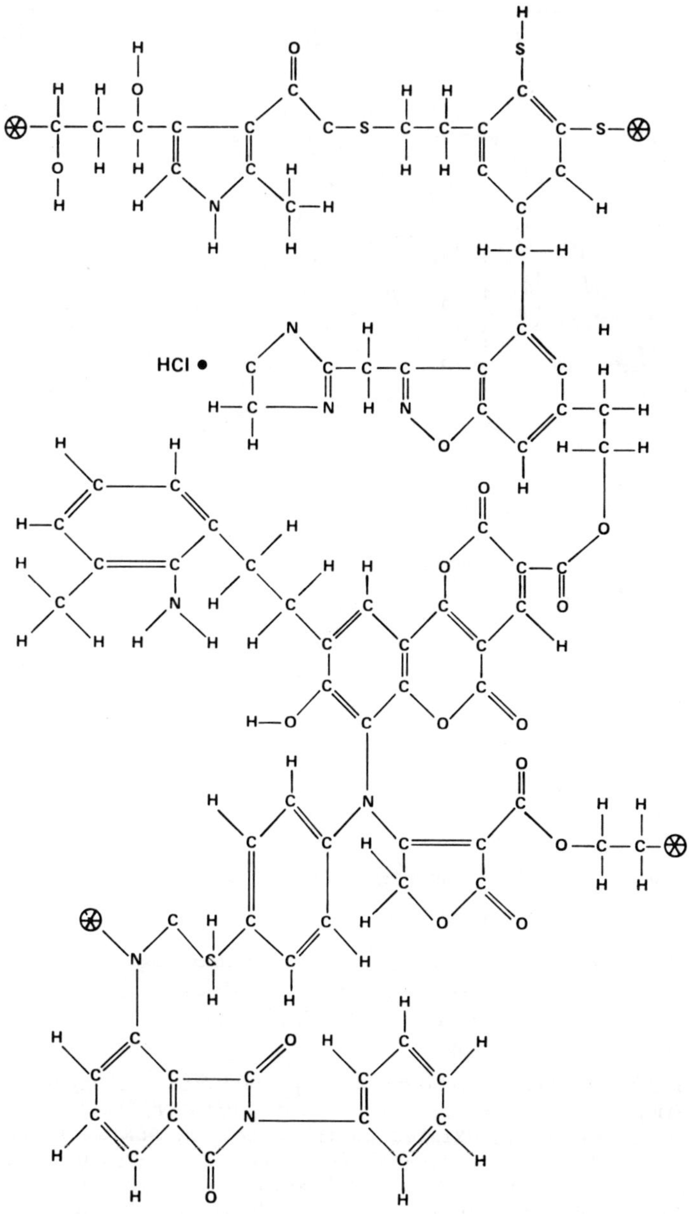

FIG. 6. Characteristic Structure of a Coal Based on Diffuse
Reflectance Infrared Spectroscopy. This is a composite structure composed of the chemical structures of the five (5)
chemical compounds that have infrared spectra most closely
matching the diffuse reflectance spectrum of coal (Fig. 5).

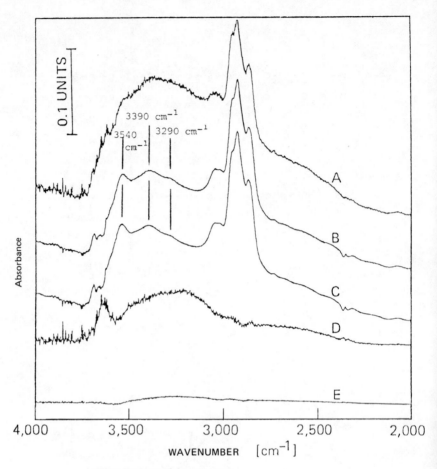

DRIFT Spectra of Wyodak Coal.

A.   27°C, 100 kPa   D.   Difference Spectrum (A-B)
B.   27°C, 1 Pa   E.   Difference Spectrum (B-C)
C.   191°C, 1 Pa

FIG. 7. Diffuse Reflectance Spectroscopic Changes Due to
Drying Coal. A. DRIS of coal powder with its full compliment
of moisture as it existed in the mine. B. DRIS of the same
area of the same samples as (A) upon exposure to 0.001 Pascal
vacuum and steady state equilibration. C. DRIS of (B) at 200
C in vacuum after 16 hours of steady state. D. DRIS difference (A-B) showing the spectroscopic features associated with
the vacuum labile moisture. E. DRIS difference (B-C) shows
that there is not a significant amount of thermally labile
hydroxyl species that was not vacuum labile at ambient
temperature.

detectable volatile loss over the temperature and pressure range of the experiment;

(5) The spectral region classically referred to as the "aromatic region" (3000 to 3100 cm^{-1}) is unquestionably made up of two components, based on the recognition of two inflection points.

(6) One can well measure the "moisture content" of the coals by an appropriate integration of the area under curves such as 7D and probably obtain better values than the classical method of measuring weight loss upon prolonged exposure to 100°C. In the DRIS method we have the further advantage of evaluating the amount of the various modes of water binding for future relation to economics and/or chemical processing. Spectrum 7E can be taken as assurance that ambient vacuum exposure is adequate to remove the water and preclude prolonged treatment required for "diffusion" out of the coal as in the classical method of "moisture" analyses;

(7) There is little doubt that there is a true band at ca 2730 cm^{-1} which may well arise from the vibration of C-H moieties of inherent aldehyde groups of the coal [2] or, alternatively be the result of combination and overtone bands of the poly-nuclear aromatic substrate [37]; and

(8) The intensity of the narrow bands (3500 to 3700 cm^{-1}) serve as specie specific measures of the concentration of water vapor in equilibrium with the coal/mineral substrate and afford a more accurate correlation than the total pressure measured by capacitance manometers and ion gauges. See the preceeding discussion of the uranium/water system.

Surface Modifications of Coals

The surface of solid coal can be made to diffusely reflect infrared radiation [28] by a single abraision to produce a matte surface as witnessed by the Spectrum A of Fig. 8. Only a trace of carbonyl oxygen is noted to give rise to the "shoulder" (inflection point) at ca. 1704 wavenumbers. This region of the sample is readily oxidized under controled conditions (400°C and 20 torr of air) to markedly alter the spectrum to that of Fig. 8B where the degree of oxidation is characterized by a virtual abscense of aliphatic hydrogen and a marked degree of oxygen incorporation into the substrate in the form of a variety of carbonyl structure (ca. 1704, 1760, 1850, etc. wavenumbers). This initial reaction zone is rather thin and can be removed by gentle abrasion to expose un-reacted substrate as noted by the resultant spectrum of Fig. 8C. The hydrocarbon bands are restored to at least the intensity of the original state. The fact that the residual carbonyl bands are retained to a slight degree may be a result of the oxygen insertion reaction preceeding the hydrogen removal reaction through the matrix of the substrate. Oxygen insertion occurs both at the methylenic sites and at carbon-carbon graphitic sites, whereas dehydrogenation occurs only at the former. Preferably, the hydro-genated component of the coal should be removed before combustion for use as chemical feedstocks and/or premium fluid fuels. Current technology uses this hydrogen component as an energy-rich

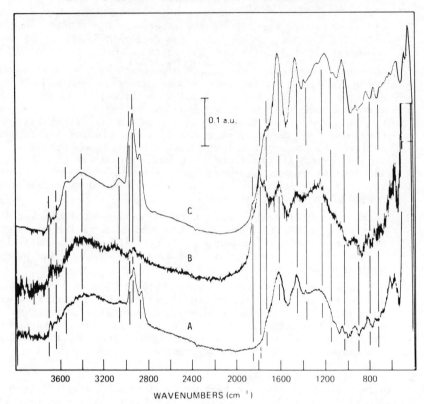

DRIFT SPECTRA: MODIFIED KENTUCKY NUMBER NINE

0.1 a.u.

C

B

A

3600   3200   2800   2400   2000   1600   1200   800

WAVENUMBERS (cm⁻¹)

FIG. 8.   Diffuse Reflectance Spectroscopy of Solid Kentucky
Number Nine: Effect of Chemical and Physical Treatment.   A.
DRIS of the abraided surface of solid Kentucky Number Nine.
B.   DRIS of the same area of the same sample as (A) after air
oxidation at 400°C.   C.   The same area of the same sample as
(B) after reabrasion to remove 10 to 50 micrometers of the
sample surface.   The signal/noise ratio is decreased markedly
by alloting more time to accrue better statistical averaing
when steady state conditions are maintained.

"kindling" to accelerate the combustion of the char component in
conventional combustion technologies.   Any gasification or liqui-
faction scheme must account for a process to utilize the residual
carbon that remains.   If we must burn coal for heat then let us do
it judiciously where tankers of fluid fuel are produced at the
power plant site and only the residual char is squandered to
produce heat and/or electricity.

SECOND DERIVATIVE SPECTROSCOPY (SECDER)

The very richness of qualitative information in the DRIS of coals
must be reckoned with in any quantitative scheme. Virtually every
band overlaps one or more of the adjacent bands so that some sort
of deconvolution scheme is required in order to evaluate the area,
shape, and/or intensity of each. Classical transmission infrared
spectroscopy of coals has the same spectral features [8,10] since
the heterogeneous nature of the environs around the vibrating
entitities generates a wide variety of force fields for a given
chemical entity. We are blessed with numerous infrared chromophores interspersed in a polynuclear aromatic matrix which in turn
is comprised of some graphitic (infrared inactive) components. For
instance, the hydrocarbon (3200 to 2700 wavenumber) band is
comprised of two aromatic and at least four aliphatic entities as
mentioned earlier. Any direct area integration can give only a
statistical "average" related to the concentration of hydrocarbons.
On the other hand, a wealth of information is available pertaining
to

(1)   the number of chemical entities (oscillating dipoles);

(2)   The type and distribution of environs (electrostatic fields);
      and

(3)   The concentration of each of these chromophores in the path of
      the diffusely reflected analytical light beam.

Infrared absorption spectra are made up of the summation of a
number of discrete bands of the Lorentzian/Gaussian shape for each
of the composite optically active oscillators which interact with
the electromagnetic radiation. Each of these bands has to be
characterized by enough parameters:

(1)   to uniquely position it on the energy (wavenumber or fre-
      quency) scale;

(2)   an intensity value (the extensive parameter related to the
      concentration of the chromophore);

(3)   a shape factor (half width at half maximum, height/width
      ratio, etc.).

One can draw upon the techniques of classical infrared spectroscopy
to visibly search out, judge the existence of, and enumerate the
number (b) of bands required to account for the peaks, shoulders,
inflections, etc. that exist in the experimental spectrum. The
modern computer that is becoming an intimate part of the spectrometer is ideally suited to carry out the task more efficiently,
more rapidly, more reproducibly, more precisely, and with less
variation due to "human error" or judgement between analyst/
researchers. A simple calculation is made to obtain the derivative
spectrum (second derivative, inverted to enhance and maintain the
"peak" character at the ordinate position corresponding to the
inflection points on the integral spectrum). In actuality, derivative spectra such as that given in Fig. 9 are really a much better

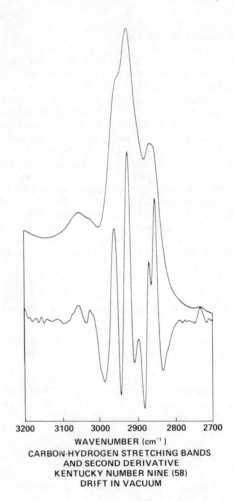

WAVENUMBER (cm⁻¹)
CARBON-HYDROGEN STRETCHING BANDS
AND SECOND DERIVATIVE
KENTUCKY NUMBER NINE (58)
DRIFT IN VACUUM

FIG. 9. Second derivative spectrum of the hydrocarbon region
of Kentucky Number Nine (K958). Upper curve: Experimentally
determined DRIS spectrum where inflection points and
"shoulders" are noted at frequencies of absorbance of infrared
radiation. Lower curve: Negative of the second derivative of
DRIS spectrum illustrates the more obvious qualitative
analyses for determining the number and position of contributing absorption bands.

means of qualitative analyses for the ultimate goal of discovering
how many species (B) of what chemical composition (wavenumbers) are
present in the given speciman. Newer spectroscopies, such as Auger
Electron Spectroscopy, AES, routinely rely on this enhanced method
to present the data to prove the presence or absence of a given
element. Furthermore, it is a simple task to program the computer

to select the wavenumber data for the local maximum intensity for
each of the B bands proven to contribute to the B inflection points
in the region of analyses. The computer is programmed to select
the data point at which the negative second derivative is a maximum
and thus we see that the closest estimate to the characteristic
wavenumber is dictated by the resolution of the spectrum (2 wavenumbers in this instance). Although the values listed in Table I
are measured to the nearest 0.01 wavenumbers, the characteristic
frequencies for each of the B inflection points actually exist
somewhere +/- 2 wavenumbers from these positions. This point is
moot anyway since the inflection points per se do not necessarily
occur at the same position (wavenumber, frequency, energy, etc.,
depending on the bent of the reader) as the centroid of the contributing bands, so an interpolation scheme does not seem warranted
at this time.

FOURIER SELF DECONVOLUTION (FOSEDE)

The development of the Fourier transform infrared spectrometers was
made possible by the computing capability, ready accesibility, very
high speed, and ingenious software for dedicated computers and
microprocessors. This equipment is now available for further data
processing. The Fast Fourier transform hardware/software can be
very beneficial as an aid in further interpretation of the data.
FTIR spectra are obtained by the Fourier transform of the interferograms and the numerical data can be stored for future presentation either in the frequency (spectral) domain or in the time
domain, based on Fourier or inverse Fourier transforms, respectively. Convolution (combining of two curves) in the spectral
domain forms a Fourier transform pair, FTP, with multiplication in
the time domain [38]. Conversely, the operation of deconvolution
(curve separation) in the spectral domain can be carried out in
three steps:

(1)   Fourier transform to the time domain;

(2)   divide the interferogram by the desired function (curve); and

(3)   subsequently retransform back to the spectral domain to form
      the deconvolved spectrum.

This operation is very useful to remove the instrument line shape
contribution from the spectrum when such an instrument line shape
is known or can be accurately calculated or predicted. For
instance, the contribution of the triangular slit function of a
grating instrument can be "stripped" from the spectrum as described
above. Fourier self-deconvolution is the mathematical extension of
the above concepts to actually reduce the apparent band width by
"deconvolving" a fraction of the inherent Lorentzian bands within
the composite envelope of the experimentally determined spectrum to
exagerate the height (at the expense of the width) of the bands
comprising the spectrum [22,28,39]. The result of Fourier self-
deconvolution of a portion of the DRIS spectrum of Kentucky Number
Nine coal is presented in Fig. 10. The degree of "spectral
enhancement" was chosen to be a maximum to allow the definition of

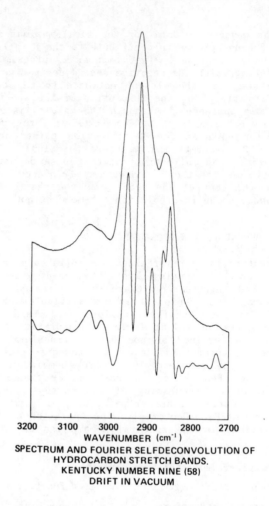

SPECTRUM AND FOURIER SELFDECONVOLUTION OF
HYDROCARBON STRETCH BANDS.
KENTUCKY NUMBER NINE (58)
DRIFT IN VACUUM

FIG. 10. Resolution enhancement by Fourier self-deconvolution
of the hydrocarbon bands of Kentucky Number Nine (K958).
Upper curve: Experimentally determined DRIS that must be used
to define the number and position of the "peaks", "shoulders",
"knees", etc. for qualitative analyses. Lower curve: Fourier
Selfdeconvolution of DRIS appears to be as useful as the
second derivative method (Fig. 9) for more readily recogniz-
ing the features in terms of "peaks" with less equivication.

the inflection points but not enough to appreciable "amplify" the
noise (the low level, somewhat periodic, or ringing features noted
in the FOSEDE of the featureless regions of the DRIS spectrum). A
simple computer program is used to assure that the degree of
"resolution" is limited to be significantly greater than that which
generates "peaks" due to the noise of the spectrum in regions where

no significant chemical information exists. In some instances the
FOSEDE or SECDER can be used to evaluate the optical interference
phenomena and obtain information related to the size of the primary
structural unit of the coal (particle size, domain size, abrasion
asperity dimensions, etc.). The reader is warned to be aware that
artifacts of this nature can be observed when an inadequate apodization (filter) function is used to calculate the initial DRIS
spectrum from its finite interferogram [27]. Most schemes for
SECDER and FOSEDE use a "smoothing" function to avoid the problems
of random noise in the enhanced spectrum. The results in Fig. 9
and Fig. 10 were not artificially "smoothed" since we utilized the
high signal/noise associated with the 25,000 scans.

Virtually all of the functions used to smooth, apodize, filter,
etc. can be classed as "filter" functions and the same nomenclature
is applicable to the FOSEDE and the SECDER operators, in the sense
that the "resolution enhancement" is in reality a "negative
filtering" (unsmoothing) effect. Other analysts have pointed out
that many of these operations can be used as smoothing or
"apodization" functions in a seminegative sense [40]. The
similarities carry over to note that the FTP of SECDER is a simple
parobolic multiplier, $(2\pi x_i)^{**}n$, (where n=2 for second derivative),
as compared to the FTP of FOSEDE which is exponential, $\exp(2\pi wx)$,
(where w is the width at halfheight). Both multipliers increase
monatonically and asymptotically to greatly emphasize the magnitude
of the interferogram at large values of x. A true positive
filtering effect (smoothing) can be effected in either case by the
proper choice of n or w. Such is not the goal of our efforts as
presented here since we utilize both techniques in the sense of a
negative filter for the express purpose of high quality determination of two values: the number of distinct bands present in a
given envelope and a first estimate (within the resolution of the
data base) of the position of the centroid of each of the bands on
the wavenumber (frequency) scale. One should not extend the
resolution enhancement beyond the point where the features of
reference regions (regions with no obvious specie related features)
becomes appreciable with respect to the related features in the
active regions of the spectrum. To this end one must be certain
that the "enhancement process" is not carried to the extreme that
one exceeds the Nyquist frequency [38] where the number of spectral
features approaches the number of data points in the same region.
The data of Figs. 9 and 10 serve to illustrate the point. In the
region between 3100 and 2800 wavenumbers there is strong evidence
for the presence of seven (7) bands spaced somewhat uniformly over
the 150 data points (300 wavenumbers at 2 wavenumber resolution) so
that there is a comfortable 21 data points for each of the bands to
be used to accurately described the height, width, and position of
each. However, if the same data were acquired at eight (8) wavenumber resolution one would have to evaluate each set of three
parameters with the aid of 5 to 6 data points for each band. Such
is not too risky, in terms of a robust solution to the fitting
problem. However, additionally more stringent "resolution
enhancement" can and will provide nonunique fitting of the data for
we now come to the case where we have more unknowns than we have

data points to describe the spectral envelope. To illustrate, if
one predicts 11 bands in the same interval with 8 wavenumber
resolution, there would be only 38 data points to provide the basis
for estimates of the 33 needed (height, width, and position)
numerical values. This situation seems trivial, but is often
overlooked by the engrossed analyst. Even the observation that the
enhanced values fall at reproducible wavenumber values is not proof
of the uniqueness of these solutions since the trends would be for
the erroneous "peaks" to tend to "protrude" between the sparse data
points which occur at approximately the same abscissa (wavenumber)
for each spectrum gathered on a given instrument. A good rule is
to acquire at least 5 data points for each curve that is imbedded
in the experimental envelope to assure good statistical definition
of the 3+ required adjustable parameters. Care should be taken
that even these points are distributed to cover the major portion
of the calculated band. Such criteria are not overly restrictive
and can be readily incorporated into a computer driven program for
the qualitative analyses of coals and related materials. In many
cases these high caliber qualitative analyses may suffice but, more
often than not, quantitative analyses are of more interest in
commerce and production. Analytical operations are somewhat remiss
if quantitative analyses are not provided for the time, energy, and
expense is minimal to allow the digital computer to process the
stored numerical spectral information the additional step and to
provide the "customer" with highly informative tabular and
graphical results. Even if the instrument and operator are
occupied during a shift, most computer/spectrometers are not
acquiring data for extended periods and the additional computations
can be carried out at this time.

MINIMUM DEVIATION COMPOSITE SUMMATIONS (MIDECS)

Accurate quantitative analyses require evaluation of the contributions of each of the composite components to the intensity of
the experimental spectrum. This is contrasted to the classical
analyses where one is more than likely analysing for the amount of
one or more chemical components in solution or in a mixture. Coal
cannot be viewed as a chemical compound, or even very accurately as
a mixture of a few chemical compounds. More fruitfully we model
coal as a continuously varying (both in time and space) composite
made up of classical organic fragments bonded together to form the
organic phase which may or may not [41] be intimately mixed with
the inorganic (mineral) constituants. The atomic ratios, interatomic arrangement, and functionality was altered over the eons by
the coalification process in the dark, hydrous anerobic confines of
the coal bed, where the dietrous of cellulosic life systems had
accrued. The heterogeneous nature of the feedstocks, the diffusive
segregation, the variation of environment (temperature, pressure,
etc.), and the vagaries of prehistory all lead us to suspect that
one can and should talk of "average" or "representative" structre
of each coal much akin to that presented in Fig. 6. Good sampling
techniques and comprehensive spectral comparison of the infrared
spectra, with all of the extensive information contained therein
[40], will allow us to determine how much variation there is for

coals of a given rank and environ. Even then we must be very cautious of assuming that coal has structure akin to a mixed polymer with a few distinct entities bonded (crosslinked) together.

The data of Figs. 9 and 10 serve as an excellent example of the composite summation process since the band level of the noise is so low by virtue of the signal averaging with the 25,000 scans for the coadded interferograms. The effect of noise will be discussed below. Our approach is to utilize the previous qualitative analyses to determine how many bands are present and to provide a first approximation as to where these bands are located on the energy (wavenumber axis. These values serve as an initial set for a reinterative process whereby the three parameters (height, width, and position) are adjusted for each of the B predetermined bands. The program is essentially one proven to described such data [42,43]. We have opted to allow the background curve to be adjusted also to give the best minimum least squares deviation between the composite summation as shown in Fig. 11. We have also allowed the band shape to vary with a fractional Gaussian/Lorentzian character to account for instrumental and other convolutions [42,43]. We note that the independent optimization of each of these 3B+3 adjustable values does provide a MIDECS that is virtually indistinguishable from the experimental spectral envelope. Perfect correlation would have resulted in a flat, horizontal, straight line for the difference curve (the top line of Fig. 11). Several tests can be cited to point out the necessary, sufficient and unique character of the MIDECS "solution" to the complex wave form of the experimental spectrum:

(1)   The quoted MIDECS curve was achieved as the convergence of the interative process with absolutely none of the 3B+3 values restrained;

(2)   The same values were obtained for each experimental wavenumber upon reconvergence after various parameters were modified to higher or lower values. The answers showed no "hysteresis" that one would expect if the technique were not sensitive to said variable;

(3)   Insertion of one or more added peak (with the 3 additional parameters) does not improve the fit. As a matter of fact the mathematical treatment is so robust that in the final convergence the additional band invariably was assigned a zero width, a zero height, and/or a wavenumber far off of the analysed scale. In other words, the mathematical treatment "rejects" the additional band;

(4)   Attempts to find MIDECS with one less band invariable leads to marked deviations from the experimental envelope with a standard deviation much greater than the noise level of DRIS;

(5)   The characteristic vibrational energies are very close to those found in pure chemical compounds [29,30] known to contain the general hydrocarbon functional groups as listed in Table 1;

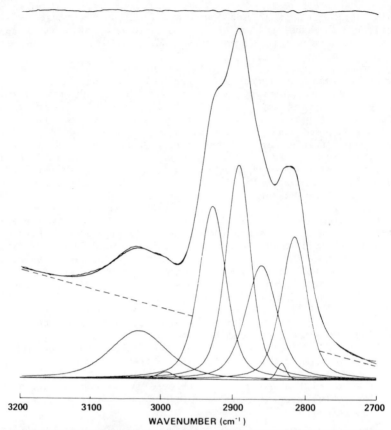

LEAST SQUARES COMPOSITE BAND SUMMATION: CARBON-HYDROGEN STRETCH
KENTUCKY NUMBER NINE (58) IN VACUUM

FIG. 11. Minimum Deviation Composite Summation Interative
Calculation Compared to the Experimental Hydrocarbon Spectrum
(K958). The MIDECS is seen to overlay the experimental
diffuse reflectance spectrum very well with only minor deviations as witnessed in the upper difference tabulation. Each
of the composite curves are the result of free convergence to
the MIDECS, with the baseline, Gaussian component, and individual band parameters (height, width, and position) all
optimized.

(6) The results of the analyses of our DRIS experiments are vir-
     tually the same as those obtained in another lab with another
     coal, and with another technique (alkali halide pellet trans-
     mission) [8];

(7) Although it is almost trivial, the SECDER and FOSEDE of the
     MIDECS are insignificantly different from the results of Figs.

TABLE 1. Characteristic Infrared Band Frequencies for Coals Based
on Various Analytical Techniques †

SECDER	FOSEDE	MIDECS	REF.[8]	REF.[29]	ASSIGNMENT
3051	3051	3051	--	3050(50)	Ar-H
3016	3016	3017	--	3025(25)	R2C=CHR & RCH=CHR
2958	2954	2958	2956	2962(10)	RCH3
2924	2924	2925	2923	2925(5)	ArCH3 & R2CH2
2896	2896	2894	2891	2900(?)	R3CH
2866	2866	2870	2864	2872(10)	RCH3 & ArCH3
2850	2850	2854	2855	2849(10)	R2CH2
2735	2735	2752	--	2740(15)	RCHO & Ar3CH

† All frequencies are reported in wavenumbers (reciprocal
centimeters) and the assignments are those used as "fingerprints"
for classical qualitative analyses.

9 and 10 for the experimentally recorded spectrum; and

(8) We have practiced enough by testing each other (and the
computer program) by supplying each other with composite bands
which were synthesized (calculated) without information as the
number of bands, position of the bands, band height, band
width, nor relative Gaussian/Lorentzian shape. For bands as
complex, or more so, as the examples shown in this report we
have been able to recalculate the parameters that were used to
compute the test envelope [44].

One must recognize some trivial problems such as the oft quoted
melding of two equal bell shaped curves to produce a summation that
is defined by its own set of 3+1 parameters. When this arises in
practice the analyst must use other information to decide which
(the large summed bands or its composite pair) meets the criteria
of necessity, sufficiency, and uniqueness for the given case. In
order to maintain an economy of postulates and sufficiency we
report the directly measured summation band. The analyses of these
synthetic envelopes is quite routine and one merely retraces the
steps outlined above:

(A) choose the number and approximate position of the bands by
SECDER and/or FOSEDE;

(B) Utilize the reiterative MIDECS to convergence to the noise
     level which was synthesized on the envelope; and

(C) Examine the deviation band both overall and locally, for dis-
     parites that are associated with inadequate background compen-
     sation and/or erroneous band computation, respectively.

The results are good and no problems arise as long as the number of
data points supplied is great enough to meet the Nyquist sampling
criteria and the "noise" band on the envelope has a rather sharp
(Gaussian, triangular, etc.) distribution about the mean (signal)
values. Flat distributions and skewed distribution of noise on the
signal tend to give rise to slow convergence and to net displacements, respectively, for the final values used to compute the
minimum least squares deviation between the calculated envelope and
the noise-free component of the synthetic envelope.

The techniques of quantitative analyses by computer aided MIDECS is
shown here to be very successful in analysing the envelope
associated with the various C-H stretching modes in solid coal
which comprises only a small portion of the DRIS data. Other work
has shown that the same technique is capable of evaluating the band
shape/size for virtually every component [2] in the entire complex
envelope. These analyses can be carried out for narrow regions, as
recommended in Ref. [8], with all 3B+3 variables optimized to give
the best correlation coefficient with respect to the experimental
segment (in deference to preselecting the 2 parameters defining the
background). These evaluated segments then serve as the initial
approximation for a global calculation where the 3 parameters of
the entire ensemble are finally adjusted by the MIDECS procedure
with respect to one overall optimized linear background. The final
result of our global analyses is similar to others [10] except that
we do not precorrect for a sloping background, do not subtract out
the mineral bands, nor do we use preselected invarient positions
(wavenumbers) for the component bands. It is only with the aid of
the rapid, accurate, and extensive numerical processing that we can
hope to defined every feature of the complex DRIS data of coal.

POWDERED COAL SAMPLES

Diffuse reflectance spectra of high quality are obtained for
samples of powdered coal as shown in Fig. 12. Thus we see that
DRIS is ideally suited for analysing the feed stuff of a commercial
furnance where the standard practice is to pulverize the coal to
the consistency of "flour" prior to injection into the firebox with
a stream of air or oxygen. In this manner efficient combustion is
achieved roughly akin to an carburized mix of liquid hydrocarbon
and air. There is not enough time to allow the larger particles to
burn by some sort of "shrinking core" mechanism. The potential of
DRIS as an online, rapid analyser is great. With adequate development and correlation studies DRIS can and will replace several
classical techniques execpt for the most demanding and exploratory
studies. DRIS can give a good measure of the carbon/hydrogen/
oxygen content, replacing cumbersome "wet chemical" methods. We

THE ISOTHERMAL OXIDATION OF COAL BY AIR AT 300 C
IS READILY MONITORED

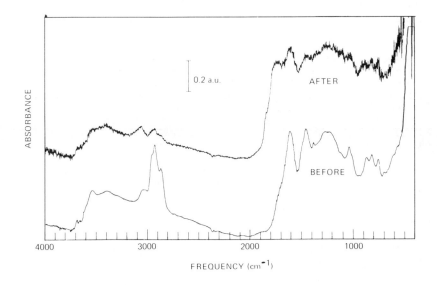

FIG. 12. Diffuse Reflectance Infrared Spectra of Powdered
Subbituminous Coal (WYODAK). Oxidation of the powder bed in
the cell within the infrared spectrometer alters the original
spectrum with intensity loss for the hydrogenous species and
gain for the oxygenated entities as noted in the upper
spectrum.

can envision that the same data can be used to evaluate the "ash
content" of the homogenized powders by calculation from the mineral
content (kaolinite, illite, silica, limestone, etc.) as deduced
from the same DRIS data for either atmospheric or vacuum spectra.
As noted earlier, a rapid facile measure of the moisture content is
obtained by comparing the original spectrum to that in vacuum. A
straight forward computer program will calculate the above values
and from theory and emperical calibrations allowing us to evaluate
virtually all of the engineering and chemical parameters of
interest. We can envision substantial savings in time and in
capital equipment. For instance, C/H/O analyses and energy content
(BTU, Joules, Calories, etc.) by calorimetry both require hours for
a single determination. The standard ASTM tests for moisture and
ash both need several hours, if not overnight, for each determination. The DRIS technique is rapid enough to provide useful information for process control before the sampled material has been
long since consumed. The speed of analyses can be increased with a
tradeoff for precision. The instrument manufacturers are now
making rugged and safe instruments that can be placed in the
hostile atmosphere of a power plant, a mine, or a coal processing
plant with the option of having the analyses controlled and moni-

tored remotely from a computer console in the operating room.

Added merits for the DRIS techniques are the measure of the
material that is to be used currently in the process without
complications associated with changes in the chemistry and
structure of coals associated with further processing (e.g.,
grinding [41] or by weathering as the analysed sample is left
exposed to air prior to use [45]). The tribulations of sending
samples off site for analyses further are enhanced due to sampling
problems, time delays, and possible confusion in record keeping.
The merits of an on site dedicated instrument are manifold.

REACTION MONITORING WITH DIFFUSE REFLECTANCE INFRARED SPECTROSCOPY
(DRIREM)

The DRIS technique is ideally suited to monitor the reactions that
occur in the uppermost regions of the powder bed as shown in the
limits of Fig. 12. The spectroscopic system can be set up to
obtain spectra of the sample at fixed or otherwise predetermined
time intervals. These stored spectra can then be analysed to
monitor the kinetics and mechanisms of the gas/solid reactions with
a degree of detail that cannot be achieved by any other single
technique. The merits of using the dedicated cell are

(A) The environmental conditions (temperature, pressure, gas
     composition, etc.) can be controlled much better than in
     natural or process situations;

(B) Many more spectra can be obtained to assure adequate detail
     with respect to temporal variations;

(C) All data are obtained for the same area of the same sample so
     that problems related to heterogeniety are mitigated;

(D) No problems related to variations due to sample preparations
     are found since all data are acquired on a single preparation;
     and

(E) In many instances one can "control" a reaction (modify rates
     and reaction paths) much less expensively than if he were to
     rely on large scale production or pilot plant equipment.

The DRIS system described here is much more than analytical tool.
It can best be viewed as a microreactor in which data can be
acquired to fit the needs of many disciplines; process engineering,
physical chemistry, organic chemistry, analytical chemistry,
combustion engineering, coal science, etc.

The changes due to oxidation are much more obvious when viewed as
difference spectra (with respect to the unreacted material in this
case) as shown by the spectra of Figs. 13, 14, and 15. These
spectra were selected from the larger record to show the qualitative changes that occur in the early, intermediate, and later
stages of oxidation. The same information exists in the original
spectra but the spectral feature of interest are difficult to
discern due to the existence of the very large feature at 1608

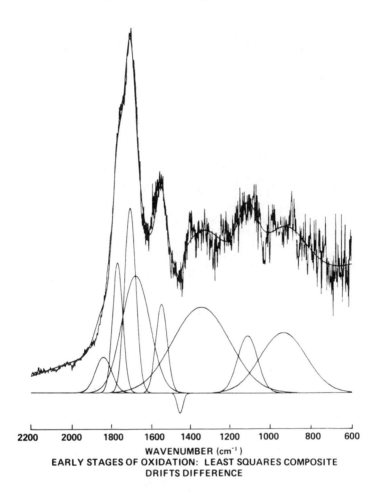

WAVENUMBER (cm⁻¹)
**EARLY STAGES OF OXIDATION: LEAST SQUARES COMPOSITE
DRIFTS DIFFERENCE**

FIG. 13. Early Stages of Oxidation of Wyodak Powder: MIDECS analysis shows that the initial oxidation involves primarily formation of aldehyde/ketone functional groups onto the surface of the coal substrate.

wavenumbers. The subtraction was carried out directly with no proportionality constant used to mitigate "background" effects so that we may note nonchemical changes (scatter, dispersion, reflection, etc.). This is to be contrasted to the technique of "adjusting" the algebraic subtraction normally used to compensate for variation between samples and for subtle differences due to sample preparation.

Following the procedure outlined above we have utilized the SECDER/FOSDEC/MIDECS mathematical analyses on each of the nine (9) difference spectra acquired in the oxidation of Wyodak powder from

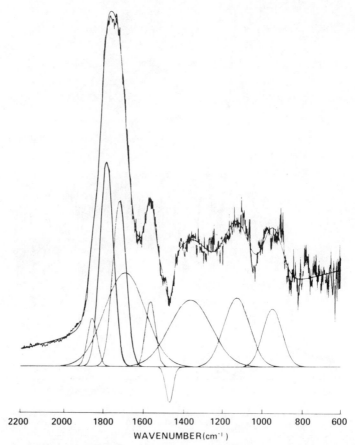

WAVENUMBER(cm⁻¹)

INTERMEDIATE OXIDATION: LEAST SQUARES COMPOSITE DRIFTS DIFFERENCE

FIG. 14.   Intermediate Stage of Oxidation of Wyodak Powder:
MIDECS analysis shows that the majority of the oxygen insertion into the surface of the coal substrate forms intermediate
degrees of oxidation in the form of esters, anhydrides, aryl
ketones, etc.

the state of Fig. 12A to that of Fig. 12B. The spectra of Figs.
13, 14, and 15 are representative of the changes wrought in the
different stages of the oxidation reaction -- each are plotted with
an adjusted ordinate to expand the maximum absorption to the same
fullscale magnitude solely to show that cursory examination of the
data in this form is an even better method of qualitative analysis
to note the variation of the relative abundance of each specie and
how they vary with the degree of reaction. The initial maximum
absorption occurs at ca. 1700 wavenumbers very characteristic of
aldo-/keto-groups in organic compounds. At the intermediate stages
of oxidation the maximum has shifted upward to ca. 1750 wavenumbers

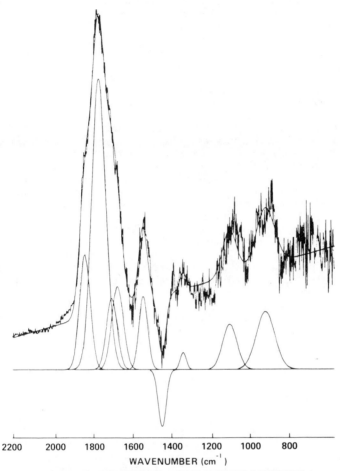

EXTENSIVE OXIDATION: LEAST SQUARES COMPOSITE
DRIFTS DIFFERENCE

FIG. 15. Extensive Oxidation of Wyodak Powder: MIDECS analysis reveal that the oxygen insertion process is progressive
and cumulative such that the highly oxygen rich functionality
(acid anhydrides, carbonate esters, etc.) is formed as the
immediate precurser for formation of carbon dioxide from the
coal substrate.

so characteristic of a more highly oxidized acido-/estero-
functionality. This same maximum prevails at more extensive oxidation and we now note the "shoulder" >1800 wavenumbers has become
quite prominent, indicating the presence of highly oxygen rich
moieties such as acid anhydrides and carbonato esters. Quantitative analyses are best obtained from the MIDECS results, but it is

worthy of note that the summation of the MIDECS values appears to be the ultimate smoothing function (in the context discussed above) as noted in the "threading" of the summation line through the noise of the experimental spectrum. We see then that, ideally, the "chemical" contribution as measured by MIDECS can be evaluated initially and the noise then discarded in contrast to the normal filtering technique whereby the "noise filter" (apodization, smoothing, etc.) is applied first and tends to decrease the apparent chemical (band) contribution also.

We also note the statistically significant "background" modification as the general increase, which is linearly inversely proportional to wavenumber (frequency). With the confidence inherent in the dedicated DRIS technique, we can be certain that this "background" alteration is due to the modification of the physical topography of the solid substrate by the oxidation process. This is to be expected due to the heterogeneous nature of the coal particles. Additional studies are underway to correlate the infrared scattering to microscopic evaluation of the solid/gas interface. This area of study may be a very promising field of research/analysis for the infrared Cassegranian microscopes currently being offered as accessories that can be utilized in both the visible and infrared modes to obtain morphological and chemical data, respectively.

The versatility of the MIDECS method is quite evident in that the negative feature at ca. 1450 wavenumbers is not only very apparent and a requisite feature of the MIDECS procedure, but also shows a very definite trend that is a direct measure of the loss of hydrocarbon from the solid substrate in the realm accessible to the incident/reflected beam, see Fig. 16. The band assignments for these studies are given in Table 2 where correlations to standard

TABLE 2. Band Assignments for Species Altered by In Situ Oxidation of Coals.†

MIDECS	REF. [8]	TENTATIVE ASSIGNMENT
1838	--	C=O Carbonate ester, anhydride, etc.
1767	1769	C=O Ester, Anhydride, etc.
1704	1712	C=O Arylalkyl ketone, aldehyde
1667	1677	C=O carboxylic acid
1548	1575	asymmetric C-C-O carboxylate salt
1455	1450	H-C-H methylene bend

† Numerical values are the characteristic frequencies expressed as wavenumbers (reciprocal centimeters) and the assignments are those used in classical qualitative infrared analyses.

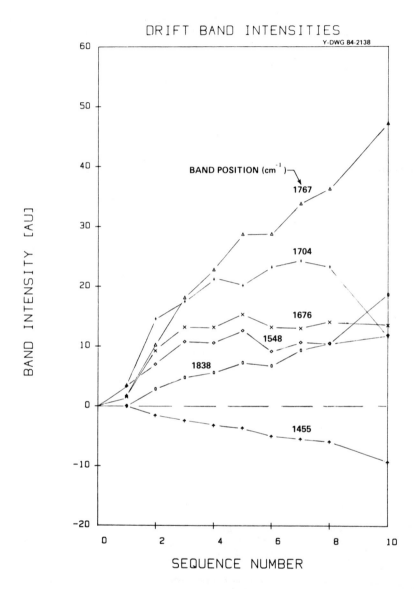

FIG. 16. Band Intensity of Carbonyls on Oxidation of Wyodak Powder: Sequential variation of diffuse reflectance infrared absorption showing the synergistic loss of hydrocarbon and the oxygen insertion to form the precursers for the production of carbon dioxide and water.

texts [29,30] and references [8,10] on other spectroscopic studies of coal are utilized. We see that the SECDER/FOSEDE/MIDECS analyses of this one portion of the DRIS spectrum provides both a measure of hydrogen loss (1450 cm^{-1}) as well as a measure of the oxygen insertion in each of four oxidation states (1677, 1704, 1767, and 1838 cm^{-1}) requisite to the formation of gaseous carbon dioxide. Concurrent analyses of the C-H vibrational modes (2700-3200 cm^{-1}) provided even greater detail with respect to speciation and sequential lability of the various hydrocarbon species as has been reported elsewhere [1].

REFERENCES

1. N. R. Smyrl and E. L. Fuller, Jr., "Chemistry and Structure of Coals: Diffuse Reflectance Infrared Fourier Transform Spectroscopy (DRIFT) Spectroscopy of Air Oxidation", in Coal and Coal Products: Analytical Characterization Techniques, edited by E. L. Fuller, Jr., A.C.S. Symposium Series 205, 133 (1982).
2. E. L. Fuller, Jr., N. R. Smyrl, R. L. Howell and C. S. Daw, "Chemistry and Structure of Coals: Evaluation of Organic Structure by Computer Aided Diffuse Reflectance Infrared Spectroscopy", in Symposium on Structure and Chemistry of Carbonaceous Materials, Prep. Div. Fuel Chem., ACS 29, 1 (1984).
3. M. P. Fuller, I. M. Hamedeh, P. R. Griffiths and D. E. Lowenhaupt, Fuel, 61, 529 (1982).
4. P. R. Griffiths and M. P. Fuller, in Advances in Infrared and Spectroscopy, edited by R. J. H. Clark and R. E. Hester, Heyden, London (1982).
5. N. R. Smyrl, E. L. Fuller, Jr. and G. L. Powell, in Analytical Spectroscopy, edited by W. S. Lyon, Anal. Chem. Symp. Series (Elsevier), 19, 357 (1984).
6. R. R. Willey, Appl. Spectrosc., 30, 593 (1976).
7. D. W. van Krevelen, "Coal", Elsevier (1961).
8. P. C. Painter, R. W. Snyder, M. Starsinic, M. M. Coleman, D. W. Kuehn and A. David, in Coal and Coal Products: Analytical Characterization Techniques, E. L. Fuller, Jr., editor, Amer. Chem. Soc. Symp. Series 205, 47 (1982).
9. H. H. Lowry, "Chemistry of Coal Utilization", Wiley (1963).
10. P. R. Solomon, D. G. Hamblen and R. M. Carangelo, in Coal and Coal Products: Analytical Characterization Techniques, E. L. Fuller, Jr., editor, Amer. Chem. Soc. Symp. Series 205, 77 (1982).
11. N. R. Smyrl, E. L. Fuller, Jr., and G. L. Powell, Appl. Spectrosc., 37, 38 (1983).
12. E. L. Fuller, Jr., N. R. Smyrl, and J. B. Condon, J. Nuclear Mater., 120, 174-194 (1984).
13. R. H. Essenhigh, "Fundamentals of Coal Combustion" in Chemistry of Coal Utilization, M. A. Elliot, ed., Wiley and Sons, 1981.
14. S. A. Yeboah, S. Wang, and P. R. Griffiths, Appl. Spectrosc., 38, 259 (1984).
15. E. L. Fuller, Jr., and N. R. Smyrl, Fuel, in press (1984).

16. P. R. Griffiths and M. P. Fuller, "Mid-infrared Spectrometry of Powdered Samples", in Advances in Infrared and Raman Spectroscopy, edited by R. J. H. Clark and R. E. Hester, Heyden and Sons, 1982.
17. W. W. Wendtlandt and H. G. Hecht, "Reflectance Spectroscopy", Interscience Publ., 1966.
18. F. A. Jenkins and H. E. White, "Fundamentals of Optics", McGraw-Hill, 1957.
19. M. J. D. Low, Appl. Opt., 6, 1503 (1967).
20. W. W. Wendtlandt, Modern Aspects of Reflectance Spectroscopy, Plenum Press, 1968.
21. P. C. Gillete and J. L. Koenig, Appl. Spectrosc., 38, 334 (1984).
22. J. K. Kauppinen, D. J. Moffatt, H. H. Mantsch, and D. G. Cameron, Anal. Chem., 53, 1454 (1981) and Appl. Opt., 20, 1866 (1981).
23. P. H. Von Cittert, Z. Phys., 69, 298 (1931).
24. W. E. Blass and G. W. Halsey, "Deconvolution of Absorption Spectra", Academic Press, (1981).
25. R. P. Young, "Computer Systems", in Infrared and Raman Spectroscopy, edited by E. G. Brame, Jr. and J. G. Grasselli, Marcel Dekker, 1977.
26. D. G. Cameron and D. J. Moffat, J. Testing and Evaluation, 12, 78 (1984).
27. P. R. Griffiths, Chemical Infrared Fourier Transform Spectroscopy, John Wiley, 1975.
28. E. L. Fuller, Jr., N. R. Smyrl, R. W. Smithwick, and C. S. Daw, Fuel Chem. ACS Preq. 28, 44 (1983).
29. N. B. Colthup, L. H. Daley and S. E. Wiberly, Introduction to Infrared and Raman Spectroscopy, Academic Press (1964).
30. L. J. Bellamy, Advances in Infrared Group Frequencies, Methuen and Co. (1968).
31. K. Nakamoto, Infrared Spectra of Inorganic and Coordination Compounds, John Wiley (1963).
32. G. C. Pimentel and A. L. McClellen, The Hydrogen Bond, Freeman and Co., 1960.
33. S. N. Vinogradov and R. H. Linnell, Hydrogen Bonding, Van Nostrand Reinhold, 1971.
34. W. C. Hamilton and J. A. Obers, Hydrogen Bonding in Solids, W. A. Benjiman Inc., 1968.
35. J. W. Larsen, "Organic Chemistry of Coal", A.C.S. Symposium Series, American Chemical Society, 1981.
36. E. L. Fuller, Jr., "Physical and Chemical Structure of Coals", in Coal Structure, edited by M. L. Gorbaty and K. Ouchi, American Chemistry Society, 1981.
37. P. Painter, M. Starsinic, E. Riesser, C. Rhoads and B. Bartges, "Concerning the Calculation of Coal Structural Parameters from Spectroscopic Data", Prepr. Div. Fuel Chem. ACS 29, 29-35, 1984.
38. R. E. Ziemer and W. H. Tranter, Principles of Communications: Systems, Modulation, and Noise, Houghton Mifflin Co., 1976.
39. J. K. Kauppinen, D. J. Moffatt, H. H. Mantsch, and D. G. Cameron, Anal. Chem., 53, 1454 (1981).
40. R. S. McDonald, Anal. Chem., 56, 349R-372R (1984).

41. E. L. Fuller, Jr., J. Colloid Interfac. Science, 75, 577
     (1980).
42. E. Suzuki, Anal. Chem., 38, 1770 (1966) and 41, 37 (1969).
43. R. W. Snyder, P. C. Painter, J. R. Havens, and J. L. Koenig,
     Appl. Spectrosc., 37, 497 (1983).
44. J. Pitha and R. N. Jones, Can. J. Chem., 44, 3031 (1966).
45. E. L. Fuller, Jr., J. Colloid and Interface Science, 89, 309
     (1982).

(*) Operated for the U.S. Department of Energy by Martin Marietta
     Energy Systems, Inc. under Contract Number DE-AC05-840R21400.

UNCONVENTIONAL APPLICATIONS OF FT-IR SPECTROMETRY TO THE ANALYSIS OF SURFACES

Peter R. Griffiths, Kenneth W. Van Every and Norman A. Wright

Department of Chemistry
University of California
Riverside, CA 92521

and

Isaam M. Hamadeh[*]

Department of Chemistry
Ohio University
Athens, OH 45701

INTRODUCTION

The characterization of molecules adsorbed on a variety of sub-strates is vital for understanding the basis of many heterogeneous catalytic reactions. Certain instrumental techniques, such as X-ray photoelectron spectrometry, Auger spectrometry, and secondary ion mass spectrometry, provide information on the elemental compo-sition at surfaces but little data from which the molecular struc-ture of adsorbed species can be derived. Vibrational spectrome-tries not only serve this purpose but often also provide clues on the way in which adsorbates are bound to surfaces. Under certain circumstances, adsorbed molecules may be present on surfaces at close to monolayer coverage, while other species, especially inter-mediates formed during heterogeneous catalytic reactions, are present at much lower coverages. At these levels, non-routine instrumentation is necessary to obtain vibrational spectra.

For the purposes of this discussion two types of surface may be distinguished. The first has a very high specific surface area -- typically several hundred square meters per gram -- while the second is a well-defined flat surface which may have a total area of less than 1 cm^2. The techniques used to study each type of surface are quite different, but in each case the advent of Fourier transform infrared (FT-IR) spectrometry has enabled significant new chemical data to be obtained. In this chapter we will review several different sampling techniques for measuring the vibrational spectra of species adsorbed on each type of surface. In each case a brief review of published work will be given and some novel developments being made in our laboratory will be described.

[*]Present Address: Monsanto Corporation
                      800 N. Lindbergh Blvd.
                      St. Louis, MO 63167

HIGH SURFACE AREA ADSORBENTS

Introduction
Samples with high surface areas are usually powdered, and may be stoichiometric or mixed oxides or metals on oxide supports. Among the techniques which may be used for measuring the vibrational spectra of species adsorbed on this type of substrate are Raman spectrometry and infrared transmission, emission, photoacoustic (PA), and diffuse reflectance (DR) spectrometry. These techniques will be discussed in varying degrees of detail below.

Raman Spectrometry
Raman Spectrometry with the advantage of requiring minimal sample preparation, allows species on the surface of powdered adsorbents to be studied directly. In addition, the intensity of Raman bands of gaseous species is very low. Thus if a band which is due to a vibrational mode of the adsorbate is observed, it is usually easily assigned to an adsorbed species rather than to the molecule in the vapor phase. A further advantage of Raman spectrometry is that metal-oxide stretching bands, which are particularly intense in the infrared spectrum of the adsorbent and often preclude measurement below 1200 cm^{-1}, are generally quite weak in the corresponding Raman spectrum so that strong Raman bands due to the adsorbate can occasionally be observed in this low wavenumber region.

On the other hand, Raman spectrometry suffers from several disadvantages with respect to the more useful sampling techniques for infrared spectrometry. First and foremost, Raman spectrometry (even resonance Raman spectrometry) is a relatively insensitive technique for the study of adsorbed species, unless the substrate is a coinage metal such as silver which can lead to the "surface enhanced Raman" effect. In these relatively rare cases, enhancement of band intensities by four or five orders of magnitude may be observed. Secondly, the intensity of background radiation due to fluorescence may be very high. The increased shot noise from the photomultiplier tube because of the fluorescent background is often enough to mask the weak Raman bands of the surface species. In general, the disadvantages of Raman spectrometry for studying the vibrational spectra of adsorbed species presently outweigh its advantages.

Infrared Transmission Spectrometry
Infrared transmission spectrometry has been used for the characterization of adsorbed molecules for over 25 years through the application of techniques originally developed by Eischens [1]. For these measurements the powdered sample is compressed into a very thin disk, the thickness of which is typically 0.1 to 0.2 mm. The adsorbent wafer may be held in a chamber where its temperature can be raised to the desired value and where the composition and pressure of the atmosphere can be controlled. Thus the adsorbent may be activated in situ, after which the chamber is evacuated, and the adsorbate introduced. Alternatively, the adsorbent can be deposited from a slurry onto the actual windows of the sample cell. In this case, thinner samples can be prepared but measurements at

high temperature are less easily performed because of thermal gradients across the window.

A variety of cells for holding catalyst disks have been described in the literature. For example a very simple cell was developed by Haaland [2] which, like several of its predecessors, had a fairly long gaseous absorption path and windows of large diameter. This this cell was designed more for studies involving the use of a high vacuum rather than a high pressure. An easily constructed cell permitting relatively high pressure (up to 2.4 MPa) and temperature (to 200°C) was described by Hicks et al. [3]. Edwards and Schrader [4] built a cell which could be taken to slightly higher pressures than this, while King [5] described a flow-cell permitting the direct measurement of heterogeneous catalytic reactions at high pressure under conditions of dynamic equilibrium.

Many reviews of the measurement of the transmission spectra of species adsorbed on pressed catalyst disks have been published, including the early books of Little [6], Hair [7], and Kiselev and Lygin [8]. The review of surface measurements by FT-IR spectrometry written by Angell [10] includes reports of much of the latest work in this area. Each of these sources gives details of many successful measurements made using pressed disks, and yet the use of this sampling technique still has several disadvantages. Perhaps the most important is the fact that the trasmittance of the wafers of many adsorbents between 3000 and 4000 cm^{-1} can be very low because of scattering. For example, Haaland [2] reported that the transmittance of a 0.2 mm thick wafer of alumina at 3000 cm^{-1} was only 0.1% despite evacuating the cell for several hours while the sample was held at 350°C and then briefly raising the temperature to 500°C in an attempt to eliminate as many hydroxyl groups as possible from the alumina surface. If even thinner disks are prepared, their fragility is increased while the number of surface sites is correspondingly reduced. In several cases these deficiencies may be unimportant, but whenever bands in the C-H, O-H, or N-H stretching region must be observed, they can completely prevent the desired data being obtained. Several workers have investigated the feasibility of using alternative approaches to circumvent these problems. Among the techniques which have been studied are infrared emission, photoacoustic and diffuse reflectance spectrometries. The advantages and drawbacks of each of these techniques will be discussed below.

Infrared Emission Spectrometry
At first glance, emission spectrometry would seem to be a good technique for characterizing the surface of adsorbents at elevated temperature since the problem of scattering, which is so deleterious for transmission measurements, is effectively absent. It is certainly true that some useful emission spectra of adsorbed species have been obtained [11] but, in the opinion of these authors, its disadvantages far outweigh its benefits. The biggest problem with infrared emission spectrometry is the generation of artifacts caused by temperature gradients [12]. It is also worth noting that the radiant energy from a blackbody emitter is only

large at high wavenumbers if the temperature of the sample is well above 500°C. Since most catalytic reactions take place at somewhat lower temperatures, the signal-to-noise ratio (SNR) of emission spectra near the important C-H stretching region is usually much too low for this technique to be used in all but a limited number of applications.

Photothermal Spectrometry

Photoacoustic spectrometry, which is by far the most common type of photothermal spectrometry to have been applied to the infrared, is sometimes thought of as a surface sensitive technique, but this is only true up to a point. The penetration depth of a beam into an isotropic sample is determined by the absorptivity at that particular wavelength. The optical absorption depth, μ_β, is defined as the reciprocal of the absorption coefficient, β, given by the equation:

$$I = I_o e^{-\beta x} \tag{1}$$

where I_o is the intensity at the surface of the sample and I is the intensity after penetrating a distance x into the sample. The absorbed radiation is converted into heat. If the input beam is modulated at a frequency, ω Hz, the effect of the heat generated x cm into the sample, s, on the temperature at the surface depends on the thermal conductivity, K_s, density, ρ_s, and specific heat, C_s, of the sample, and the modulation frequency, ω.

The thermal penetration depth, μ_s, is a measure of the depth from which a temperature increase in the bulk of the sample can be sensed at the surface. For the type of samples most commonly encountered in infrared spectrometry, Rosencwaig and Gersho [13] have calculated the thermal penetration depth as:

$$\mu_s = \left(\frac{K_s}{\pi \omega \rho_s C_s} \right)^{1/2} \tag{2}$$

For molecules with a fairly small absorption coefficient, μ_β is large and the effective penetration depth is given approximately by μ_s. Since μ_s varies with $\omega^{-\frac{1}{2}}$, the modulation frequency of PA spectra measured on an FT-IR spectrometer can be varied by changing the scan speed of the interferometer. Conceptually at least, this property leads to a method of depth profiling. The faster the scan speed of the interferometer, the closer to the surface of the sample is the region to be probed.

The modulation frequency, ω, imposed by a Michelson interferometer scanning with a mechanical velocity, V cm s^{-1}, is given by:

$$\omega = 2V\bar{\nu} \text{ Hz} \tag{3}$$

where $\bar{\nu}$ is the wavenumber of the radiation. V usually can take values between about 0.01 and 3 cm s^{-1} on modern FT-IR spectrometers, but the highest scan speeds are rarely used in PA/FT-IR measurements because of the very poor SNR of the spectra. In addition at very low scan speeds many interferometers become unstable. Thus, the typical range of mirror velocities used in PA/FT-IR measurements is from about 0.02 to 0.2 cm s^{-1}. In view of the dependence of μ_s on w^{-2}, it can be seen that the effective penetration depth can be varied by little more than a factor of three in depth profiling experiments.

Typical values of μ_s are on the order of 5 μm. Therefore not only is the PA signal due to the surface species observed but also the signal from at least the next 10^3 molecular layers (for an isotropic sample). Only when the surface/bulk ratio is very high, e.g. for highly porous adsorbents with surface areas greater than 100 m^2 g^{-1}, should good surface sensitive be achieved. Of course, this criterion is equally applicable to measurements made by transmission, emission, and diffuse reflectance spectrometry, but the SNR of transmission and DR measurements is significantly better than that of PA spectrometry.

For monolayer coverage on adsorbents of high surface area, adsorbate molecules can be present at concentrations as great as one or two percent by weight. Under these circumstances, adequate PA spectra of species adsorbed on substrates of this type can still be measured, since now the "surface" comprises a significant part of the "bulk". Molecules adsorbed at a substantially incomplete surface coverage cannot generally be observed because of the poor SNR of PA spectra. This is certainly not the case for diffuse reflectance spectrometry, which will be discussed in the next section. A comparison of a single-scan DR spectrum of a sample containing CO adsorbed on rhodium supported on γ-alumina with the PA spectrum of the same sample measured by averaging 1024 scans is shown in Fig. 1. It is also worthwhile noting that the scan-speed of the interferometer was four times faster for the DR spectrum compared with the PA spectrum.

In certain cases, the observation of bands due to weakly held surface species can be made even more difficult because of spectral interferences by the presence of adsorbate molecules present in the vapor phase. In this respect it should be noted that the PA effect is much more efficient for species in the vapor phase than for bulk surface species (whose temperature changes have to be transferred to the gas phase). An example of this effect was observed in our laboratory [14]. In an early attempt to measure the PA spectrum of adsorbed pyridine, a slurry of silica gel in a solution of pyridine in CCl$_4$ was prepared. After equilibrium the majority of the solution was removed by filtration. The silica residue was subsequently heated to about 80°C for 2 hours and then subjected to a vacuum for a further 30 minutes to remove all weakly held species. The sample was then transferred to a PA cell and flushed with nitrogen for a short time. In spite of all these precautions, the

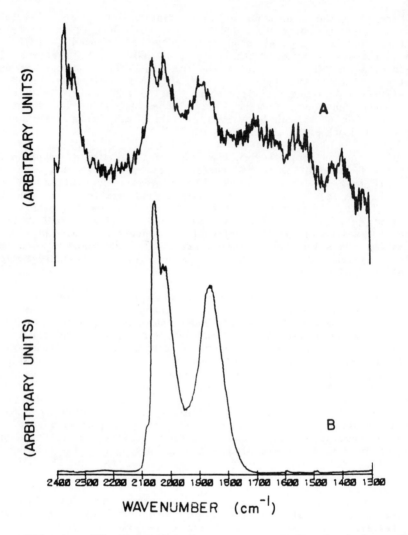

FIG. 1. (A) Photoacoustic spectrum of CO adsorbed on 5% Rh/Al$_2$O$_3$ which had been reduced in flowing H$_2$ at 200°C. CO was passed over the catalyst for 5 min at 1 mL/min after which N$_2$ was passed over the catalyst for a further 5 minutes. 1024 scans at a mirror velocity of 1.58 mm/s were averaged for both sample and reference spectra. (B) DR spectrum (in Kubelka-Munk format) of the same catalyst treated in the same manner as for Fig. 1A, except that after exposure to CO, N$_2$ was passed over the catalyst for 25 minutes. 1 scan at a mirror velocity of 6.33 m/s was used for sample and reference spectra, i.e., a reduction in time compared with the PA spectrum of a factor of 8000.

dominant feature in the spectrum was the 798 cm^{-1} band due to CCl_4 in the vapor phase, see Fig. 2.

To avoid this effect, pyridine was adsorbed onto silica gel directly from the vapor phase. The sample was placed in the PA cell and flushed with N_2 for 15 minutes to remove any weakly held pyridine. Nevertheless, the presence of a strong Q branch in the measured spectrum again indicated that bands due to adsorbed pyridine were being masked by the corresponding band from pyridine in the vapor phase.

In summary, we believe that in spite of several claims to the contrary, PA/FT-IR spectrometry is not a particularly powerful technique for studying adsorbed species because of the intrinsically low SNR of PA spectra, the rather large thermal penetration depth, and severe interferences from the vapor phase.

An alternative technique to PA spectrometry for measuring FT-IR photothermal spectra involves the deflection of a laser beam passing close to the surface of a sample when the atmosphere above the sample is heated. Beam deflection occurs by the same mechanism that causes the observation of mirages in hot weather. In beam deflection photothermal (BDPT) FT-IR spectrometry, the atmosphere above the sample is heated by temperature changes due to radiationless transitions of vibrational excitations in exactly the same manner as in PA spectrometry. Instead of the generation of an acoustic signal, however, the temperature modulations of the atmosphere lead to corresponding modulations in the refractive index of the gas, and hence a modulated deflection of the laser beam by the mirage effect. This modulated beam deflection is monitored with an optical detector.

Low and his coworkers have described the construction of a BDPT/FT-IR spectrometer [15-17]. Although the development of this instrument is still quite recent, the SNR of the measured spectra is at least equivalent to that of PA spectra. The biggest advantage of BDPT over PA spectrometry for studying surface species is the fact that it is unnecessary to hold the sample in a small enclosed cell of constant volume. Interference of BDPT spectra by species in the vapor phase which are slowly desorbing from the sample is therefore far less severe. In addition massive samples can be studied, whereas PA spectrometry is restricted to small samples which fit inside the cell [18].

One possible advantage of both PA and BDPT spectrometry is found for spectral regions where the absorption is very strong, i.e., where $\mu_\beta < \mu_s$. In this case the relative contribution of species near the surface is much greater than when $\mu_\beta > \mu_s$. For example, Low et al. [19] have reported the measurement of CO adsorbed on a catalyst compared of 50% nickel on carbon. This type of adsorbent is particularly difficult to study by any other technique, including transmission and diffuse reflectance FT-IR spectrometry.

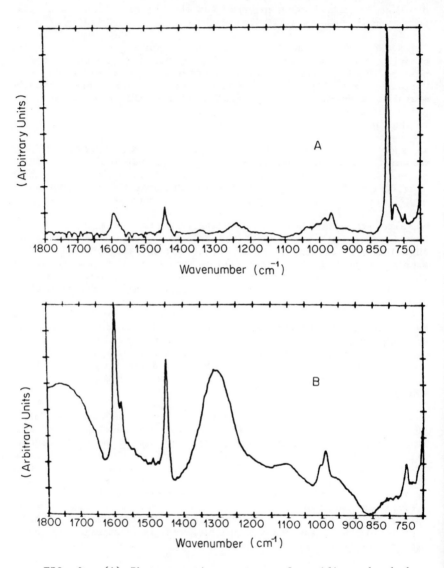

FIG. 2. (A) Photoacoustic spectrum of pyridine adsorbed on silica from CCl_4 solution. Note both the good compensation of the SiO_2 bands but the appearance of an intense feature at 798 cm^{-1} due to CCl_4 vapor. (B) Diffuse reflectance spectrum of the same sample. Note the good discrimination against vapor-phase species and the excellent SNR, but the poor baseline flatness due to inadequate compensation of the intense SiO_2 absorptions.

Diffuse Reflectance Spectrometry

Diffuse reflectance infrared spectrometry is not a new technique for measuring the spectra of adsorbed species on powdered adsorbents. In 1964, Körtum and Delfs [20] reported measurements of the spectra of ethylene and hydrogen cyanide adsorbed on metal oxides using a hemiellipsoidal reflectometer and a grating spectrometer. In 1979, Niwa et al. also described measurement of adsorbed species by DR infrared spectrometry [21] using an optical configuration which permitted the discrimination of radiation emitted by the hot sample. In this work, the adsorption of pyridine on HY zeolites was studied. Although both of these studies demonstrated the feasibility of DR infrared spectrometry for surface studies, in neither case was the SNR sufficiently high to justify great optimism that DR spectrometry would become a useful technique for the study of adsorbed species.

Even early measurements of DR spectra of bulk samples made using an FT-IR spectrometer showed little indication of the potential of this technique. For example, a special-purpose slow-scanning FT-IR spectrometer containing an integrating sphere yielded spectra with a very poor SNR [22]. This state of affairs changed in 1978, when Fuller and Griffiths [23] described a simple ellipsoidal reflectometer designed to work at high efficiency in conjunction with a commercial rapid-scanning interferometer and mercury cadmium telluride (MCT) detector. Later the same authors showed that the SNR of DR spectra of very dilute dispersion of absorbing materials in a non-absorbing matrix (such as KBr) measured using this device was significantly better than that of transmission spectra of the same samples pressed into a transparent disk [24]. Through the use of a simple cell held at the sample focus, this group was able to demonstrate the measurement of DR spectra of CO adsorbed on supported metal catalysts at low coverage at surprisingly high sensitivity.

Rapid development of DR accessories for FT-IR and even grating spectrometers ensued and now these accessories are available from a variety of vendors. At least two (Harrick Scientific and Barnes-Spectratech) make controlled atmosphere chambers in which powdered adsorbents can be held at elevated temperature, activated in situ where appropriate, and treated with the desired adsorbate. Recently we described an optical system for measuring DR spectra of samples in a controlled atmosphere which is so efficient that under certain conditions it is necessary to insert a screen in the infrared beam to decrease the energy at the MCT detector so that the analog-to-digital converter (ADC) is not overloaded and the detector response remains linear [26]. The DR spectra shown in the next section were all measured using this device.

It is interesting to note that when the sample is held at elevated temperature (<200°C) the d.c. energy emitted by the sample and the hot cell has the effect of reducing the response of an MCT detector. Indeed unless some effort is made to reduce the amount of radiant energy emitted from the cell body which reaches the detector, the signal from an MCT detector can sometimes become less than the signal from a triglycine sulfate (TGS) pyroelectic bolometer, which usually has a much poorer sensitivity than MCT detectors. In the device described by Hamadeh et al. [25], a ring which is in thermal contact with the water-cooled body of the cell was fitted over the sample, so that only radiation which was either diffusely reflected or emitted from the sample is permitted to reach the detector.

For absorbates with one or more strongly absorbing bands adsorbed on substrates with a high surface area (<200 m^2 g^{-1}) we believe that spectral features can be observed by diffuse reflectance FT-IR spectrometry at surface coverages as low as 0.01%. The following example is designed to illustrate the power of DR infrared spectrometry for studying both the activation of catalysts and the adsorption of molecules on relatively complex surfaces.

THE CO/Rh/Al$_2$O$_3$ SYSTEM

Ever since the early work of Eischens [1], the adsorption of carbon monoxide on a variety of supported metal catalysts has been studied by many spectroscopists. Perhaps the most interesting metal to be studied is rhodium, and the interpretation of the spectra of CO adsorbed on supported rhodium catalysts is still the subject of much speculation. Four bands are easily distinguished in spectra of CO on Rh/Al$_2$O$_3$. The two bands at about 2030 and 2100 cm^{-1} were assigned to the antisymmetric and symmetric stretching modes, respectively, of the geminal dicarbonyl species below (Species I). The band near 2070 cm^{-1} was assigned to CO linearly bonded to rhodium (Species II, below), and the broad band near 1860 cm^{-1} was assigned to the bridged species, III [26].

I   II   III

It was also believed that the gem-dicarbonyl (I) was only formed on rhodium atoms which were isolated on the alumina surface while species II and III were formed on rhodium atoms which are present in clusters on the surface.

Later several additional species were postulated, involving not only zero-valent rhodium, but also rhodium in the I, II and III oxidation states. These have been summarized by Rice et al. [27], and their conclusions appear in Table 1. Despite the number of species shown in this table, there are still several unexplained species appearing in spectra of CO adsorbed on Rh/Al$_2$O$_3$, especially at low surface coverage where the improved SNR of spectra measured

TABLE 1. CO/Rh/Al$_2$O$_3$ Species Proposed by Rice et al. [27]

Frequency Range (cm^{-1})	Variation with Coverage	Site Distribution	Oxidation State	Proposed Structure
2136	No	Atomic	III	RhCl$_3$(CO)·2H$_2$O
2116-2120	No	Atomic	II	OC $\diagdown$ O $\diagup$ Rh
2096-2102 2022-2032	No	Atomic	I	OC $\diagdown$ CO $\diagup$ Rh
2080-2100	?	?	I	O–C–Rh
2042-2076	Yes	Clusters	0	O–C–Rh
2000-2020	?	Clusters	I	O=C bridging Rh⋯Rh
1900-1920	Yes	Clusters	0	OC–Rh, C(=O), Rh–CO
1845-1875	Yes	Clusters	0	O=C bridging Rh⋯Rh

by DR spectrometry increases one's confidence in their existence.
We therefore initiated a program to study this system by DR FT-IR
spectrometry, and some of the preliminary results and conclusions
are reported below.

Rhodium catalysts are usually prepared by preparing a slurry of
rhodium (III) salt, such as $RhCl_3 \cdot 3H_2O$ or $Rh(NO_3)_3$, and the support
material, allowing it to dry, and reducing it with hydrogen at
elevated temperature. For this work a catalyst with a nominal
composition of 1% Rh on Al_2O_3 was purchased from Alpha Products
(Danvers, MA). The DR spectrum of this catalyst as received is
shown in Fig. 3A. This spectrum indicates the presence of a small
amount of a saturated hydrocarbon by the C-H stretching bands
between 3000 and 2800 cm^{-1}. From the singlet near 2350 cm^{-1} some
CO_2 appear to be physically adsorbed on the support. (Chemisorbed
CO_2 absorbs at a much lower wavenumber [6]). Subjecting the
catalyst to a vacuum at ambient temperature (Figure 3B) or to an
argon purge at 350°C (Fig. 3C) does not significantly decrease the
intensity of these bands, although the amount of moisture is
certainly decreased, as evidenced by the reduction in intensity of
the O-H stretching bands near 3400 cm^{-1}. Even passing hydrogen gas
through the catalyst does not reduce the intensity of the C-H or
CO_2 bands (Fig. 3D), suggesting that the hydrocarbon and CO_2
molecules causing these bands are trapped in the pores of the
support rather than being absorbed on an accessible region of the
surface.

An examination of the region below 1800 cm^{-1} is even more instruc-
tive. A strong sharp band at 1380 cm^{-1} is clearly observed in the
spectrum of the catalyst as received. This band is almost certainly assignable to nitrate ion remaining after reduction of
$Rh(NO_3)_3$ in the preparation of the catalyst by the manufacturer.
The intensity of this band is not significantly reduced after
evacuation at room temperature but decreases somewhat on increasing
the temperature (Fig. 3C), presumably because of decomposition in
the nitrate:

$$2Rh(NO_3)_3 \rightarrow Rh_2O_3 + 6NO_2 + 3/2 \ O_2$$

A slow decomposition of the oxide to rhodium metal at this temperature is also possible. On reducing the temperature to 300°C and
passing H_2 through the sample, the nitrate band is rapidly eliminated from the spectrum, suggesting (but not proving) the complete
reduction of Rh^{+3} to the zerovalent state.

Two other features in the fingerprint region are also reduced in
intensity during this treatment sequence, while two new bands
appear. The two features that disappear are the band located at
1602 cm^{-1} and the shoulder near 1640 cm^{-1}. The 1640 cm^{-1} band is
presumably assignable to weakly bound water, and its disappearance
should closely parallel the reduction in intensity of the broad O-H
stretching band centered near 3550 cm^{-1}. The 1602 cm^{-1} band is
possible associated with more tightly bound water molecules, which
probably also give rise to the more discrete spectral structure on

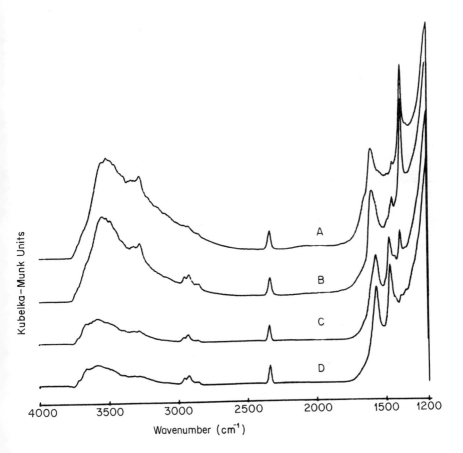

FIG. 3.   DR spectrum of a 1% Rh/Al$_2$O$_3$ catalyst, (A) as received; (B) after evacuation for 4 hr at ambient temperature;
(C) after heating under flowing argon for 4 hr at 350°C; (D)
after treatment under flowing H$_2$ for two hours at 300°C.

the long wavelength tail of the 3550 cm^{-1} water band (near 3300
cm^{-1}).

Two bands, at 1570 and 1470 cm^{-1}, increase in intensity during this
activation process.   The origin of these bands is a bit of a
mystery.   Similar bands have been observed when supported metal
catalysts have been treated with CO, and are usually assigned to a
carbonate or chemisorbed CO$_2$ speices.   Such an assignment does not
seem very likely in this case, since no carbon-containing species
has been added to this sample. The intensity of the band due to
CO$_2$ trapped in the pores of the alumina does not decrease throughout this treatment so that it may be presumed that the 1570 and
1470 cm^{-1} bands are not caused by a reaction product of this

trapped CO_2. Most surface carbonates show a splitting of the 1440 cm^{-1} antisymmetric stretching band with one component shifting to higher wavenumber and the other to lower wavenumber [6]. However in this work, both components are observed above 1440 cm^{-1}. The frequencies of these bands are similar to those of the symmetric and antisymmetric stretching modes of carboxylate ions, but no source of carboxylic acids in this work is apparent. Thus the assignment of the 1570 and 1470 cm^{-1} bands must be left as an open question. We are now preparing rhodium catalysts at various loading levels and on several different supports in an attempt to obtain an accurate assignment for these bands.

When CO is admitted to the catalyst activated as in Fig. 3D, several of the bands listed in Table 1 are observed, see Figs. 4 and 5. When the equilibrium pressure of CO is very low (9 millitorr) the strongest band in the spectrum is at 2038 cm^{-1}, which is lower than is usually observed for the linearly bonded species (II). In view of the absence of a band near 2100 cm^{-1}, this band is more logically assigned to Species II than the anti-symmetric stretching mode of Species I, because of the weakness of the symmetric stretching mode of Species I around 2100 cm^{-1}. The 2038 cm^{-1} band is quite asymmetric, and a shoulder at lower wavenumber ($\sim$2000 cm^{-1}) is readily apparent. In addition a weak feature at 2096 cm^{-1} can also be observed. The intensities of the two bands due to Species I are usually approximately equal, so that if the 2096 cm^{-1} band is the symmetric stretching mode of Species I, the antisymmetric band would be of too low intensity to be observed under the more intense broad band centered near 2038 cm^{-1}.

Thus we believe that the first CO molecules to be adsorbed at low coverage attach to the most active surface sites giving the strongest Rh-C bonds for Species II. Under this circumstance, the C-O band would be weaker and absorb at lower wavenumber than adsorbates on less active sites. The shoulder near 2000 cm^{-1} is assigned to the species shown in Table 1 as -Rh-Rh-, with each Rh atom in the +1 oxidation state. As the pressure of CO is increased, this band becomes less apparent as it is masked by the rapid growth of spectral features to higher wavenumber. This interpretation relies on the logical assumption that there are very few sites where two rhodium atoms in the +1 oxidation states are nearest neighbors.

At low CO coverage, Fig. 4, the dominant band between 2100 and 2000 cm^{-1} is centered at 2039 cm^{-1}. As the pressure of CO is increased this band shifts first to lower wavenumber and then to higher wavenumber and a second feature starts to become apparent as a shoulder at higher wavenumber. At CO pressure above 5 torr, this feature (at 2057 cm^{-1}) becomes dominant, see Fig. 5. Besides these bands, a well resolved feature at 2097 cm^{-1} also increases in intensity with increasing CO pressure. The frequency of this band does not vary with coverage.

The initial reduction in frequency, from 2039 cm^{-1}, of the band assigned above to tightly bound Species II molecules is followed by

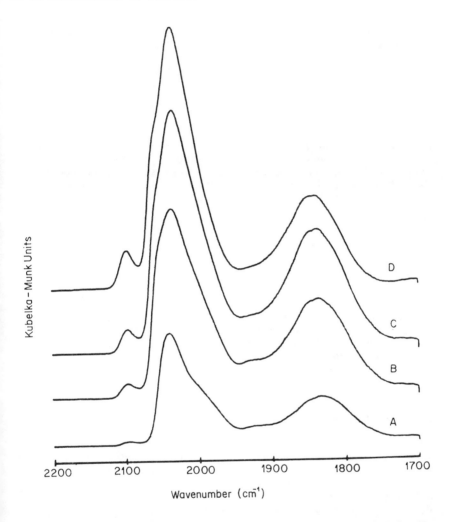

FIG. 4. DR spectra of CO in 1% Rh/Al$_2$CO$_3$ activated by H$_2$
reduction at 300°C for 5 hr at low equilibrium pressures of
CO, (A) 0.009 torr; (B) 0.015 torr; (C) 0.026 torr; (D) 0.094
torr.

an increase to about 2039 cm^{-1} as the CO pressure is increased.
This behavior suggests that after the first Species II surface site
is occupied, a second site with an even stronger Rh-C bond becomes
occupied (giving rise to the 2031 cm^{-1} band). The two sites might
well be edges, steps, or kinks on the surface of the Rh rafts. At
higher CO pressure, the 2097 and 2039 cm^{-1} bands due to Species I
grow at approximately the same rate, and the band at 2057 cm^{-1}
increases in intensity at a somewhat faster rate. We believe the
2057 cm^{-1} band is also due to Species II molecules. The fact that

the C-O bond strength is greater than that of the Species II
molecules adsorbed at low CO coverages suggests that the Rh-C bond
is somewhat weaker, i.e., that these sites are less active. In
addition, the number of sites of the type occupied at low surface
coverage. In combination these two observations lead us to believe
that the 2057 cm-1 band is associated with Species II bonding at
smooth regions of the rhodium rafts, somewhat akin to the (III)
surface of crystalline rhodium.

The well-known feature near 1860 cm-1 assigned to bridge-bonded CO
is very obvious in these spectra. Our data confirm that the wavenumber of this band increases with increasing surface coverage.
However we see little or no evidence for any spectral feature
assignable to the feature shown in Table 1 as

This species, if it indeed exists, should give rise to at least two
absorption bands since these are two chemically different types of
CO molecules in this structure. It is likely that the addition of
another bridge-bonded or linearly bonded CO molecule near a bridge-
bonded CO would cause electron density to be withdrawn from the
Rh-C bond of the (previously isolated) bridge-bonded species. In
this case the C-O bond would be strengthened and the C-O stretching
bond would be shifted to higher wavenumber, in a similar manner as
the mechanism postulated above to account for the shift of the
bands assigned to Species II.

It may also be noted that at intermediate surface coverages (e.g.
around 0.2 torr), there appear to be several unresolved components
contributing to the very broad bridge-bonded CO band. We have
attempted to further resolve these bands by Fourier self-
deconvolution [28], but with little success.

This rather brief study has been discussed at some length because
it gives a good idea of the wealth of information which can be
obtained from DR spectra of species adsorbed on high surface area
adsorbents. This high information content is mainly due to the
very high SNR of the spectra, and also in part to the good wavenumber reproducibility of FT-IR spectra. We expect to see many
more reports of investigations of heterogeneous catalysts using DR
FT-IR spectrometry in the near future.

LOW SURFACE AREA ADSORBENTS

Compared to the relatively simple problem of studying adsorbates on
adsorbents with a high surface area, measurement of the vibrational
spectrum of even a full monomolecular layer adsorbed on a flat

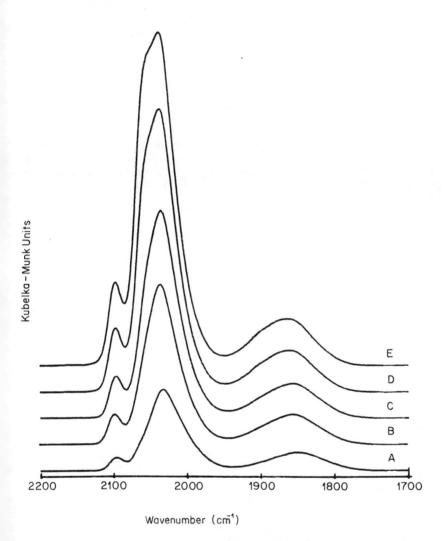

FIG. 5. DR spctra of CO on 1% Rh/Al_2O_3 at equilibrium pressures above 0.1 torr, (A) 0.204 torr; (B) 0.502 torr; (C) 1.26 torr; (D) 3.50 torr; (E) 8.42 torr.

surface can present a formidable task. In general neither emission nor photoacoustic spectrometery has the sensitivity to permit the study of species adsorbed at monolayer coverage on flat metallic surfaces, and transmission and diffuse reflectance spectrometry are, of course, completely inappropriate.

The most sensitive method for characterizing adsorbed species by measurement of the vibrational spectrum is electron energy loss spectrometry (EELS). This technique has the capability of measuring the spectra of species present at less than 1% coverage but has two important disadvantages. Firstly EELS is a form of electron spectrometry and therefore can only be performed under a high vacuum. Secondly spectra can only be measured at about 40 cm^{-1} resolution.

Essentially the only technique which is appplicable to surface measurements at reasonably high vaccum conditions is infrared reflection-absorption spectrometry (IRRAS). It has always been obvious that for maximum IRRAS sensitivity the angle of incidence of the beam should be as great (i.e. as close to grazing) as possible. Many of the recent advances in this field were made possible, however, by the important work of Greenler, who showed that the absorption of grazing incidence radiation polarized parallel to the surface by materials on the surface is much greater than for perpendicularly polarized radiation [29,30].

Even with incidence angles greater than 70° the absorption of parallel polarized radiation by species present at monolayer coverage on surfaces rarely exceeds 1%, and is usually much less. Despite the excellent performance of modern FT-IR spectrometers, they still have difficulty in measuring very weak bands (even with sensitive detectors such as MCT) because of the limited dynamic range of the ADC. The very high intensity of interferograms in the region of the centerburst is so large relative to the miniscule modulations caused by adsorption bands of adsorbed species that even a 16-bit ADC may be insufficient for the characterization of surface species on flat substrates.

Ideally, therefore, one would like to suppress the centerburst while retaining the weak modulation due to the absorption bands. A few early attempts were made to suppress the centerburst in IRRAS by dual-beam, or optical subtraction, FT-IR spectrometry [31,32]. Although the results of these experiments were promising, the fact that the interferometer must usually be removed from the optical bench supplied by the spectrometer manufacturer and then installed in special-purpose optics has restricted the acceptance of this technique.

An alternative, and more easily implemented, technique for suppressing the centerburst in IRRAS measurements has involved the use of a photoelastic modulator (PEM). PEM's consist of an isotropic crystal, most commonly zinc selenide, which is sinusoidally expanded and compressed at high frequency by a piezoelectric transducer. Plane polarized radiation is passed to the PEM. When the amplitude of modulation of the ZnSe plate has a certain magnitude, the emerging beam is alternately converted to left and right circularly polarized radiation at a modulation frequency equal to that of the PEM.

This behavior has been applied to the measurement of vibrational circular dichroism spectra using an FT-IR spectrometer by Nafie and others [33,34]. For this work the modulation frequency of the PEM must be significantly higher than the highest modulation frequency imposed by the interferometer (see eq. (3)). With no sample in the beam path, there is no difference between the spectrum of the instrument background measured with left and right circularly polarized radiation. Thus if a lock-in amplifier (LIA) is used to observe the signal from the detector at the modulation frequency of the PEM, no output is measured. If, on the other hand, a chiral molecule is present in the beam, a weak interferogram is observed at the output of the LIA because of the difference in absorption of the sample to left and right circularly polarized radiation.

If plane polarized radiation is input to the PEM and the modulation amplitude is doubled from the previous case, the plane of polarization is rotated by 90° at each extreme of the expansion/compression cycle. Thus if a dichroic sample is placed between the PEM and the detector, an interferogram caused by the difference in absorption by the sample of parallel and perpendicular polarized radiation will be observed at the output of the LIA [34]. In this case the first harmonic of the modulation frequency of the PEM must be used as the reference for the LIA.

This effect was applied by Dowrey and Marcott [35] to the study of species on the surface of a flat copper plate by grazing incidence polarization modulation FT-IR spectrometry. The modulation frequency of the PEM in this work was about 70 kHz, much higher than the highest Fourier frequency, ω, imposed by the interferometer (which was about 1 kHz) but was still readily detected by an MCT detector. Indeed the specific detectivity (D*) of MCT detectors is at its maximum value between 10 and 100 kHz. Further reports of the characterization of surface species by polarization modulation IRRAS using an FT-IR spectrometer have been published by Golden et al. [36,37]. This technique appears to be applicable to the measurement of the IRRAS spectra of species on the surface of flat substrates only when an MCT or other detector with a good response at high modulation frequency can be used.

Because of the cut-off wavelength of MCT detectors, however, measurements of this type cannot be made much below 500 cm^{-1}. Thus only intramolecular vibrations can be studied and not the very important bands due to the substrate-adsorbate stretching mode. For CO adsorbed on a variety of transition metals, this mode typically gives rise to an absorption band between 350 and 500 cm^{-1}. The best detector for high sensitivity measurements in this region is the liquid helium cooled gallium-doped germanium bolometer, the sensitivity of which is at least an order of magnitude greater than that of a narrow-range MCT detector. Because of this exceptionally high sensitivity, low resolution FT-IR measurements made with a Ga:Ge bolometer in the region around 400 cm^{-1} can also be limited by the dynamic range of the ADC.

Polarization modulation measurements of a surface species have been performed in the mid-infrared using a bolometer in conjunction with either variable-wavelength filters or monochromators [38,39].

These experiments did not involve the use of a PEM, but instead used a rotating polarizer. The reason for using a rotating polarizer instead of a PEM was that the response time of the bolometer is slow, whereas the modulation frequency of most PEM's is on the order of 100 kHz. Typically the D^* of a Ga:Ge bolometer operating at 1 kHz is about 80% of its D^* at d.c., and its sensitivity falls off rapidly for higher modulation frequencies.

The application of polarization modulation techniques for IRRAS measurements of adsorbed species on flat surfaces using a bolometer and an FT-IR spectrometer therefore presents a fundamental problem. The slowest velocity at which most commercial interferometers can be driven is about 0.05 mm s^{-1}, but as mentioned earlier in this chapter, at this speed the drive is usually much less stable than at higher mirror velocities. Were a 100 Hz rotating polarizer to be used, the Fourier frequency, ω, of the highest wavenumber in the spectral range of interest (say 1000 cm^{-1}) must be reduced well below 100 Hz, say to 20 Hz. From eq. (3) it can be seen that the velocity of the moving mirror must be reduced below 20/2×1000 cm s^{-1}, i.e. 0.1 mm s^{-1}. However, at this velocity interferograms usually become quite unstable.

To complete this review of unconventional applications of FT-IR spectrometry for surface analysis, we propose an alternative method for polarization modulation for IRRAS measurements in the transition region between the mid- and far infrared. It can be seen from the discussion above that a higher polarization modulation frequency than 100 Hz is desirable. Ideally this frequency should be between 500 and 1000 Hz to permit efficient separation from the Fourier frequencies, ω, imposed by an interferometer scanning fast enough to avoid instabilities but still at a low enough frequency that the bolometer is operating near its peak D^*. One way of achieving a polarization modulation frequency of 1 kHz is to pass a plane polarized beam through two PEM's each operating under conditions that would create polarization modulation, but at frequencies which are different by 1 kHz. The beat frequency could then be used as the polarization modulation frequency.

At the time of this writing, no demonstration of surface analysis using this technique has been reported. Nevertheless we recently demonstrated the feasibility of beating two ZnSe PEM's together to create a difference frequency of 1 kHz. In this investigation, the fundamental frequencies of the two PEM's were 70.925 and 70.425 kHz, so that their first harmonics were 141.85 and 140.85 kHz, respectively. Interferograms were generated using a Nicolet 60-SX spectrometer with a Ge:KBr beamsplitter. The mirror velocity was set at 0.29 mm s^{-1} (VEL = 4). A low pass optical filter with a 1600 cm^{-1} cut-off was placed in the beam, so that the useful wavenumber range was 1600 to 700 cm^{-1}. The highest Fourier frequency was therefore 2×0.029×1600, i.e., 93 Hz. Thus the first harmonic frequency of both PEM's, their beat frequency, and the Fourier frequencies are all separated by at least a decade.

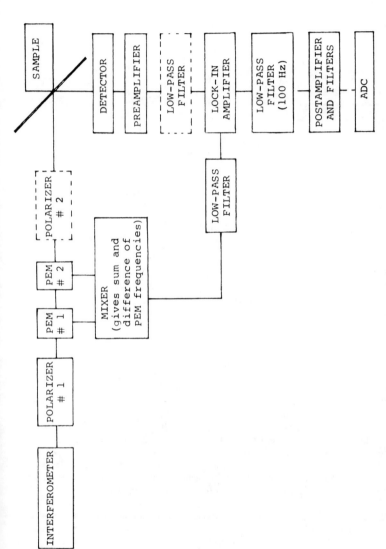

FIG. 6. Block diagram of the triple modulation system designed for polarization modulation at 1 kHz. Polarizer #2 was inserted in the beam for testing the concept, but is removed when species are adsorbed on the sample mirror. The low-pass filter between the preamplifier and LIA was present for the preliminary experiments but has since been removed.

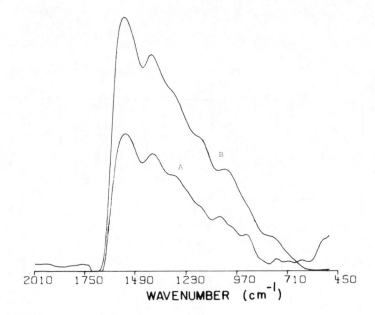

FIG. 7. (A) Single-beam background spectrum measured using
the triple modulation scheme with two polarizers, one on each
side of the two PEM's. (B) Single-beam spectrum measured
using the FT-IR spectrometer in its conventional single-
modulation mode. A DTGS detector was used for both measurements.

To study the feasibility of using the beat frequency for polarization modulation, a second polarizer was placed after both PEM's.
The polarizers should therefore be "crossed" and "in line" once per
beat cycle. The signal from a TGS detector is passed through a
low-pass filter to remove the 141 and 142 kHz components of the
signal (which are barely detected by the TGS detector in any
event). The signal is then input to a lock-in amplifier whose
reference is the beat frequency. The output of the LIA is
amplified and processed with the standard high- and low-pass
filters of the Nicolet spectrometer and then digitized and stored
in the data system. This is shown schematically in Fig. 6.

The Fourier transform of the resulting interferograms is shown as
Fig. 7A. This spectrum has the same shape as the conventional
single-beam spectrum which is reproduced as Fig. 7B. This is, we
believe, the first successful "triple modulation" FT-IR spectrum to
be reported. The fact that this result is in line with our predictions indicates that the approach outlined above for IRRAS
measurements near 400 cm^{-1} using a Ga:Ge bolometer is indeed
feasible. Work is progressing in our laboratory towards this goal.

SUMMARY

In this chapter we started with a description of some relatively routine techniques for the characterization of adsorbed species by FT-IR spectrometry and finished by describing some very unconventional work which is at a very early stage. FT-IR spectrometry has already led to some important measurements in the field of surface chemistry, and we believe that much more work will be reported in the future. We look forward to participating further in this exciting field of study.

ACKNOWLEDGMENTS

Some of the work described in this paper was supported by the U.S. Department of Energy under Grant No. DE-FG22-82PC50797. However, any opinions, findings, conclusions or recommendations expressed herein are those of the authors and do not necessarily reflect the views of DOE. The purchase of one of the PEM's by the 3M Corporation is gratefully acknowledged. Finally the lock-in amplifier was provided by the National Center for Biomedical Infrared Spectroscopy of Battelle's Columbus Laboratories with funds granted by the National Institutes of Health.

REFERENCES

1. R. P. Eischens and W. A. Pliskin, Adv. Catal., 10, 1 (1958).
2. D. M. Haaland, Surface Sci., 102, 405 (1981).
3. R. F. Hicks, C. S. Kellner, B. J. Savatsky, W. C. Hecker, and A. T. Bell, J. Catal., 71, 216 (1981).
4. J. F. Edwards and G. L. Schrader, Appl. Spectrosc., 35, 559 (1981).
5. S. T. King, Appl. Spectrosc., 34, 632 (1980).
6. L. H. Little, "Infrared Spectra of Adsorbed Species", Academic Press, NY (1967).
7. M. L. Hair, "Infrared Spectroscopy in Surface Chemistry", Marcel Dekker, NY (1967).
8. A. V. Kiselev and V. I. Lygin, "Infrared Spectra of Surface Compounds", Wiley, NY (1975).
9. A. T. Bell and M. L. Hair, "Vibrational Spectroscopies for Adsorbed Species", A.C.S. Symp. Ser. 137, Washington, D.C. (1980).
10. C. L. Angell, Ch. 1 in "Fourier Transform Infrared Spectroscopy", Vol. 3 (J. R. Ferraro and L. J. Basile, eds.), Academic Press, NY (1982).
11. P. C. M. van Woerkom and R. L. de Groot, Appl. Opt., 21, 3114 (1982).
12. P. R. Griffiths, Appl. Spectrosc., 26, 73 (1972).
13. A. Rosencwaig and A. Gersho, J. Appl. Phys., 47, 64 (1976).
14. P. R. Griffiths, S. A. Yeboah, I. M. Hamadeh, P. J. Duff, W. J. Yang, and K. W. Van Every, in "Recent Advances in Analytical Spectroscopy", (K. Fuwa, ed.), pp. 291-298, Pergamon Press, Oxford (1982).

15. M. J. D. Low and G. A. Parodi, Chem. Biomed. Environm. Instrum., 11, 265 (1982).
16. M. J. D. Low and M. Lacroix, Infrared Phys., 22, 139 (1982).
17. M. J. D. Low, M. Lacroix, and C. Morterra, Appl. Spectrosc., 36, 582 (1982).
18. M. J. D. Low, M. Lacroix, and C. Morterra, Spectrosc. Lett., 15, 57 (1982).
19. M. J. D. Low, C. Morterra, A. G. Severdia, and M. Lacroix, Appl. Surface Sci., 13, 429 (1982).
20. G. Körtum and H. Delfs, Spectrochim. Acta, 20, 405 (1964).
21. M. Niwa, T. Hattori, M. Takahashi, K. Shirai, M. Watanabe, and Y. Murakami, Anal. Chem., 51, 46 (1979).
22. R. R. Willey, Appl. Spectrosc., 30, 593 (1976).
23. M. P. Fuller and P. R. Griffiths, Anal. Chem., 50, 1906 (1978).
24. M. P. Fuller and P. R. Griffiths, Appl. Spectrosc., 34, 533 (1980).
25. I. M. Hamadeh, D. King, and P. R. Griffiths, J. Catal., 88, 264 (1984).
26. A. C. Yang and C. W. Garland, J. Phys. Chem., 61, 1504 (1957).
27. C. A. Rice, S. D. Worley, C. W. Curtis, J. A. Guin, and A. R. Tarrer, J. Chem. Phys., 74, 6487 (1981).
28. J. K. Kauppinen, D. J. Moffatt, H. H. Mantsch, and D. G. Cameron, Appl. Spectrosc., 35, 271 (1981).
29. R. G. Greenler, J. Chem. Phys., 44, 310 (1966).
30. R. G. Greenler, J. Chem. Phys., 50, 1963 (1969).
31. M. J. D. Low and H. Mark, J. Paint Technol., 43 (553), 31 (1981).
32. G. J. Kemeny and P. R. Griffiths, Appl. Spectrosc., 35, 128 (1981).
33. L. A. Nafie and M. Diem, Appl. Spectrosc., 33, 130 (1979).
34. L. A. Nafie and D. W. Vidrine, Ch. 3 in "Fourier Transform Infrared Spectroscopy", Vol. 3 (J. R. Ferraro and L. J. Basile, eds.), Academic Press, NY (1982).
35. A. E. Dowrey and C. Marcott, Appl. Spectrosc., 36, 414 (1982).
36. W. G. Golden and D. D. Saperstein, J. Electron Spec., 30, 43 (1983).
37. W. G. Golden, D. D. Saperstein, M. W. Severson, and J. Overend, J. Phys. Chem., 88, 574 (1984).
38. H. Pfnur, D. Menzel, F. M. Hoffman, A. Ortega, and A. M. Bradshaw, Surface Sci., 93, 431 (1980).
39. F. M. Hoffman, Surface Sci. Reports, 3, 107 (1983).

APPLICATIONS OF VIBRATIONAL SPECTROSCOPY TO THE STUDY OF PESTICIDES

Kathryn S. Kalasinsky

Mississippi State Chemical Laboratory
Mississippi State University
Mississippi State, MS 39762

INTRODUCTION

The major thrust of research for the Mississippi State Chemical Laboratory (MSCL) lies in the area of pesticide chemistry. Throughout this work we have found that vibrational spectroscopy is an invaluable tool. Its applications are widespread and include gas chromatography-Fourier transform infrared (GC/FT-IR), high performance liquid chromatography-FT-IR (LC/FT-IR), resonance Raman spectroscopy, attenuated total reflectance (ATR) and diffuse reflectance infrared Fourier transform (DRIFT) spectroscopy, and library searching techniques.

The following will describe some of the applications that the MSCL has found in these areas of vibrational spectroscopy.

GC/FT-IR

Gas chromatography-Fourier transform infrared (GC/FT-IR) certainly has more applications to pesticide chemistry [1] than do the rest of the vibrational spectroscopy techniques, and therefore, this discussion will be further subdivided into different topics. We have found GC/FT-IR useful for degradation analysis, isomer separations, reaction monitoring, contaminant detection, complex mixture identification, and in the process, we have determined the optimal conditions for pesticide analysis by GC/FT-IR. In essence, we have found GC/FT-IR to be useful in every aspect of pesticide chemistry where GC alone had traditionally been used. The additional information available in an infrared spectrum is a bonus which is important to any analysis.

Degradation Analysis
One of the most important factors to consider when formulating a pesticide bait is the capability to control the degradation process of the toxicant. This must be done while still maintaining (1) control of the target insect, (2) acceptability to the target insect, and (3) the capability for efficient distribution [2]. The portion of the bait which has not been consumed by the target insect must degrade rapidly in the environment. In order to

277

develop a bait with these attributes, numerous formulations of a
toxicant must be tested in both laboratory and field situations.

GC/FT-IR analysis of laboratory-degraded baits provides the separation and identification of the degradation products without requiring the traditional, time-consuming separations by column and
liquid chromatography on amounts of material large enough for
conventional, dispersive infrared analysis [3]. The amounts of
material recovered from field-degraded baits in actual environmental situations are typically even smaller, and GC/FT-IR can
provide the analysis of the degradation products while eliminating
a great deal of tedious work.

Other analytical techniques, such as electron capture gas chromatography (EC/GC) and gas chromatography-mass spectrometry (GC/MS),
are more sensitive than GC/FT-IR but they cannot always provide the
positive identification which is needed [4]. Many of the degradation products are derivatives and geometric isomers, and infrared
spectra are the only data which are unique for each one.

An example of the GC/FT-IR analysis of a laboratory degraded mirex
bait is shown in Figs. 1 and 2. Mirex (dodecachloropentacyclo-
[5.3.0.0.^{2,6}0.^{3,9}0.^{4,8}]decane) is a pesticide which is very
effective in combatting the imported fire ant [5-8]. Figure 1
shows the gas chromatogram and average absorbance plots
(Chemigram®) over several different frequency ranges (as a function
of time) for the bait formulation which was determined to be most
suitable for the environment [9,10]. Figure 2 shows the infrared
spectra of the four major decomposition products [11] obtained from
this GC/FT-IR experiment.

Isomer Separation
Once the most effective bait formulation has been determined, it is
necessary to undertake indepth studies of each degradation product
[12-14], and large quantities of each component must be prepared in
the laboratory. Oftentimes laboratory preparations do not exhibit
the same stereospecificities as do the field degradations. All
geometric isomers occur in most preparations, whereas one isomer
may dominate in actual field cases. In general, these isomers are
very difficult to separate, and spectral subtraction has been
useful and necessary [1]. An example of this is shown in Figs. 3,
4, and 5. Figure 3 is the gas chromatogram and average absorbance
plots over specified frequency regions of the laboratory preparation of the 5,10-dihydrogen derivative of mirex. (One of the
degradation products shown in Figs. 1 and 2.) Both the anti and
syn isomers were present although they were inseparable by GC.
Only the anti isomer was strongly evident in the field degradation
products. Figure 4 is a stacked plot of the infrared data
collected during the elution of the GC peak and gives a real-time
representation. The only major difference in the infrared spectra
of these two isomers is that the peak at 1100 cm^{-1} in the anti
isomer shifts to 1120 cm^{-1} in the syn isomer. Note the peaks at
these two frequencies in the stacked plot. The infrared data
collected during the first part of the GC peak elution were co-

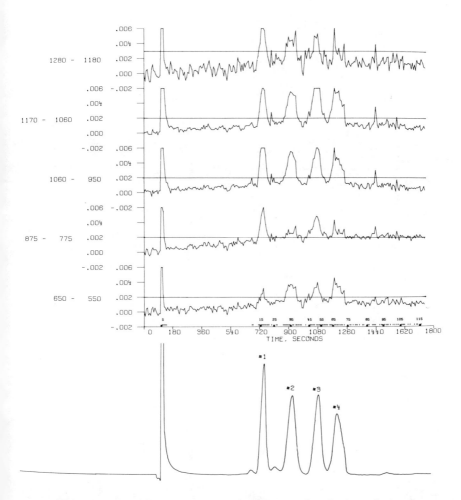

FIG. 1. Gas chromatogram and Chemigram® of mirex bait formulation degradation products. (Used by permission, Ref. [1].)

added to generate the spectrum of the <u>anti</u> isomer shown in Fig. 5,
and the infrared data during the latter part of the GC peak elution
were co-added to generate the spectrum of a mixture of the <u>anti</u> and
<u>syn</u> isomers (also shown in Fig. 5). Subtracting one spectrum from
the other produced the spectrum of the pure <u>syn</u> isomer (again shown
in Fig. 5). This was the first time that spectral data of the pure
<u>syn</u> isomer were obtained. Data from more recent experiments have
confirmed these original conclusions and spectra.

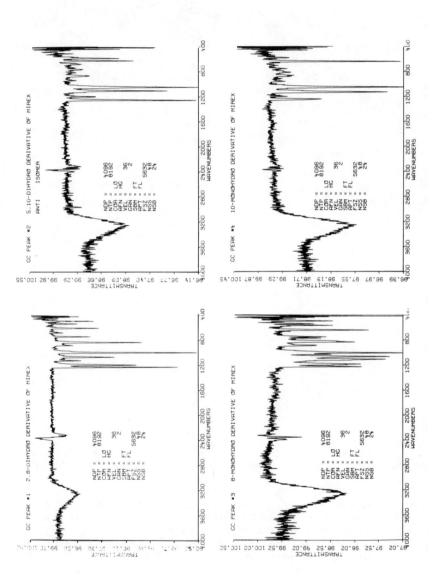

FIG. 2.   Infrared spectra of mirex bait formulation degradation products.   (Used by permission, Ref. [1].)

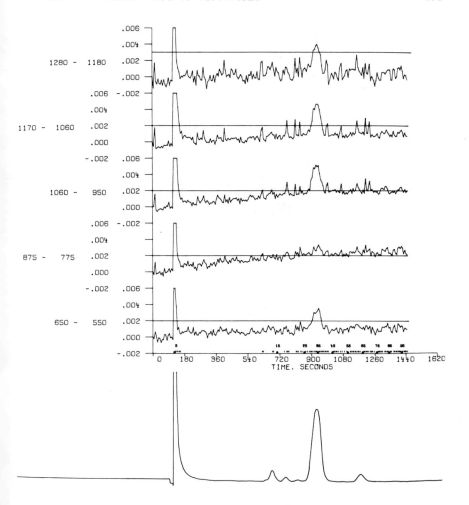

FIG. 3.   Gas chromatogram and Chemigram® of products of 5,10-
dihydrogen mirex preparation.   (Used by permission, Ref. [1].)

Reaction Monitoring

GC/FT-IR has also been used to monitor long-term reactions in
pesticide chemistry [1].   Many of the sample preparations take
several days (or weeks) for the reaction to proceed to the desired
products.   Aliquots of the reaction material are removed at regular
intervals, and the point at which the products of interest are at a
maximum concentration can be determined by GC or GC/FT-IR.   The
advantage of using GC/FT-IR is that the remainder of the reaction
products or by-products can also be monitored.   An example is shown
in Figs. 6 and 7 where a reaction for the preparation of the
8,10-dihydrogen derivatives of mirex was being monitored.   The
fourth and fifth peaks of the gas chromatogram and average

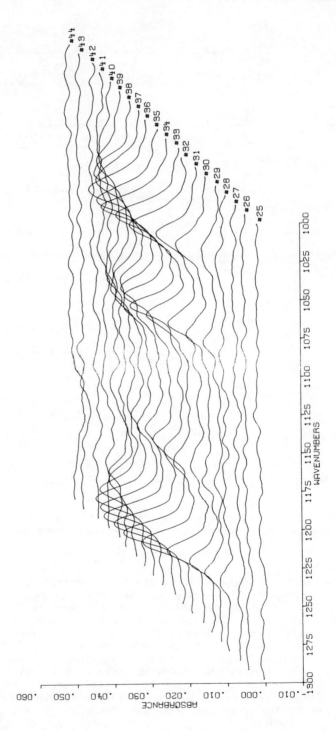

FIG. 4. Real-time display of infrared spectra from GC/FT-IR experiment of the products of the 5,10-dihydrogen mirex preparation. (Used by permission, Ref. [1].)

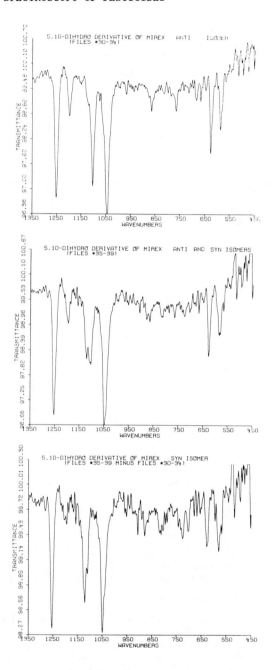

FIG. 5. Infrared spectra obtained from GC/FT-IR experiment of the products of the 5,10-dihydrogen mirex preparation. (Used by permission, Ref. [1].)

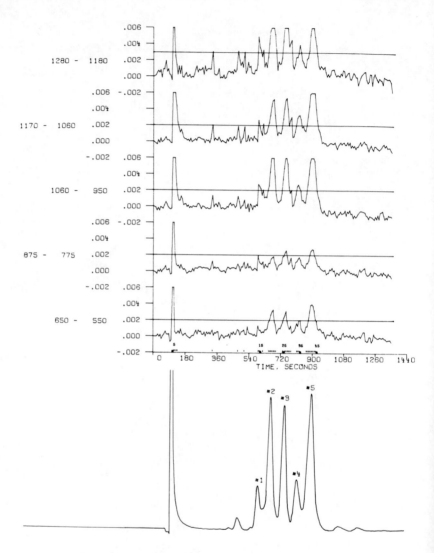

FIG. 6. Gas chromatogram and Chemigram® of products of 8,10-
dihydrogen mirex preparation. (Used by permission, Ref. [1].)

absorbance plots of Fig. 6 are due to the two isomers of 8,10-
dihydrogen mirex which were the desired products. The first and
second peaks were identified as being due to two of the trihydrogen
derivatives which had previously been prepared, but the third peak
represented a new, unidentified derivative. It was later
determined that this was one of the trihydrogen derivatives for
which previous attempts at preparation had failed. Had this reaction been monitored only by GC, that compound would have been

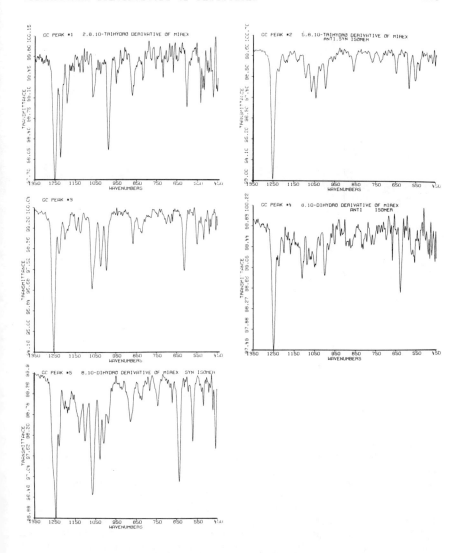

FIG. 7. Infrared spectra of the products of the 8,10-dihydro-
gen mirex preparation. (Used by permission, Ref. [1].)

discarded in the sample clean-up without a knowledge of its
identity. The infrared spectra of all five GC peaks collected
during this GC/FT-IR run are shown in Fig. 7.

Contaminant Detection
Another major phase of work to be considered before a pesticide can
be marketed is a biological study on a rat population to determine
the LD/50 and carcinogenity of the pesticide and each of its degra-

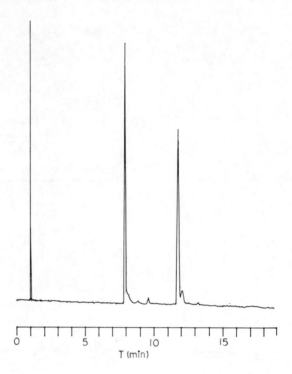

FIG. 8. Electron capture gas chromatogram of the mirex degradation products used in a carcinogenity study.

dation products. It is very important that only pure materials be
used so that an assurance is made that none of the effects come
from by-products. One such example compares two studies of one of
the degradation products of mirex -- one in which the derivative
was shown to be carcinogenic and another in which it was shown not
to be. Purity checks made by the standard method of EC/GC (shown
in Fig. 8) indicated that the degradation products used in the
former study were pure. Further checks made using GC/FT-IR showed
the products to be impure and identified the contaminant as a known
carcinogen. The gas chromatogram from the GC/FT-IR experiment is
shown in Fig. 9, and the infrared spectrum of the contaminant is
shown in Fig. 10. In this case, the problem lay in the high
sensitivity of the electron capture detector for halogenated
(chlorinated) compounds and its low sensitivity for the organo-
phosphorus compound. The infrared detector, on the other hand, has
numerous frequency "windows" through which to view a wider variety
of compounds.

Complex Mixture Identification
Oftentimes in the analysis of pesticides, one is confronted with
complex mixtures of related compounds. Such is the case with DDT,

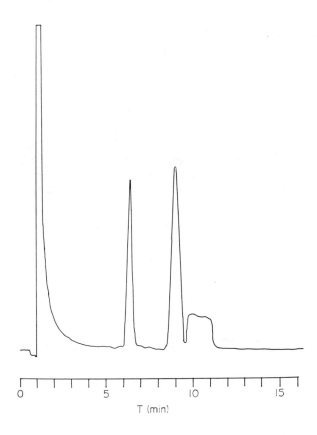

FIG. 9. Thermal conductivity gas chromatogram from a GC/FT-IR experiment of the mirex degradation products used in a carcinogenity study.

DDE, DDD, and their isomers. Capillary GC/FT-IR can separate and identify each component as shown in Figs. 11 and 12. Three different types of chromatographic displays from the computer data (Fig. 11) as well as full infrared spectra and identification of all seven compounds (Fig. 12) are available. EC/GC and GC/MS can also separate these compounds with better detection limits, but many of the isomers are indistinguishable by these methods.

Optimal Conditions for Pesticide Studies
In the past ten years, GC/FT-IR has made great strides from the initial prototypes to current-day optimized systems. Little developmental work remains. Virtually any GC/FT-IR system on the market today would be generally suited for any type study; however, if an older system is to be used or if a new system is to be developed, there are several aspects which one must be aware of, especially if pesticide studies are to be considered.

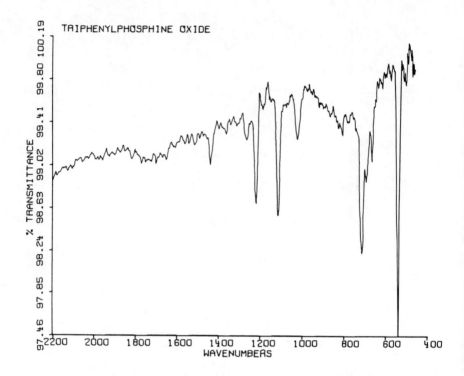

FIG. 10. Infrared spectrum of the contaminant found by GC/FT-IR in the mirex degradation products used in a carcinogenity test.

If a conventional GC detector is to be used in-line with a GC/FT-IR system, it should be placed after the lightpipe to avoid any dead volume or sample loss before the infrared analysis. GC detectors are not necessary in GC/FT-IR experiments since the computerized data can be used to generate several different chromatographic displays (as shown previously in Fig. 11). The transfer line used between the GC column and the infrared lightpipe should be glass-lined stainless steel or fused silica, as short as possible, and of an inner diameter that closely matches that of the GC column. Many current systems run the GC column inside of the transfer line to a point very close to the lightpipe.

The columns best-suited for pesticide studies have been found to be the wide-bore fused silica columns with heavy loadings. Many pesticides have very low infrared absorptivities, and relatively large quantities must be injected for infrared detection. The fittings used should be zero-dead-volume (ZDV) stainless steel, but it is important that the amount of metal surfaces exposed to the eluents be minimized since many pesticides decompose readily in the presence of hot metals. We have found that a good test is to

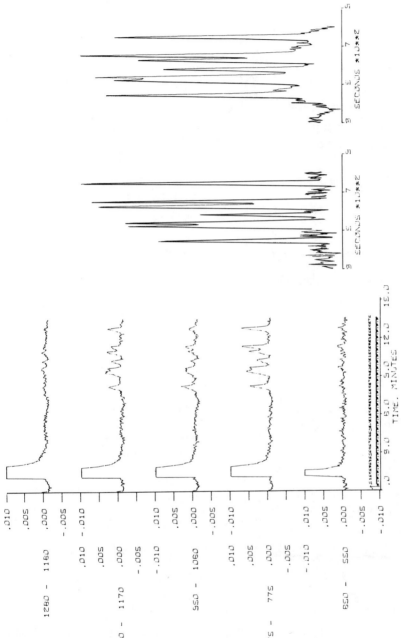

FIG. 11. Chemigram®, Gram-Schmidt reconstructed chromatogram, and infrared absorbance chromatogram of a seven component mixture of isomers of DDE, DDD, and DDT.

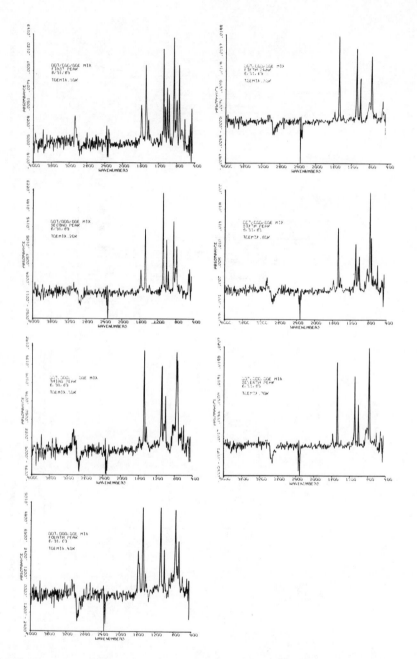

FIG. 12. Infrared spectra of the seven component mixture of the isomers of DDE, DDD, and DDT. The GC peaks were identified as follows: 1, o,p-DDE; 1, p,p-DDE; 3, o,p-DDD; 4, m,p-DDD; 5, p,p-DDD; 6, o,p-DDT; 7, p,p-DDT.

inject a mixture of DDT and DDE and look for signs of decomposition in the resulting spectra.

The lightpipe dimensions depend upon many factors including the choice of column size, but for most studies, the best lightpipe size is 1-2 mm inner diameter and 10-20 cm length. The lightpipe should be optically aligned at the temperature at which the samples are to be run. Optical throughput is a function of the lightpipe temperature because of thermal expansion which takes place.

Rapid-scanning should be used, if available, and co-added scans should be dumped to disk storage approximately every second to maintain chromatographic resolution. Detection limits for the GC/FT-IR experiments for pesticides are routinely in the mid-nanogram range. Although GC/MS detection limits can be in the picogram range, the identification of similar compounds by this method is sometimes ambiguous.

LC/FT-IR

GC/FT-IR analysis is applicable to many degraded pesticide products; however, there are also many high-molecular-weight, ionic pesticide products which decompose in a gas chromatograph if indeed they are volatile enough to elute. High performance liquid chromatography must be employed with these compounds. Unlike GC/FT-IR, LC/FT-IR is a relatively new technique with unrefined experimental designs and procedures. To address the applications which arose in pesticide chemistry, a new system was developed [15].

Previously, there had been two main techniques for LC/FT-IR analysis with both being applicable to normal phase separations. One system employed a flow-through cell [16,17] and the other utilized a solvent elimination system with subsequent analysis of the solute by diffuse reflectance infrared Fourier transform (DRIFT) spectroscopy [18,19]. However, neither system was readily applicable to aqueous reverse phase HPLC which constitutes the majority of present-day LC analyses. In our experiments, a modification of the DRIFT technique to deal with the problem of water elimination was undertaken [20,21].

The modified LC/IR interface for regular bore columns consists of a sample concentrator, a moving "sample train" which contains a halide salt substrate applicable for the diffuse reflectance analysis, and a nebulizer which sprays eluent onto the KCl substrate. For microbore columns, the volume of solvent is greatly reduced, which allows the solvent concentrator to be eliminated.

The substrate "train" is a length of metal with a rack gear mounted on the bottom for stepper motor control. The top side can be configured with small compartments spaced closely together or as an uninterrupted trough which yields a more continuous flow system. A novel method of forming the substrate was devised to enhance the sensitivity of the technique and create a sturdier surface [22]. A KCl-methanol slurry is made and packed into the compartments, then the surface is scraped to remove excess slurry. In the case of the compartmentalized train, each individual compartment is then compressed by a small diameter rod. For the trough, the entire surface is compressed using a rectangular block. In both cases the methanol is allowed to evaporate leaving a hard, uniform surface that is an excellent diffuse reflectance substrate.

For the removal of water, a post-column organic reactant (2,2-dimethoxypropane) is pumped into the eluent stream before sample deposition. The reaction products (methanol and acetone) [23,24] and solvents are volatile and are evaporated:

$$\underset{\underset{O-CH_3}{|}}{\overset{\overset{O-CH_3}{|}}{H_3C-C-CH_3}} + H_2O \overset{H^+}{\rightarrow} 2CH_3OH + H_3C-\overset{\overset{O}{||}}{C}-CH_3$$

This system is compact, applicable to "on-the-flow" systems with FT-IR analysis and can provide a reconstructed chromatogram from the infrared data. It also allows for signal averaging of a particular fraction of interest and for recovery of fractions for alternate analysis. A diagram of the microbore LC/FT-IR interface with either the continuous trough or compartmentalized train is shown in Fig. 13.

The pesticide study which triggered the need and subsequent development of this system involved the use of photooxidizing dyes for control of manure-breeding insects. It was found that Rose Bengal and Erythrosin B worked very well in the control of flies in chicken houses by simply spraying the dyes onto the manure beds. The flies that breed in the manure ingest the dye. The young flies have translucent bodies, and when they fly out into the sunlight, the dyes photooxidize and the flies literally explode!

Although these dyes are registered food dyes and are consumed by humans every day, the degradation products must still be known before a full pesticide registration can be obtained. This is where LC/FT-IR came into use.

A liquid chromatogram of the two dyes using ultraviolet detection and a water/methanol solvent system is shown in Fig. 14. The corresponding infrared reconstructed chromatogram from the LC/FT-IR data is shown in Fig. 15, and the infrared spectra collected from the LC elution are shown in Fig. 16. A comparison of the infrared spectrum collected from the water/methanol solvent by this method and one collected from a film cast from a methanol solution is shown in Fig. 17. It is clear that the LC/FT-IR interface is capable of removing the water from the solvent system. A real-time display of the infrared data collected during LC elution is shown in Fig. 18 with Erythrosin B in 100% methanol.

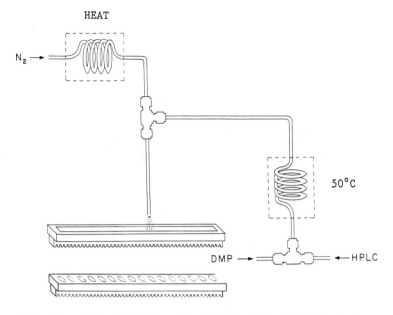

FIG. 13. Schematic diagram showing the essential features of the microbore HPLC/FT-IR interface. (Used by permission, Ref. [15].)

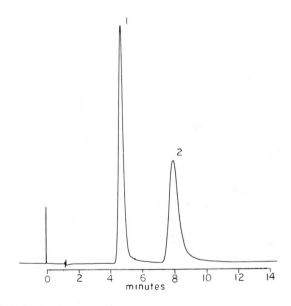

FIG. 14. Liquid chromatogram (UV detection) of Erythrosin B and Rose Bengal.

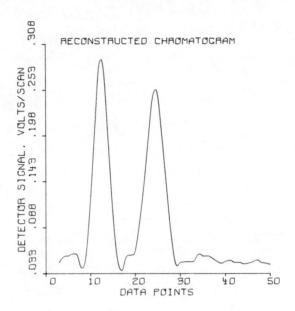

FIG. 15.   Infrared reconstructed chromatogram from the LC/FT-
IR data of Erythrosin B and Rose Bengal.

RESONANCE RAMAN SPECTROSCOPY

As mentioned in an earlier section, the pesticide formulation is as
important as the pesticide itself.   The formulations can alter the
photochemistry of the degradation process of particular pesticides.
The combination of halocarbons and amines has been found to form
charge transfer complexes which exhibit UV or visible absorption
bands which are different from the absorption bands of the individual components [25,26].   The object is to find a charge
transfer complex which absorbs as much of the spectrum of sunlight
as possible to aid in photodecomposition of the pesticide.
Resonance Raman spectra can be used to determine the association
constant for the complexation in the ground state by measuring the
relative intensities of appropriate peaks [27,28].   By using this
method, the more suitable mixtures and their concentrations can be
narrowed down without laboratory and field tests.   Such work was
conducted in our laboratory on DDT, mirex, and dibromomirex with
aliphatic amines [29].   The final formulation marketed for mirex
was just such a complex [9,10].

DIFFUSE REFLECTANCE

Diffuse reflectance infrared Fourier Transform (DRIFT) spectroscopy
has already been discussed as a useful tool for pesticide chemistry
inasmuch as it is used for LC/FT-IR analysis; however, it also has

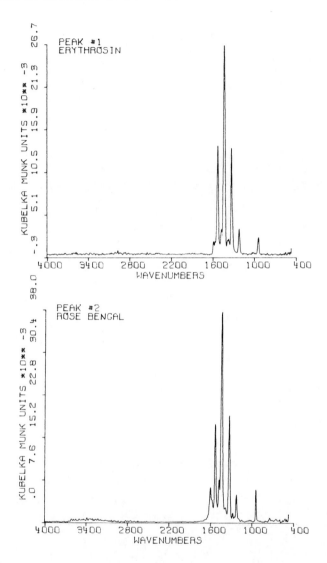

FIG. 16. Infrared spectra collected from LC elution of Erythrosin B and Rose Bengal.

other applications. One of the necessary requirements for an effective formulation is the capability for efficient distribution. In other words, an appropriate carrier must be found for the pesticide. In some instances, polymeric beads have been used, and in these cases the pesticide coated on these carriers can be studied by using DRIFT spectroscopy without an initial extraction procedure. Spectra of the pesticides and their formulations

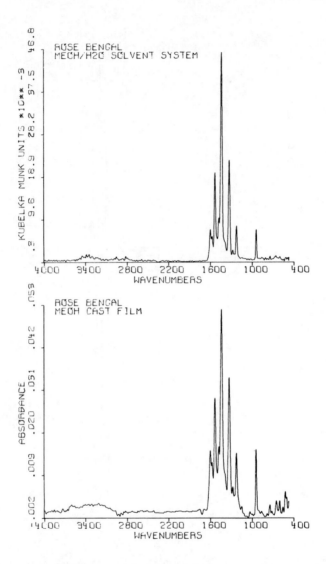

FIG. 17. Infrared spectra of Rose Bengal from a water/metha-
nol solvent system and from a methanol cast film.

studied by this method cannot be shown due to their proprietary
nature.

ATTENUATED TOTAL REFLECTANCE

Attenuated total reflectance (ATR) is one of the most useful tools
in analyzing industrial samples, but it has also found its way into

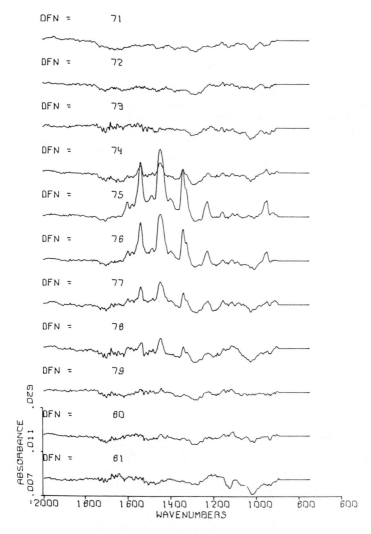

FIG. 18. Real-time display of infrared spectra from an LC/FT-IR run of Erythrosin B in methanol.

the study of pesticides. For example, a field of cotton had been sprayed with a pesticide which remained on the crop after picking. The cotton was washed, and the remaining pesticide was monitored by ATR infrared spectroscopy until the toxicant was removed.

LIBRARY SEARCHING

Spectral library searching is a powerful tool and has found its way into many analytical techniques. The fundamentals of infrared library searching have been discussed previously [30], so an application other than the obvious ones will be discussed here.

Every time an infrared spectrum is obtained on an FT-IR (or any other computerized) system, the data should be added to a spectral library regardless of whether or not an identification has been made. The data should then be fully catalogued as to the source of the material and other pertinent information. Multiple occurrences of unknowns from different sources can sometimes be used to identify a sample. Such was the case on several occasions during the years that mirex was being studied in the MSCL. By obtaining the spectrum of the same, but unknown, mirex derivative from various reaction pathways, the unknown derivative was identified by cross-matching the possibilities from reaction kinetics.

Clearly this approach will work best when the samples of interest comprise a limited, but recurring, population. Fortunately, however, this is often the case for the individual analyst working on a specific project.

EQUIPMENT

The work described herein was performed using the following various pieces of instrumentation: Nicolet 7199 Fourier transform interferometer, Varian 3700 gas chromatograph, Waters M-45 liquid chromatography pumps with a Waters 680 automated controller, Spex Ramalog DUV Raman spectrometer equipped with Spectra-Physics 171 argon ion and 375 dye lasers, Harrick DRA-SN5 diffuse reflectance attachment, and a Perkin-Elmer 186-0055 attenuated total reflectance accessory.

ACKNOWLEDGMENTS

The work described above represents the efforts of many skilled individuals who should not go unrecognized. It spans many years and briefly touches on many different research projects. Those who made major contributions are: Dr. Earl G. Alley, Dr. Victor F. Kalasinsky, Mr. James A. Smooter Smith, Ms. Donna Kaye Cassell, Mr. James D. Cain, Dr. Soraj Pechsiri, and Dr. Kashinath Nag.

It is a pleasure also to thank the Mississippi Imported Fire Ant Authority for the funds to purchase the Nicolet FT-IR and Spex laser-Raman systems.

REFERENCES

1.   K. S. Kalasinsky, J. Chromatogr. Sci., 21, 246 (1983).
2.   K. Nag, M. S. Thesis, Mississippi State University, 1977.
3.   K. S. Kalasinsky, Proc. Soc. Phot., 289, 156 (1981).

4. E. G. Alley and B. R. Layton, in "Mass Spectrometry and NMR Spectroscopy in Pesticide Chemistry," R. Hague and F. J. Niros, eds., Plenum Press, New York, 1974.
5. E. G. Alley, J. Environ. Qual., 2, 52 (1973).
6. E. G. Alley, presented at the Annual Meeting of Mississippi Entomology Society, Mississippi State University, Mississippi State, MS, November, 1978.
7. E. G. Alley, presented at the American Farm Bureau Forest Commodities Seminar, Birmingham, AL, July, 1979.
8. E. G. Alley, J. L. Smathers, and R. R. Ingram, presented at the Annual Imported Fire Ant Research Conference, Gainesville, FL, March, 1980.
9. E. G. Alley, presented at the Annual Imported Fire Ant Research Conference, Raleigh, NC, April, 1978.
10. E. G. Alley, presented at the Annual Imported Fire Ant Research Conference, Louisiana State University, Baton Rouge, LA, May, 1979.
11. B. R. Layton, presented at the Annual Imported Fire Ant Research Conference, Raleigh, NC, April, 1978.
12. J. L. Smathers, presented at the Annual Imported Fire Ant Research Conference, Louisiana State University, Baton Rouge, LA, May, 1979.
13. E. G. Alley, B. R. Layton, J. Eubanks, J. L. Smathes, R. R. Ingram, and K. S. Kalasinsky, presented at the Second Chemical Congress of the North American Continent, Las Vegas, NV, August, 1980.
14. E. G. Alley, presented at the Mississippi Section ACS Midwinter Chemical Symposium, Mississippi State, MS, 1981.
15. K. S. Kalasinsky, J. A. S. Smith, and V. F. Kalasinsky, Anal. Chem., submitted for publication.
16. D. W. Vidrine and D. R. Mattson, Appl. Spectrosc., 32, 502 (1978).
17. D. W. Vidrine, J. Chromatogr. Sci., 17, 477 (1979).
18. D. T. Kuehl and P. R. Griffiths, Anal. Chem., 52, 1394 (1980); J. Chromatogr. Sci., 17, 471 (1979).
19. C. M. Conroy and P. R. Griffiths, Pittsburgh Conference on Analytical Chemistry and Applied Spectroscopy, Atlantic City, NJ, March, 1984, paper 658.
20. K. S. Kalasinsky and V. F. Kalasinsky, Pittsburgh Conference on Analytical Chemistry and Applied Spectroscopy, Atlantic City, NJ, March, 1982, paper 30; March, 1983, paper 357.
21. K. S. Kalasinsky, J. A. S. Smith, and V. F. Kalasinsky, Pittsburgh Conference on Analytical Chemistry and Applied Spectroscopy, Atlantic City, NJ, March, 1984, paper 659.
22. V. F. Kalasinsky, J. A. S. Smith, and K. S. Kalasinsky, Appl. Spectrosc., 39, 000 (1985).
23. D. S. Erley, Anal. Chem., 29, 1564; F. E. Critchfield and E. R. Bishop, Anal. Chem., 33, 1034 (1961).
24. R. A. Bredeweg, L. D. Rothman, and C. D. Pfeiffer, Anal. Chem., 51, 2061 (1979).
25. R. Foster, "Organic Charge-Transfer Complexes," Academic Press, New York, 1969, pp. 94-103, 126-155, 276-282.
26. J. Yarwood, "Spectroscopy and Structure of Molecular Complexes," Plenum Press, New York, 1973, pp. 120-208.

27. G. Maes and T. F. Hyuskens, J. Am. Chem. Soc., $\underline{82}$, 2391 (1978).
28. K. H. Michaellan, K. E. Rieckhoff, and E. M. Voigt, J. Phys. Chem., $\underline{81}$, 1489 (1978).
29. K. Nag, Ph.D. Dissertation, Mississippi State University, 1979.
30. S. R. Lowry and D. A. Huppler, Anal. Chem., $\underline{53}$, 889 (1981).

TIME RESOLVED STUDIES VIA FT-IR OF POLYMERS DURING STRETCHING AND
RELAXATION

J. A. Graham*, R. M. Hammaker and W. G. Fateley

Chemistry Department
Willard Hall
Kansas State University
Manhattan, Kansas 66506

INTRODUCTION

Background
The advantages of Fourier transform infrared (FT-IR) instrumentation have been well documented [1-3]. These include the multiplex
(Fellgett) advantage [4], high throughput or etendue (Jacquinot)
advantage [5], and accurate wavenumber (cm^{-1}) determination [6].
Largely through the use of powerful minicomputers and small He/Ne
lasers the present day commmercial FT-IR spectrometer has become
the fast and reliable infrared tool that has long been considered
to be the dispersive infrared instrument. There are major differences, however. The FT-IR spectrometer is faster, extremely
sensitive, more precise and accurate.

The FT-IR spectrometer is capable of performing experiments that
were not possible with the dispersive infrared instrument. Some of
these experiments include sensitive sample handling accessories
such as diffuse reflectance attachments, photoacoustic cells, and
solid and liquid attenuated total reflectance cells. Other areas
of exploration have led to detection of single monolayers of
materials adsorbed on surfaces [7]. Perhaps one of the more
exciting fields enhanced by FT-IR instrumentation has been the area
of time resolved spectroscopy (TRS), also known as stroboscopic
interferometry. This exciting area of spectroscopy allows for the
observation of dynamic processes on a relatively short time scale.

One of the goals of the infrared spectroscopist is fast spectral
acquisition. Due to inherent limitations caused by the nature of
infrared radiation and the instrumentation necessary to measure
infrared radiation, the infrared spectroscopist is plagued with
difficulties not experienced with the faster methods employed by UV
and visible spectroscopists. Dispersive infrared spectrometers
have been developed to the point where a spectrum with adequate
resolution (for example, 6 cm^{-1} resolution at 1000 cm^{-1}, degrading

*Hercules Inc., Research Center, Wilmington, Delaware 19894

to 10 cm-^{1} at 3000 cm-^{1} from a Perkin-Elmer 1330) can be acquired in three minutes. A sample can seriously decompose or undergo some physical change during the course of spectral acquisition, resulting in infrared spectra which are not indicative of the state of the sample at any given time.

Parallel development of FT-IR spectrometers has enabled acquisition of high resolution spectra in very short time. For instance, 2 cm-^{1} resolution spectra can be acquired on an IBM model 98 vacuum FT-IR spectrometer in approximately one sec. In general, interferometers obtain spectral data much faster than dispersive spectrometers, because of the multiplex advantage whereby all the information contained in the range of detector response is obtained in one measurement. The speed advantage enjoyed by interferometry is particularly noticeable for the examination of large spectral ranges. The interferogram must be mathematically transformed from a measure of detector response vs. optical displacement in Fourier space to a measure of spectral intensity vs. wavenumber (cm-^{1}) in the spectral domain. The Fourier transformation requires a certain amount of computational space and time, but it occurs after the interferogram has been obtained, and does not prolong the acquisition time. A spectroscopist can thus monitor a kinetic process using FT-IR, provided the spectral sampling interval is shorter than the interval during which the change occurs [8].

What happens when the infrared spectroscopist wishes to examine the spectrum of a sample that is undergoing change at a subsecond rate? One approach, using dispersive instrumentation, is not to scan, but to set the grating in one position to monitor one spectral element, and chop the radiation at a rate ranging from the normal 13 Hz up to several thousand Hz [9]. Although this technique has yielded important information, the limitations to this method are obvious. Prior knowledge of which bands will undergo changes is necessary, and extraordinary reproducibility must be attained to determine whether spectral changes occur at the same time for each band.

Extremely fast-scanning dispersive instruments have been developed for the study of transient species, but these instruments have had limited commercial development. Pimentel and his co-workers [10] designed a spectrometer with a Littrow mirror which rotated at approximately 10,000 rpm and enabled a spectral acquisition rate of 10 cm-^{1}/sec and a resolution of approximately 20 cm-^{1}. They used a zinc-doped germanium detector with a response time of approximately 10 nsec, which is inherently faster than the detector's impedance response time. Further improvements were reported by Pimentel in 1968 [11], but this kind of instrument must be regarded as an extraordinary solution to an extraordinary problem.

One practical alternative involves the use of FT-IR instrumentation to perform time resolved spectroscopy (TRS). TRS, a sort of stroboscopic spectroscopy, is limited in that only reversible systems may be investigated without using complicated and expensive flowthrough systems. Thus, degradation and nonreversible reactions are beyond the capabilities of this technique. Nevertheless, there are many reversible chemical and physical changes that can be observed using TRS. Useful insights about reversible reactions may be gleaned from TRS, and it is a technique that could prove to have many applications.

History
TRS using FT-IR has been applied to two very different systems with varying degrees of success: systems examining the reactions generated by flash photolysis or electron bombardment, and systems examining phase transitions of stretched polymer films. One of the first systems examined [12] was a reversible reaction generating NO, N_2 and NO_2 from a mixture of O_2 and N_2 bombarded by a 32 kV electron beam. The information obtained was collected with a spectral resolution of 10 cm^{-1} and a temporal resolution of 0.5 ms. Improvements in instrumentation allowed excitation-relaxation studies [13] of CO_2 and CO to be performed with improved spectral and temporal resolution. The spectral resolution obtained was 2 cm^{-1} with a temporal resolution of 0.1 ms.

Closely following these studies was a study [14] of the reversible UV photodecomposition of acetone. It was later demonstrated [15] that these results were incorrectly interpreted owing to the generation of spectral artifacts. These artifacts were produced by concentration fluctuations in the sealed sample cell. It might also be noted that the quantum yield from the UV source was too low to achieve the reported results.

It should be noted here that the interpretations of the generation of these spectral artifacts is not quite correct. Garrison et al. described a test whereby spectral artifacts were artificially produced. The procedure involved computer sorting ten cosine waves, single frequency interferograms, all having identical frequencies but varying amplitudes. The variation in amplitude simulated a fluctuation in concentration. The result shown in Fig. 1 depicts an irregular waveform whose Fourier transform would produce two or more spectral frequencies. If a concentration change, such as that experienced by Mantz et al., was produced during a TRS experiment, the concentration change should be a continuous one and reproducible with every pulse of the flash lamp. The result, in the sorted interferograms, would be 10 interferograms of unaltered frequency with a cyclicly varying amplitude. The results, i.e., would be similar to starting conditions in Garrison's et al. test. The test described by these authors implies that an uncontrolled change in concentration has occurred. It is, therefore, postulated that the experiment performed by Mantz et al. and reproduced by Garrison et al. must have experienced some difficulties due to one or a combination of the following events: 1) incorrect timing of data acquisition, 2) incorrect firing of the flashlamp, or 3) not allowing sufficient time for the recombination of the photolysis products.

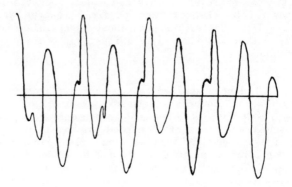

FIG. 1. Illustration of an irregular waveform produced by
computer sorting ten cosine waves of varying amplitudes. Used
by permission, Ref. [14].

The TRS study of polymer changes induced by mechanical stretching
has also been attempted. Stretched polymer TRS studies [16,17] of
isotactic polypropylene were begun in 1977, and recent results
[18,19] were published elsewhere. These studies showed that isotactic polypropylene undergoes reversible changes in the methyl
deformation region. Two infrared absorption bands, located at 1457
and 1378 cm^{-1}, were used to follow the deformation of the polymer.
The data were presented as an intensity ratio I_{1378}/I_{1457}, hereafter denoted as R, plotted against file number. Each successive
file or data point represented 1.14 ms. The ratio R followed
fairly closely the input stimulus function, which was a square
wave. In addition, a background test was performed to illustrate
that the observed changes are not due to cyclic noise or systematic
errors.

Similar encouraging TRS studies are being undertaken elsewhere
[20,21]. Polymers examined include low density polyethylene (PE),
poly(butyleneterephthalate) (PBT), and an ethylene-vinyl acetate
copolymer. Time resolved experiments performed on the low density
PE indicated that the dichroic ratio of the CH_2 segments in the
polymer lagged the input strain function by 70° or approximately
400 ms. It should also be noted that the dichroic ratio first
decreased, reached a minimum, increased to its original value and
then increased to a higher value before returning to rest near the
starting value (see Fig. 2). This indicates that a pseudo relaxation state may have been achieved. These results were obtained
with a temporal resolution of approximately 112 ms.

The sample of PBT was examined for reversible conformation changes
in the 1000 cm^{-1} to 900 cm^{-1} region of the infrared spectrum, an α
to β crystalline form change. In an attempt to increase the signal
to noise (S/N) ratio and decrease the time required for computerized data sorting, uniform groups of data were sorted in a block as

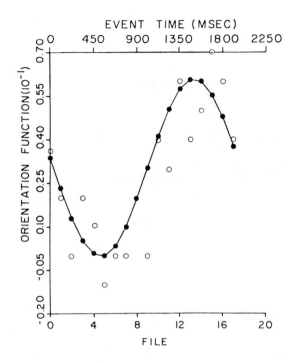

FIG. 2.   Dichroic ratio measured for the polyethylene methyl
bending.   Used   by   permission   from   Lasch   et   al.,   Appl.
Spectrosc., <u>38</u>, 343 (1984).

if they occurred at the same sample event time.   The data was
acquired at a temporal resolution of 122 ms.   One consequence of
sorting a group of data in a block is that small changes in the
baseline of the interferogram, due to modulation of the infrared
light passing through the sample, will show up as a sawtooth
function in the sorted interferograms.   The authors corrected this
by leveling out the discontinuities.   Further data has been
recorded and will be reported in a later section.

It should be apparent from references to both types of systems,
flash photolysis and oscillatory deformations, that improper
application of TRS can lead to artifacts which can easily be
interpreted as data.   The following sections will explain the
mechanism of TRS, examine the pitfalls which underlie the system,
and suggest an approach to eliminate spurious results.

THEORY

Data Collection
In order to understand how interferograms are used in TRS, it is useful to understand interferograms and their relationship to spectra. There is no one-to-one correspondence between points in the interferogram and points in the spectrum; rather, each point in the interferogram contributes to every point in the spectrum. The points in the interferogram are transformed into points in the spectrum using the Fourier transform algorithm,

$$B(\tilde{\nu}) = 2\int_{o}^{\infty} I(\delta) \cos (2\pi\tilde{\nu}\delta) \, d\delta.$$

The other half of the Fourier transform pair,

$$I(\delta) = \int_{o}^{\infty} B(\tilde{\nu}) \cos (2\pi\tilde{\nu}\delta) \, d\tilde{\nu},$$

is known as the inverse Fourier transform, and can be used to transform a spectrum into an interferogram, although this is less common. Phase error corrections and apodization functions complete these transforms. They are beyond the scope of this article; for further information the reader is directed to books by Bell [1] and Chamberlain [2].

The Fourier transform of monochromatic radiation, such as that from a laser, is a cosine wave, so a laser signal is used to determine when to collect data points for the interferogram at discrete, very precise intervals of optical retardation. The laser interferometer is coupled with the infrared interferometer, so any deviation from constant mirror velocity is matched by a corresponding deviation in collection rate, and precision from data point collection is maintained. The laser signal is frequency doubled by the Michelson interferometer, with data point collection occurring at every instance of zero amplitude and interpolated data collection at the maxima and the minima of the frequency doubled laser cosine wave. The He/Ne laser common to most interferometers has a wavelength of 632.8 nm, resulting in data point collection every 0.0791 μm of optical retardation. An interferogram which has a data point collected every 0.0791 μm enables transformation to a spectrum with a high wavenumber limit of 15,800 cm^{-1} without encountering spectral features known as foldback artifacts. When the spectral region of interest has a high wavenumber limit of 3950 cm^{-1}, a data point need only be taken at every fourth possible position. Selectively omitting data points is known as undersampling; in the example above the undersampling ratio (UDR) is 4. Figure 3 depicts a scheme in which data points are collected for a static system using a UDR of 2. A typical interferogram might consist of approximately 2K of digitized points, which appear as a continuous curve when plotted.

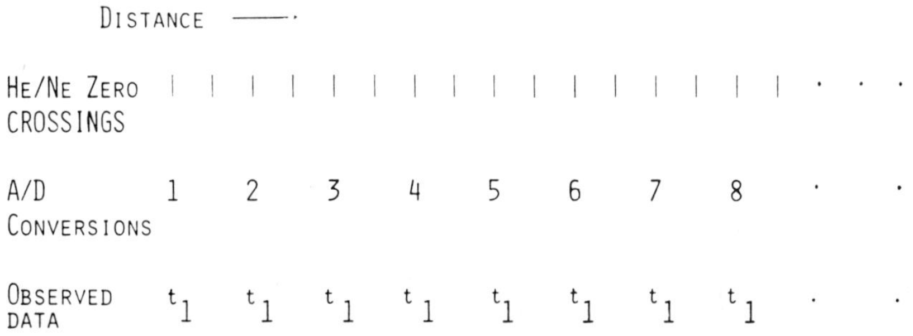

DISTANCE ————·

HE/NE ZERO | | | | | | | | | | | | | | | | · · ·
CROSSINGS

A/D   1   2   3   4   5   6   7   8   ·   ·
CONVERSIONS

OBSERVED t_1 t_1 t_1 t_1 t_1 t_1 t_1 t_1 ·   ·
DATA

FIG. 3.   Scheme for data point collection for a static
stretching system.

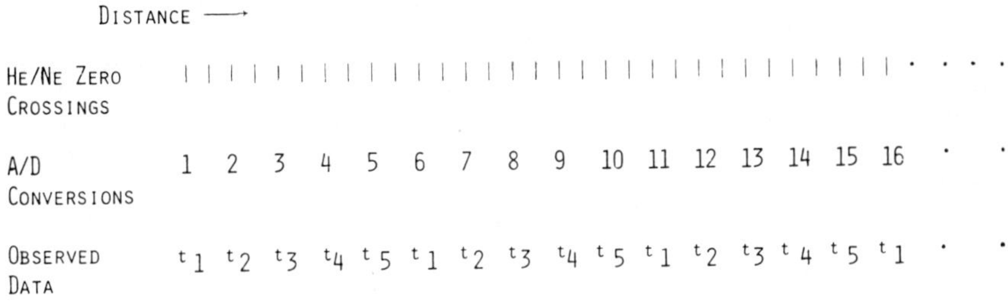

DISTANCE ——→

HE/NE ZERO | | | | | | | | | | | | | | | | | | | | | | | | | | | | | | | · · · ·
CROSSINGS

A/D   1 2 3 4 5 6 7 8 9 10 11 12 13 14 15 16   ·   ·
CONVERSIONS

OBSERVED t_1 t_2 t_3 t_4 t_5 t_1 t_2 t_3 t_4 t_5 t_1 t_2 t_3 t_4 t_5 t_1 ·   ·
DATA

FIG. 4.   Scheme for data point collection for a dynamic
stretching system with UDR=10, and important information may
be missing from many of the files.

When rapid reversible changes are observed using TRS, data collection becomes more complicated. The data point collection must be
synchronized with the event cycle to within instrumental limits so
the experimenter knows the stage of the cycle for each data point.
In the scheme depicted in Fig. 4, the reversible reaction is cycled
many times during the acquisition of a single interferogram, and 5
data points are collected during each event cycle. Thus, only
certain points correspond to a particular stage during the event
cycle. For instance, in Fig. 4 only points 1, 6, 11, 16,..., $5n+1$
correspond to points in the optical interferogram collected at
instant t_1 in the event cycle. The UDR for the collected interferogram is 2, and points acquired at t_1 of the event cycle occur
at only 1 out of every 5 points, resulting in an effective UDR of
10 for the theoretical interferogram for t_1. If physical strobing
is used to mask all radiation except that occurring during a given
time, for example, t_2, as depicted in Fig. 5, an interferogram with
a UDR of 10 is the result. In either case, the UDR may be large

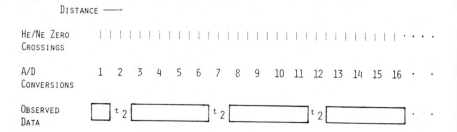

FIG. 5. Scheme for data point collection for a physically strobed system.

enough to restrict the high wavenumber limit and data are needlessly discarded.

The interesting solution to this problem makes use of the precision of the laser and the computational power of the computer. Once the reaction cycle has been synchronized with the collection of data points, it is a simple matter to offset data collection by one data point for each scan. Each filled interferogram is still meaningless in and of itself, because its points have been collected at different intervals during the cycle. However, as demonstrated in Fig. 6, a data point has been collected at every sampling point for each time interval of the cycle. As can be seen from Fig. 6, t_1 has been acquired at A/D conversion 1 in file 1, A/D conversion 2 in file 2, and so on. Here t_2 has been acquired at A/D conversion 2 in file 1, A/D conversion 3 in file 2. None of the data had been wasted because all data are pertinent to the experiment.

The next step is to construct meaningful interferograms from the composite interferogram. New files are created, taking pt. 1 from file 1, pt. 2 from file 2, continuing this sorting until a file is created solely out of points of the interferogram which occur at t_1 of the cycle. Similar files are created to create interferograms corresponding to other times of the cycle.

A typical interferogram consists of about 2048 pts., the first 296 of which are used for phase correction, so it should be noted that Figs. 3-6 depict only a portion of data collection. The number of interferograms needed is inversely proportional to the number of event cycles that occur during each interferogram. If the system undergoes 1 event cycle per interferogram, each interval will be 1/2048 of the time required for 1 scan, and 2058 files will have to be acquired and sorted. Similarly, as the system goes through more cycles per scan, the duration of each interval is decreased and fewer files need to be filed, theoretically to the point where a cycle only lasts 2 data points, and two files are collected for sorting.

DISTANCE ⟶

HE/NE ZERO | | | | | | | | | | | | | | | | | | | | · · · ·
CROSSINGS

A/D   1 2 3 4 5 6 7 8 9 10 11 12 13 14 15 16 17 18 19 20 · · · ·
CONVERSIONS

OBSERVED
DATA

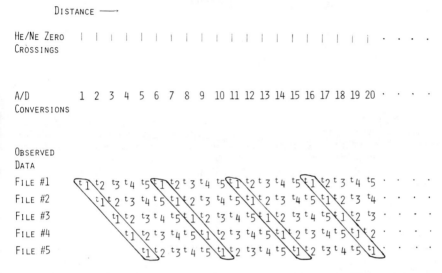

FILE #1
FILE #2
FILE #3
FILE #4
FILE #5

FIG. 6. Scheme for data point collection for TRS, with an indication of how many points are to be sorted.

Various parameters, such as mirror velocity and optical retardation, can be adjusted to obtain the desired time interval and resolution for each experiment. While a change in resolution will not affect the number of files to be collected, it will affect the number of data points per file. An increase in resolution will result in longer data collection and sorting times.

The variables undersampling ratio (UDR) and mirror velocity may be used to define the repetition period, time/data point, or data collection period (DCP). Equation (1) shows how these variables are related.

$$\text{Mirror Velocity} = \frac{(\text{UDR}) \ (\text{distance between data points})}{(\text{DCP})} \qquad (1)$$

The distance between the data points is defined as the actual distance traveled by the moving mirror and is dependent on the type of interferometer as well as the number of data points per cycle of the He/Ne laser cosine waveform. As described earlier, the distance traveled between data points for an interferometer with four zero crossings per He/Ne laser cycle is 0.0791 μm. It is possible, therefore, to calculate the data collection period DCP. By varying the mirror velocity and/or the UDR, it is possible to select the time resolution that is desired. The investigator should keep in mind that the free spectral range (FSR) is affected by the UDR and serious problems with foldback artifacts may occur [2]. The relationship between the FSR and the UDR is given by:

$$\text{free spectral range (FSR)} = \frac{15800.8}{\text{(UDR)}}, \text{ cm}^{-1} \qquad (2)$$

where 15800.8 cm^{-1} is the He/Ne reference laser frequency. If the UDR can not be set such that foldback artifacts will not occur, a filtering of radiation outside the defined FSR must be performed. This may be accomplished by a practical combination of the following: optical filters, beamsplitter range, detector response, and source emissivity.

In conjunction with selecting the data collection period (DCP), the investigator must also select the period of the event cycle. This is an important parameter and should be chosen carefully such that sufficient time is allowed between the beginning and ending of the sample event. If the event cycle period is too short, the perturbation mechanism will begin before the sample has had time to reach equilibrium. If the event cycle period is too long, data collection will be wasted between the time the system reaches equilibrium and the time the new perturbation cycle begins. The optimum event cycle period should be chosen such that approximately 10% of the data points collected per event cycle occur after sample equilibrium has been reached. In this way, no collected data is wasted and sufficient time is allotted for the system to reach equilibrium.

The number of files to be collected (FTC) is dependent on two parameters: the total number of data points per event cycle (NDC) and the time delay or offset (TDO) between the start of the perturbation cycle in successive interferograms. The exact relationship is given by :

$$\text{FTC} = \frac{\text{(NDC)}}{\text{(TDO)}} \qquad (3)$$

where both NDC and TDO are expressed in units of data points. In most cases the parameter TDO is set equal to one data point. The result of setting TDO equal to a value greater than one is twofold. First, the total number of files to be collected is reduced inversely proportional to the value of TDO. Second, the UDR of the sorted file is increased proportionally to the value of TDO. Therefore, a reduction in the number of files to be collected due to a time delay greater than one data point is obtainable only at the expense of the final UDR and hence the free spectral range. In addition, it should be noted that the odd numbered files will have data point collections occurring at every other zero crossing as compared to the even numbered files. This should not have a serious effect on the outcome of the experiment.

The main variables for characterizing a TRS experiment are: (UDR) undersampling ratio, (NDC) number of data points per cycle, (TDO) time delay or offset, and mirror velocity. Once these parameters are established, the TRS experiment is essentially defined. These parameters will affect the time resolution, cycle event period, free spectral range, total data collection time, and the time required for data sorting.

The Spectral Data Sorting Process
After acquiring the spectral data, each interferogram is useless as
a measure of the spectral information as a function of time. If
the collected interferograms were to be Fourier transformed as
collected, each spectral file would appear nearly identical. In
addition, there would be no representation of time as each spectral
file would represent an average of all transient species during the
event cycle. As described earlier, data point one in file one
represents t_1 at mirror position one, while data point two in file
two also represents t_1 but at mirror position two, and so on. If a
sorting routine is performed on the collected data such that a new
interferogram is constructed whereby all data points in the interferogram are of the same t_n, then useful time information may be
obtained.

A newly constructed interferogram consisting entirely of data
points representing t_1, may be obtained by copying the data points
from the collected data, diagonally, to a new file. A new file
representing each t_n may be created in a similar manner until the
number of new files is equal to the number of data points in the
event cycle. The number of data points to be sorted may be
enormous and will require a fairly large amount of computer space
and time to perform. As an example, if the UDR is set equal to 8
with NDC set equal to 50 and the resolution is two, the number of
points to be sorted is calculated as follows:

$$\text{number of points in interferogram (NDI)} = \frac{(OPR)}{(4)(UDR)(0.0791\mu m)} \quad (4)$$

where OPR is the optical retardation of the moving mirror in μm.
The value of OPR may be calculated by taking one half of the reciprocal of the resolution. This value must further be divided by
two for a Genzel type interferometer. If, as in this example, the
resolution is 2 cm^{-1}, then the optical retardation is 0.5 cm.
Therefore, the NDI would be given by:

$$NDI = \frac{(5000 \text{ mm})}{(4)(8)(0.0791\mu m)} = 2048 \text{ data points} \quad (5)$$

The number of interferograms to collect (NIC) is equal to NDC, the
number of data points per event cycle. The number of data points
to sort is given by:

$$\text{Number of data points to sort} = (NDC)(NDI) \quad (6)$$

and in this example the value of NDC is 50; thus, the number of
data points to sort is given by:

$$\text{Number of data points to sort (NDS)} = (50)(2048) \quad (7)$$
$$NDS = 102,400 \text{ data points.}$$

Computational time increases roughly as the square of the number of
files: while five files might require 4 hours of sorting, 100
files might take up to 16 hours. It is useful to reduce the computations to a minimum for systems with limited computational power.

EXPERIMENTAL PARAMETERS

Polymer Films Examined

One of the first polymer films examined was isotactic poly(propylene) supplied by John Rabolt [22]. The film examined most extensively was annealed poly(butyleneteraphthalate), four GT,
supplied by Jack Koenig [23].

Instrumental Parameters

All experiments were performed on an IBM instruments Model 98
Fourier transform infrared spectrometer. The instrument was
operated with a Ge/KBr beamsplitter and a heated silicon carbide
source. The transmitted energy was measured with an Infrared
Associates, liquid nitrogen cooled, narrow band Hg-Cd-Te photo-
conductive detector. The usable frequency range extended from 4500
cm^{-1} to 815 cm^{-1}. All sections of the instrument were operated in
a vacuum except the sample chamber, which was separated from the
rest of the instrument by an accessory sampling box. The box was
constructed of aluminum on four sides and the bottom. The top of
the box was fitted with a one inch sheet of plexiglass. The two
sides perpendicular to the beam path were fitted with KBr windows
to allow the infrared radiation to pass through the box. In all
experiments, except where indicated, dry nitrogen was flushed
through the box to reduce the amount of water vapor present.

The majority of the instrumental parameters was under computer
control. Most parameters remained unchanged throughout all
experiments performed. Deviations will be noted where necessary.
Beside each parameter a three-letter mnemonic may be found in
parentheses. Because a reference to a three-letter mnemonic is
much easier than the complete name, all further references to
parameters will be made using the three-letter mnemonic.

Stretcher Interface

The synchronization of the film stretcher with the data collection
of the computer is provided by the timing box [24]. In order for
the timing box to operate properly two diagnostic signals must be
acquired from the computer. One of these signals is called "XA'S"
and represents the trigger pulses for the computer to collect the
data. These pulses are in reality the maxima, minima and inflection points for the cosine wave as detected from the reference
interferometer with a He-Ne laser as the source. The "XA'S" signal
was acquired from the front panel of the operator console. A
rotary switch, Y'OUT, was set for "XA'S" and is then connected to
the back of the timing box with a coaxial cable having BNC
connectors on both ends. The second required signal is a reset
pulse that is used to signal the collection of data on the next
forward going scan of the moving mirror. This signal was acquired
from the rear of the electronic chassis which is directly above the
disk drive. A coaxial cable with BNC connectors was used to
connect the signal at the rear of the electronic chassis to the
rear of the timing box.

The front of the timing box has four sets of switch registers that are used to program the box. The switch register marked Pulse Max is used to indicate the number of data collection pulses per stretching cycle. The switch register marked Trial Max is used to indicate the number of interferograms to be collected and should have a value that is one unit greater than the actual number of interferograms to be collected. The switch register marked Delay Inc is used to indicate the number of data points by which the start of the stretching process should be delayed for each successive interferogram. The switch register marked N/C is used to represent the number of co-additions of interferograms. The front of the box also has two toggle switches. One is for the AC power to the box and the other, labeled ARM, is used to activate the counters just prior to performing the experiment. The output signal from the timing box was acquired from a BNC connector on the front panel.

The output of the timing box was connected to a variable pulse amplifier (VPA) using a coaxial cable with BNC connectors. The VPA is a homemade device designed and constructed by D. Honigs and J. Mowla [25]. The front of the VPA has two connectors marked input and output. In addition, the front panel also contains a toggle switch that turns the unit on and off. On the left side of the VPA there is an adjustable potentiometer that controls the output pulse width. Two colored wires exiting from the rear panel were connected to a five volt power supply. In our experiments the five volt supply in the timing box was used.

The output signal from the VPA was connected to a Hewlett-Packard model 467A power amplifier/supply. A Telequipment oscilloscope type D61A was connected to the output of the model 467A amplifier.

The time base on the oscilloscope was set at five msec for all experiments. The trigger mode was set to channel one, and the channel one voltage control was set at five volts per division. The trigger level was placed at the manual position and adjusted for proper triggering during a trial experiment.

The output of the model 467A power amplifier was also connected to the input of a Harrison 6824 A power supply-amplifier, Hewlett-Packard. The meter was monitored in the DC volts position and the mode switch was set in the amplifier position. The output of the amplifier was connected to the film stretcher by a coaxial cable with the BNC connectors on both ends. The amplifier setting was adjusted by the variable gain knob and was usually set to provide approximately 40 volts on the voltage meter. The final adjustment was set to achieve the desired percentage stretch of the polymer film during a trial experiment.

The solenoid used in the film stretcher was a Sporlan MB9S2 solenoid valve designed to operate at 240 volts and 50-60 Hertz. A two inch screw was glued to the end of the solenoid plunger and provided a means for attaching the movable set of jaws to the plunger. A spring located behind the plunger was also removed so

FIG. 7. View of a polymer stretching device.

that as little resistance as possible was applied in the stretching direction.

Figure 7 shows the solenoid mounted in an aluminum housing. A stationary set of jaws were attached to one side of the aluminum housing by an adjustable bolt. As previously described the other set of jaws was attached to the solenoid plunger. Both the stationary and the movable jaw assemblies were constructed of polycarbonate plastic.

Film Alignment
To mount the polymer film in the jaw assemblies, the removable parts of both sets of jaws were first disassembled. The polymer film was placed on top of the stationary jaw. The film dimensions were kept small enough so as not to interfere with the polycarbonate screw holes. After proper positioning of the film, the top section of the stationary jaw was carefully placed on top of the film and secured with two screws. Care was exercised not to disturb the film's position while securing the top part of the jaw assembly.

The movable jaw was then positioned at the desired distance from the stationary assembly. A six mm thick aluminum spacer was used to separate the two jaw assemblies in all experiments performed. The movable jaw was, therefore, placed slightly closer than six mm away from the fixed jaw. The loose end of the polymer film was then secured by the movable jaw assembly in the same manner as previously described for the stationary jaw assembly. This procedure provided a method for controlling the amount of pre-stretch on the polymer film.

At this point, the polymer film was inspected for any wrinkles that may have been caused by the mounting procedure. It is extremely important that the film be as flat and wrinkle free as possible. Any deviation from this criteria may lead to excessive modulation of the transmitted infrared radiation which in turn may generate spectral artifacts. If any wrinkles were found the entire mounting procedure was repeated until the problem was remedied.

Next the aluminum spacer was inserted between the movable and the stationary jaw assemblies. This step determines the amount of pre-stretch or pre-strain that is forced on the polymer film. If the amount of pre-stretch was not correct then the polymer film was repositioned in one of the jaw assemblies until the correct amount was obtained. Next the spacer was taped down in order to prevent it from moving during the experiment. Care was taken to insure that the tape did not restrict in any fashion the movement of the movable jaw assembly.

After securing the polymer film in the stretcher assembly, the position of the film with respect to the infrared beam was examined. This was done by placing the stretcher assembly inside the accessory box at the infrared beam position and visually inspecting the beam on the polymer film. The infrared beam was observed by dialing in the near-infrared CaF_2 beamsplitter and turning off the room lighting. The jaw assemblies could then be moved to the correct position and once again secured in place.

In order to prevent vibrations from the stretcher housing from being transmitted to the optics bench, foam packing material was used as an insulator. This material was placed below the stretcher housing as well as on both sides. For correct vertical positioning of the stretcher housing the appropriate number of foam strips were used. The optical filter holder was placed immediately after the stretcher housing and was firmly secured to the accessory sampling box by a sheet metal screw. The filter used was purchased from Wilks Scientific and transmitted infrared radiation from 1650 cm^{-1} down to at least 800 cm^{-1}.

The plexiglass top was then positioned on the accessory sampling box and the electrical leads from the solenoid were passed through the top. A dry nitrogen purge was effected by passing the purge hose through an opening in the plexiglass top into the accessory sampling box. The remainder of the space in the cover opening was filled with foam packing material to provide a mixing barrier against the atmosphere.

Data Collection

To prepare for data collection it is essential that all electrical connections and parameters be checked for their proper configuration. The parameter FLS was set equal to the file name of the first data collection file. This name must consist of four letters followed by three numbers only. A good example would be:

FLS = ACBB000.

TABLE 1.   Standard Procedure for Data Collection

    I.   FLS = first file name (ACBB000)

   II.   Press ESC key

  III.   Enter the following commands

   IV.   SLP 51 carriage return (for collection of 50 data files)

    V.   CLS carriage return

   VI.   INC FLS 1 carriage return

  VII.   ELP carriage return

 VIII.   Turn on all electrical components including the AC switch on the timing box: the arm switch should not be on at this time.

   IX.   Type #

    X.   Activate arm switch on timing box immediately

   XI.   Adjust for proper stretching cycle by rotating the knob on the left side of the variable pulse amplifier (VPA) while observing the event on the oscilloscope

  XII.   Adjust the signal amplification to achieve the required percentage of dynamic stretching

 XIII.   When everything is ready, or if a problem exists, turn the arm switch off and type CTRL Q 1

Table 1 is a list of commands and procedures that were used to prepare for data collection.

In all experiments performed the outline of the procedure in Table 1 was followed. After determining that everything was ready for data collection, the parameter FLS was set equal to the first file name. A single scan data collection was then performed on the static system. The file name was then incremented by one number and steps eight through ten from Table 1 were executed to begin the stretching cycle. After approximately 12 minutes the data collection as well as the stretcher motion stopped automatically. At this time a second single scan data collection of the static system was acquired. It was not necessary to increment the filename because the data collection loop had already done so. Immediately following the experiment the stretcher assembly was removed and a single scan data collection was performed. This served as the collection of a single beam background spectrum which was used to correct all the sample spectra for detector response, source emissivity, etc.

If a second experiment was to be performed for the purpose of the co-addition of interferograms, then the filename was changed and the entire procedure repeated. Upon completion of any experiment the filename was changed and the timing box was disconnected from the front panel of the FT-IR. These two steps were taken to prevent accidental loss of data and/or any possible damage from occurring to the timing box.

Background Collection

The procedure for collecting a background experiment was similar to that described for the dynamic experiment in the previous section. Some differences do exist, however. The electronic hardware as well as the stretcher interface were set up and programmed the same as for the dynamic experiment. The film alignment, however, is quite different. First, a suitable polymer film was positioned in the stretcher jaw assemblies as previously described. Next, a second polymer film was attached to the outside of the stretcher housing. The second film was placed so that it was a few millimeters above the film located in the jaw assembly. At this point the stretcher was placed into the spectrometer sample chamber and positioned so that the infrared beam passed through the top static film only. This was easily achieved by placing one or two less foam strips underneath the stretcher assembly while it was being positioned. The stretcher should be aligned so that the focus of the infrared beam is located approximately at the center of the top film.

The stretcher solenoid was connected to the interface hardware in the usual manner. In the trial run the percentage stretch of the bottom film was adjusted to equal that of the film used in the actual dynamic stretching experiment. At this time the top film was observed for any movement or flexing. If movement was detected, then steps were taken to prevent it as this could lead to spectral artifacts. The rest of the experiment was performed according to the procedure previously discussed.

COMPUTER TREATMENT OF DATA

Sorting Program

Once the data for an experiment has been collected, it necessarily follows that it would be processed into time resolved spectra. The first step is to translate the data from its time independent form to its more useful time dependent form. The theory behind this transformation is found in an earlier section.

To execute the program the computer was first directed to the monitor mode. The first step in executing the program was to call the pascal program by typing:

RUN PASCAL IBMTRS2.CODE

followed by a carriage return. The computer responded with the
title of the program followed by a command. The appropriate data
was input followed by a carriage return. The first command is:

ENTER NUMBER OF DATA FILES TO BE SORTED

where the number entered represented the number of data points per
stretching cycle. The computer then responded with:

ENTER NAME, EXTENSION OF FIRST DATA FILE

where the filename represented the name of the first input file and
the word extension represented the extension of the filename as
well as the disk it was stored on. An example of a valid response
might have been:

ACBB001.DATA=D2

where data was the extension and =D2 tells the pascal program that
the data file was stored on disk number two. Therefore it was
important that all data files pertaining to a specific experiment
were stored on the same disk. The last command printed by the
computer was:

ENTER NAME,EXTENSION OF FIRST SORTED FILE

where the filename represented the name of the first output file.
The operation normally lasted about four and one half hours for 50
data files that had a resolution of 2 cm^{-1}. The time required for
processing increases by the square of the number of data points to
be processed.

Peak-Pick and Plotting Programs
After the spectral data sorting process was completed, the time
resolved interferograms were then Fourier transformed in the conventional manner. In order to facilitate the compilation of the
massive amount of data that was generated during an experiment, a
computer program was written to pick the intensities of the
important peaks.

Basically the program picked the data point with the highest
absorbance value over a limited frequency range. The frequency
range was controlled by the program operator. Because a great
majority of the data collected was for one particular film, the
program was written to accommodate the most commonly picked peaks
in the spectrum of that film.

The operation of the program was similar to that described for the
spectral data sorting program found in the previous section. The
first four lines were similar to the spectral data sorting program
and required the operator to input the number of data files to be

handled, the name and extension of the first data file for input and output, and the name and extension of the file to be used as a reference file for background correction.

The next prompt output by the program asked for the input of the number of sample peaks. Either 4 or 7 was entered, where the numbers represented the number of peaks to be measured.

After inputting the required information, the computer would then ratio the single beam sample data points against the single beam background data points. Next the program computed the intensities of the peaks selected and corrected them for background absorbance.

As the peak intensities were calculated, their values were directed to the printer along with the frequency position from which they were obtained. When the program had finished calculating all of the peak intensities, it stored them in order of decreasing cm^{-1} for each time resolved spectrum. The name of this new file was that which was previously specified as the output file.

In order to plot the data in a meaningful manner, a second program was written to separate the data into files that contained only the peak intensities for a selected frequency. The program asked for the name of the file containing the packed data that was generated by the peak-picking program. The name of the output file was also required. The program then responded by telling the operator the number of spectra as well as the number of peaks per spectrum that reside in the packed file. Next the operator was asked to input whether the output file was to be a ratio file or an absorbance file. A ratio file consisted of points made up of the ratio of two spectral peak intensities. An absorbance file consisted of points made up of individual spectral peak intensities.

After a plot file had been created using the pascal program johnplot, the file was then plotted using the standard plot functions found in the FT-IR data collection program ATS. The plot output was automatically directed to the digital plotter connected to the computer.

RESULTS

Experiments
For the purpose of simplicity, the results cited in this section have been separated into three separate experiments. The first experiment involves the infrared spectra of the static and dynamic stretching of isotactic polypropylene. The second experiment deals with the infrared spectra of statically and dynamically stretched poly(butyleneterephthalate) (PBT). In addition, this experiment also involves time resolved spectroscopic (TRS) experiments performed at various oscillatory periods, i.e., stretching frequencies. One part of the experiment also shows the effects of co-adding multiple experiments of the sample, collected using identical parameters. The third experiment also shows the effects

of averaging multiple experiments, but obtained in a different manner. In addition, the experiment was performed at a high level of pre-stretch on the PBT sample.

Figure 8 shows the effects that the moving sample has on the modulation of the infrared light as seen in an interferogram. These interferograms were collected for a butyl-latex rubber sample. The sample stretching period was 57 msec with a spectral resolution of 2 cm^{-1}.

One of the interferograms was collected just prior to performing a time resolved experiment. The rest of the data were collected during a time resolved experiment and represent files one, ten, twenty, and thirty out of a total of fifty files per stretching cycle. Table 2 is a listing of the instrumental parameters used during this and other time resolved experiments. Table 3 is a listing of the time resolved parameters used during most of the time resolved experiments performed. Table 4 lists the optical hardware, source, beamsplitter, detector, optical filter, and infrared window materials that were used in most of the time resolved experiments.

Experiment Number One

Figure 9 represents the infrared spectra collected during a time resolved experiment on isotactic polypropylene (IPP). The instrumental and time resolved parameters used are the same as those given in Tables 2 and 3. The hardware components used were the same as those listed in Table 4. The sample was mounted in the stretcher with a prestretch of five percent. The amount of dynamic stretching was approximately five percent. The stretching cycle was 56 msec or 50 data points per stretching cycle. The two spectra in Fig. 9 represent the infrared spectrum of the IPP sample at points one and eighteen in the event cycle. Points one and eighteen represent a time of 1.12 and 20.16 msec after the beginning of the event cycle, respectively. The spectra shown are plotted from 0 to 1.6 in absorbance units. The spectral range shown extends from 1525 cm^{-1} to 1425 cm^{-1}. Figures 10 through 12 are individual time resolved spectra of the polymer from six different times throughout the even cycle.

The absorbance ratio A_{1378}/A_{1457}, hereafter denoted R, was calculated for the 1378 and 1457 cm^{-1} methyl deformation bands for all 50 files. A plot of R versus file number for the time resolved stretching experiment is shown in Fig. 13. Points 0 and 51 represent the static absorbance of the infrared band immediately before and after the performance of the time resolved experiment. Points 1 through 50 represent the dynamic absorbance of the infrared band during the time resolved experiment. Figure 14 shows the plot of R versus file number for the time resolved background collection experiment. Points 0 through 51 represent the same time relationship as those described earlier for Fig. 13.

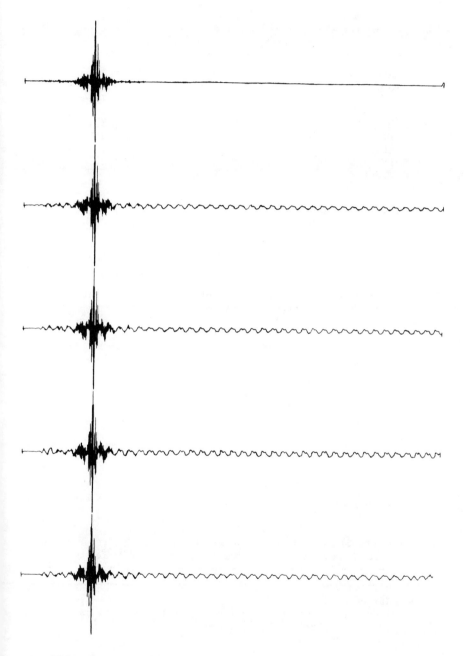

FIG. 8. Interferograms depicting the effect of a moving sample on the modulation of the infrared light: (a) prior to TRS collection, (b) file 30, (c) file 20, (d) file 10.

TABLE 2. Instrumental Parameters Used During Most of the Time
          Resolved Experiments

APF = BX	OPF = 4
APT = 1	PIP = 512
BMS = 6	PTS = 1024
DTC = MI	RES = 2
HFQ = 1975.0	SRC = MI
HTF = 6	SSP = 3
LFQ = 0.0	VEL = 0
LPF = 3	ZFF = 2
NSS = 1	

TABLE 3. Time Resolved Parameters Used During Most of the Time
          Resolved Experiments Performed

Pulse Max	50
Trial Max	51
Delay Increment	1
N/C	1

TABLE 4. Optical Hardware and Experimental System Conditions Used
          for Most of the Time Resolved Experiments Performed

silicon carbide source
KBr beamsplitter
liquid nitrogen cooled MCT detector (4000 - 815 cm^{-1})
optical filter (1650 - 400 cm^{-1})
KBr windows

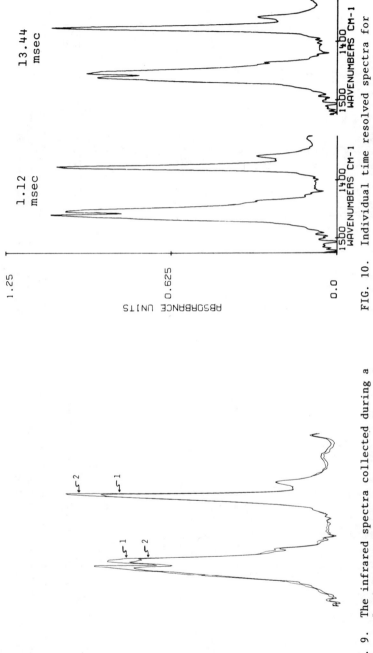

FIG. 10. Individual time resolved spectra for isotactic polypropylene representing files 1.12 and 13.44 msec after the stretching cycle began.

FIG. 9. The infrared spectra collected during a time resolved experiment on isotactic polypropylene from 1525 cm⁻¹ to 1425 cm⁻¹: (1) 1.12 msec after the stretching process began, and (2) 22.40 msec after the stretching process began.

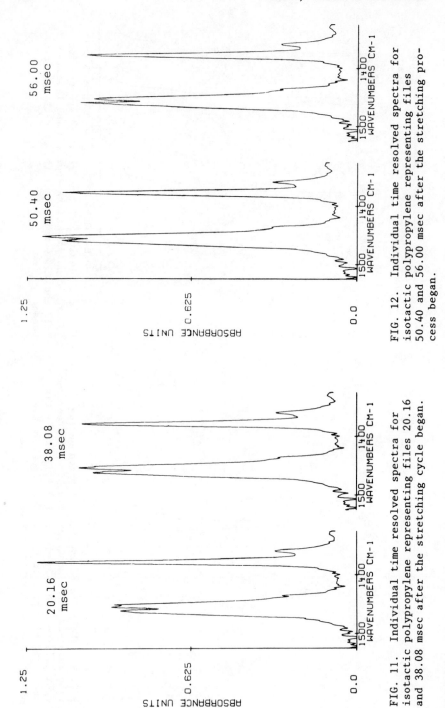

FIG. 11. Individual time resolved spectra for
isotactic polypropylene representing files 20.16
and 38.08 msec after the stretching cycle began.

FIG. 12. Individual time resolved spectra for
isotactic polypropylene representing files
50.40 and 56.00 msec after the stretching process began.

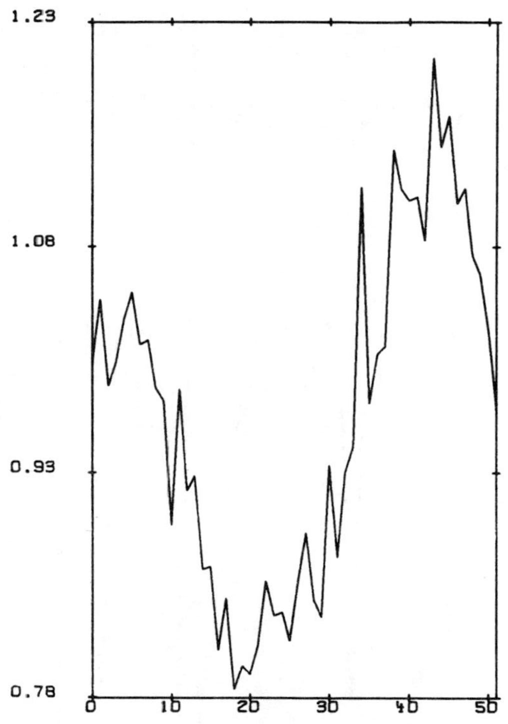

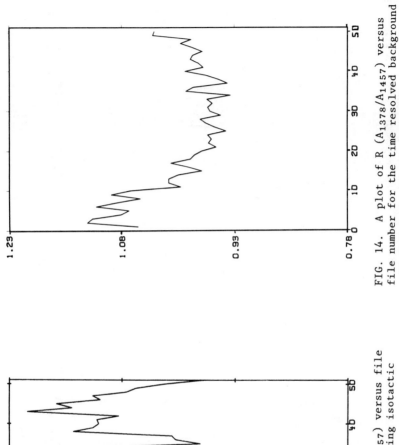

FIG. 13. A plot of R (A_{1378}/A_{1457}) versus file number for the TRS experiment using isotactic polypropylene as the sample.

FIG. 14. A plot of R (A_{1378}/A_{1457}) versus file number for the time resolved background collection experiment using IPP as the sample.

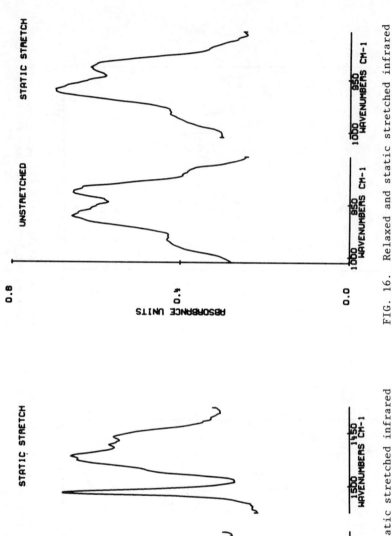

FIG. 15. Relaxed and static stretched infrared spectra of PBT shown for the 1525 to 1425 cm⁻¹ region.

FIG. 16. Relaxed and static stretched infrared spectra of PBT shown for the 1005 to 905 cm⁻¹ region.

Experiment Number Two
Figures 15 and 16 are relaxed and static stretched infrared spectra
of PBT for two regions of the infrared spectrum. The spectrum of
the unstretched sample represents the state of the PBT during a
relaxed state. The spectrum of the static stretched sample represents the state of the PBT at maximum elongation, approximately ten
to twelve percent. The two regions of the infrared spectrum shown
are 1525 to 1425 cm^{-1} and 1005 to 905 cm^{-1}.

Four separate time resolved experiments were performed to complete
this section of the work performed. Three sequential time resolved
experiments were performed using PBT. The fourth experiment was
the collection of background time resolved data using the PBT
sample. The outline of the procedure used has been described in
the experimental section. The instrumental and time resolved
parameters used were identical to those listed in Tables 2 and 3.
The hardware components used were the same as those listed in Table
4. The sample was mounted in the stretcher with a prestretch of
three to four percent. The dynamic stretch was approximately six
percent.

Files for the three sequential sample collection experiments were
averaged and stored prior to computer sorting of the data. Files 0
and 52 were collected just prior to and immediately after the
operation of the TRS experiment and were not involved in the
computer sorting of the data. Files 1 through 50 were collected
during the TRS experiment and were used during the computer sorting
of the data.

The infrared spectrum shown in Fig. 17 is that of the averaged data
from the three sequential experiments. The file number is one, and
the time is 1.14 msec after the beginning of the event cycle. The
spectrum is plotted from 1650 to 800 cm^{-1} and 0.0 to 1.0 absorbance
units.

Four individual TRS spectra are displayed in Figs. 18 and 19. The
data was plotted from 1525 to 1425 cm^{-1} on the ordinate scale and
from 0.0 to 0.65 absorbance units on the abscissa scale. The time
values shown are 1.14 and 20.52 msec for Fig. 18, and 41.04 and
55.86 msec for Fig. 19. These time values correspond to files 1,
18, 36, and 49 out of a total of 50 for the event cycle.

The remainder of the data for experiment number two will be displayed as absorbance or ratio plots. An absorbance plot is a plot
of the infrared absorbance of a designated band versus the file
number, each file represents 1.14 msec after the beginning of the
event cycle, of the band in the event cycle. A ratio plot is a
plot of the ratio of the absorbance of two infrared bands versus
the file number of those bands in the event cycle.

The absorbance ratio A_{1459}/A_{1475}, hereafter denoted R, was
calculated for the 1459 and 1475 cm^{-1} methyl deformation bands for
all fifty files. A plot of R versus file number for the time
resolved stretching experiment using polybutyleneterephthalate as

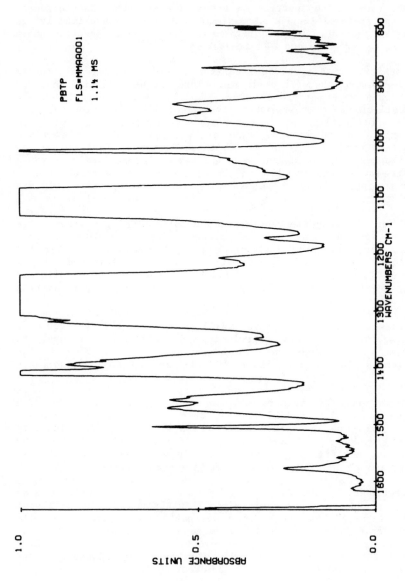

FIG. 17. Infrared spectra of PBT file 1 (1.12 sec) of a TRS experiment. Spectra shown are the average of three sequential experiments.

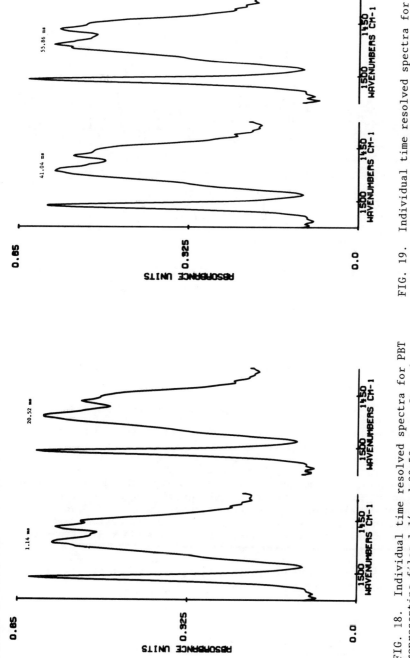

FIG. 18. Individual time resolved spectra for PBT representing files 1.14 and 20.52 msec after the stretching cycle began.

FIG. 19. Individual time resolved spectra for PBT representing files 41.04 and 55.86 msec after the stretching cycle began.

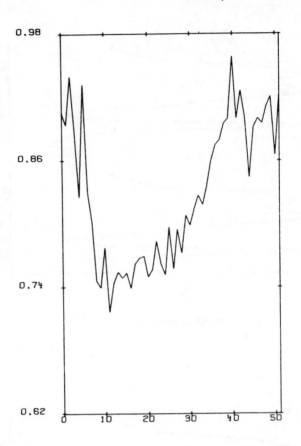

FIG. 20. A plot of R (A_{1459}/A_{1475}) versus file number for the time resolved stretching experiment using polybutylene terephthalate as a sample. The part shown is the average of three sequential experiments.

the sample is shown in Fig. 20. The plot of R versus file number as shown in Fig. 20 is the result of the average data for three sequential experiments. Figure 21 shows the plot of R versus file number for the time resolved background collection experiment. Again, files 0 through 51 represent the same time relationship as those described earlier for Fig. 13.

An entire TRS experiment is shown in Figs. 22 and 23, 1545 to 1425 cm^{-1}, using PBT as the sample.

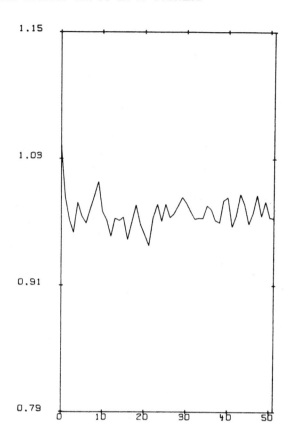

FIG. 21. A plot of R (A_{1459}/A_{1475}) versus file number for the time resolved background experiment using PBT as a sample.

CONCLUSION

The results of stretching-relaxation studies have been reported in this paper. We conclude that this is a very good method to follow the molecular structural changes in polymer molecules in time frames down to 10 μs.

The interpretation of the molecular structural changes observed here will be the subject of a future paper.

ACKNOWLEDGMENT

This work was supported in part by NSF Grant CHE-81-09570. We greatly appreciate this support.

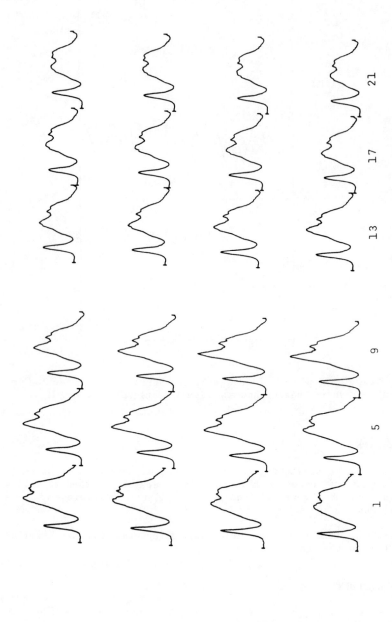

FIG. 22. Individual infrared spectra from a TRS experiment with files 1 through 24 depicted from bottom to top and left to right.

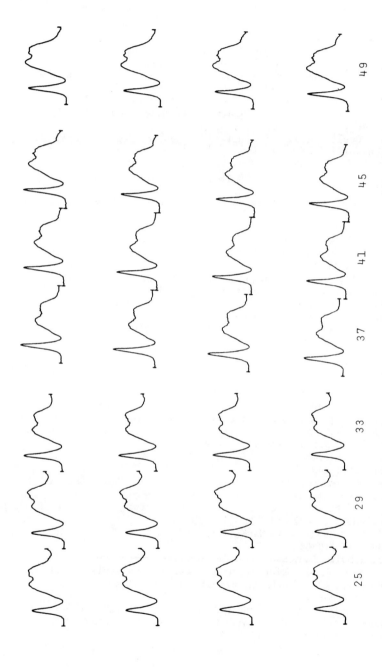

FIG. 23. Individual infrared spectra from a TRS experiment with files 25 through 50 depicted from bottom to top and left to right. The two remaining files at the top right hand corner are the infrared spectra of PBT collected immediately before and after the TRS experiments.

REFERENCES

1. R. J. Bell, Introductory Fourier Transform Spectroscopy, Academic Press, New York (1972).
2. J. Chamberlain, The Principles of Interferometric Spectroscopy, John Wiley and Sons, New York (1979).
3. P. R. Griffiths, Chemical Infrared Fourier Transform Spectroscopy, John Wiley and Sons, New York (1975).
4. P. Fellgett, Thesis (University of Cambridge), 1951.
5. P. Jacquinot and P. Doufour, Rech. J. du C.N.R.S., 6, 91 (1948).
6. G. Guelachvilli, Dissertation, Doctor of Science, Physics (University of Paris, South), 1972.
7. W. G. Golden, Polymer Preprints, 25, 158 (1984).
8. H. W. Siesler, J. Polym. Sci., Polym. Lett. Ed., 21, 99 (1983).
9. I. Noda, A. E. Dowrey, and M. Curtis, J. Polym. Sci., Polym. Lett. Ed., 21, 99 (1983).
10. K. C. Herr and G. C. Pimentel, Appl. Opt., 4, 25 (1965).
11. G. C. Pimentel, Appl. Opt., 7, 2155 (1968).
12. R. E. Murphey, F. Cook, and H. Sakai, H., J. Opt. Soc. Am., 65, 600 (1975).
13. H. Sakai and R. E. Murphy, Appl. Opt., 17, 1342 (1978).
14. A. W. Mantz, Appl. Spectrosc., 30, 459 (1976).
15. A. A. Garrison, R. A. Crocombe, G. Mamantov, and J. A. deHaseth, Appl. Spectrosc., 34, 399 (1980).
16. W. G. Fateley and J. L. Koenig, J. Polym. Sci., Polym. Lett. Ed., 20, 445 (1982).
17. D. E. Honigs, R. M. Hammaker, W. G. Fateley, and J. L. Koenig, in "Vibrational Spectra and Structure", Vol. 11, (J. R. Durig, ed.), Elsevier Scientific Publishing Company, Amsterdam (1973), pp. 231-239. (N. B., Figure 10 on p. 233 is missing half of the vertical lines in the He/Ne Zero Crossings row. If a vertical line is inserted between each line, Fig. 10 will correctly demonstrate the UDR of 2, and will be consistent with the discussion on pp. 233, 234.
18. J. A. Graham, W. M. Grim, III, and W. G. Fateley, J. Mol. Struct., 113, 311 (1984).
19. W. M. Grim, III, J. A. Graham, R. M. Hammaker, and W. G. Fateley, Am. Lab., 16, 22 (1984).
20. D. J. Bucchell, J. E. Lasch, R. J. Farris, and S. L. Hsu, Polymer, 23, 965 (1982).
21. S. E. Molis, W. J. MacKnight, and S. L. Hsu, Appl. Spectrosc., 38, 529 (1984).
22. J. Rabolt, IBM, San Jose, CA, private communication.
23. J. Koenig, Division of Macromolecular Molecules, Case-Western Reserve, Cleveland, OH, private communication.
24. Our own design.
25. D. Honigs and J. Mowla, formerly of the Department of Chemistry, Kansas State University, Manhattan, KS, private communication.

CHEMICAL UTILITY OF LOW FREQUENCY SPECTRAL DATA

J. R. Durig and J. F. Sullivan
Department of Chemistry
University of South Carolina
Columbia, SC 29208

INTRODUCTION

Beginning in the late fifties there has been an increasing number of vibrational investigations in the far infrared spectral region. In this spectral region one observes heavy atom vibrations, skeletal bending modes of organic molecules, metal ligand stretching and bending motions, the lattice modes of molecular crystals, and a number of interesting effects in semiconductors. While quite a few of the early dedicated researchers overcame the many difficulties associated with spectroscopic work in the far infrared spectral region, the true potential of this spectral area is still not fully realized.

It is interesting to note that work in far infrared spectroscopy began just before the turn of the century and Rubens and his co-workers published about eighty papers of spectral studies in the far infrared region from the early 1890's to 1910. Palik [1], utilizing the definition of the far infrared region as being between 25 and 1000μ (400 to 10 cm^{-1}), compiled a bibliography from 1892 to 1960 and there are approximately 350 papers published during this time. However, since 1960 there has been considerably more research activity in this spectral region due mainly to the development of the interferometer which helped to overcome some of the myriad of problems which plagued researchers. This spectral region suffers from the lack of suitable optical materials, low energy sources, poor detectors, and spectral interference from the pure rotational spectrum of water. It should be noted that Rubens and Wood [2] published the first true infrared interferogram in 1911 although the interferometer was more similar to a Fabry-Perot type device than the more familiar Michelson design which is used today. However, these early workers suffered the same major problem that Michelson faced which was the lack of computational resources with which to transform the interferogram into a spectrum. The structural simplicity of the interferometer compared to a dispersive spectrometer was more than offset by the unintelligibility of the output data.

In the fifties, Strong, Gebbie, Loewenstein, Vanasse and others showed that the advantages claimed for interferometry by Fellgett [3] and Jacquinot and Dufour [4] could be realized and by this time it was possible to calculate the necessary Fourier transforms albeit it was still a laborious and time consuming task. The key developments after this did not come from the spectroscopists but from the availability of inexpensive but powerful minicomputers coupled with the publication of the Cooley-Tukey algorithm for fast Fourier transforms. Gebbie [5] has commented that computers, like photographic plates, are wonderfully sophisticated and essential spectroscopic tools that, fortunately, the spectroscopist did not have to invent himself! Thus, by the late sixties one could pur- chase commercial far infrared interferometers which made this spectral region more accessible to a larger number of chemists. However, it should be noted that interferometric infrared spec- troscopy was first developed for use in the far infrared spectral region and only relatively recently has its range extended to cover the mid-infrared spectral range. In view of this, we shall briefly examine why interferometers are so suited for investigations in the far infrared spectral region.

The major problem involved in far infrared spectroscopy is the relatively poor signal-to-noise ratio. The primary reason for this problem is because there are really no good energy sources for this region which have high output over a broad enough range of fre- quencies. The commonly used blackbody sources, such as Nernst glowers or globars, all suffer from the same limitation. The energy distribution for such a source at 1000°C contains less than one hundredth of one percent of the total energy in the region below 100 cm^{-1}, with the peak output near 3000 cm^{-1}. Not only is there very little radiant power of the desired frequency, there are tremendous amounts of energy, relatively speaking, in unwanted shorter wavelengths. Increasing the source temperature only marginally increases the energy available at long wavelengths, but dramatically increases the high-frequency output. If the detection system employed merely responds to the total incident flux, this high-frequency energy must be prevented from reaching the detector and "swamping out" the desired signals. For the spectral region below 100 cm^{-1} most interferometers have a high pressure mercury lamp where the effective temperature of the plasma is about 10,000°C but this source has its maximum at about 45 cm^{-1} and the quartz envelope provides the radiation above about 80 cm^{-1}.

The overall sensitivity to far infrared radiation is further reduced due to the lack of efficient detection devices. The photo- multiplier tubes and many of the semiconductor devices used so successfully in other spectral regions have no counterparts available for the long wavelength regions and generally one must rely on thermal devices such as bolometers. Unless cooled to cryogenic temperatures, such devices must operate with high back- ground noise levels due to random thermal events within the detector itself. Recently developed detectors of the doped semi- conductor variety show promise of being more efficient and quieter than any previously known, but they require cryogenic cooling.

While spectroscopists will continue to receive newer and better detectors from research in related fields, the margin for improvement is slim, and a major breakthrough in far infrared detection is not expected. Therefore, one must make maximum use of the available energy in the far infrared spectral region and it is this aspect of interferometric methods which made it so attractive for studies in this frequency range.

It seems appropriate to add a few comments with regard to the realizability of Fellgett's and Jacquinot's advantages in the region below 500 cm^{-1}. Since the spectral range is smaller, the number of multiplexed spectral elements is correspondingly smaller. For a range of 450 cm^{-1} at 1 cm^{-1} resolution, Fellgett's advantage is only about a factor of 20. However, as the resolution is increased, the relative gain is increased, so that for the same region at 0.25 cm^{-1} resolution a factor of 40 is obtained. While not as striking as in the mid-infrared region, this factor of 40 is still significant. It has been pointed out by Griffiths [6] that the throughput of a far infrared interferometer is more a function of the f number of the source foreoptics than of the interferometer itself and hence one accrues little advantage from the throughput of the interferometer versus that of a dispersive instrument.

The greatest superiority of interferometers in the far infrared spectral region versus the earlier grating instruments is due to the ease with which effective stray light is reduced. There are two distinct ways in which this problem affects a grating monochromator. The most important is the fact that when radiation diffracted in one order of a grating reaches the exit slits, radiation in other orders is also present. To achieve good spectral coverage of the far infrared, several gratings of different ruling widths are necessary and, for each, the various orders are quite close together. To overcome this problem it is necessary to introduce several narrow-band transmission or reflection filters to separate the desired energy. Generally, these filters not only fail to totally eliminate the undesired radiation, but also significantly attenuate the desired radiation as well. Another type of stray light problem arises if nondispersed light manages to be reflected around the monochromator and fall upon the detector. There is no way to distinguish this flux from the desired flux, and the problem can be severe if the detector has a broad response range in view of the large amounts of high intensity, short wavelength light emitted by high-temperature blackbody sources.

With an interferometer, stray light of the second type which has not been modulated by the interferometer appears at the detector as a constant dc offset and, as such, has no effect on the computed spectrum provided it is of insufficient intensity to overload the detector. While there is no direct correspondence to overlapping orders in an interferometer, stray light which has been modulated is also easily handled. If the interferogram is sampled at sufficiently high frequency, energy below this Nyquist frequency will be properly accounted for in the computed spectrum. To prevent aliasing of higher frequency components, it is only necessary to

include one suitable filter (such as black polyethylene) to sufficiently attenuate the undesired radiation. If the interferometer
is of the rapid scan type, another attractive mode of filtering is
available. For such instruments, the detector signal fluctuates at
audio frequencies. Since the Fourier components due to high frequency radiation vary most rapidly, it is possible to eliminate
them from the interferogram by the inclusion of a simple audio
band-pass filter. This has the distinct advantage of being applied
after the detector on an electrical signal which is easily
amplified, and thus the optical throughput is not impaired. These
methods can reduce the effect of stray light to a negligible
amount. Also, the overall simplicity of an interferometric spectrometer as compared to a dispersive spectrometer is also an asset.
To achieve reasonably high resolution with an interferometer, the
basic design is only slightly more complicated, whereas one must
increase the size of the gratings and other optical components in a
dispersive instrument if reasonable throughput is to be maintained
at narrow slitwidths. The final advantage is that one is forced to
acquire the final spectrum in some computer readable form where it
is possible to use the computer for post-processing of the spectrum
to reduce noise, tabulate peaks, and subtract out the effect of
spectrally interfering species such as water vapor. Therefore, for
spectral studies below 200 cm^{-1} an interferometer is almost a
necessity and if relatively high resolution gas phase data is
desired then the optical bench needs to be evacuable to prevent the
spectra from being degraded by the pure rotational spectrum of
water vapor. For a more complete description of the development of
far infrared interferometry the reader should consult reference [7]
and articles cited therein.

GAS PHASE STUDIES

Initial applications of Fourier transform interferometry in the far
infrared spectral region were mainly investigations of pure
rotational spectra of light molecules, studies of skeletal bending
modes and heavy atom vibrations, and the determination of spectra
for molecules in the solid state. It is reasonably easy to see why
these studies were the initial ones when one considers the information obtained. For most molecules the energy level spacings
between pure rotational energy levels are quite small and transitions between these levels occur in the microwave region. However,
if the molecule is sufficiently "light", these transitions can
occur in the far infrared. The molecular constants which can be
derived from pure rotational spectra are especially important for
small molecules, since they can be used to model more complex
molecules. Although most of the fundamental vibrations for many
molecules occur in the mid-infrared region, normal modes involving
heavy atoms such as halogens and skeletal deformations occur in the
far infrared. It is these low-lying modes which contribute the
largest amount to the thermodynamic functions of a molecule. Also,
the determination of a molecular force field through normal coordinate analysis is impossible without frequencies for all the
normal modes. When such calculations are carried out, it is
desirable to have frequencies from the gaseous phase rather than

frequencies from the Raman spectrum of the liquid from which many of the earlier data were taken. Also it should be noted that the low frequency modes are generally the most sensitive to changes in the conformation or structure of a molecule.

When a molecule is studied in the solid state, interactions between molecules become quite important, and one obtains information on intermolecular forces and molecular dynamics of the crystal lattices. Also the determination of bulk optical properties such as reflectivity and refractive indices has also been an important use of far infrared data. Finally, the observation of the electronic phenomena peculiar to the solid state of certain minerals and metals has led to the development and verification of fundamental theories of the interaction of matter with electro-magnetic fields. Therefore, these early studies provided the chemists and physicists with data which in many cases was unique and could not be obtained from other scientific studies.

For the most part, more recent studies in the far infrared spectral region have not included studies on the pure rotation of small molecules, but have included molecules which have low frequency anharmonic vibrations such as internal torsions, small ring inversions, and bending modes of quasilinear molecules, as well as the studies of molecules with heavy atoms and solids. For the studies of anharmonic vibrations, it is usually necessary to obtain the spectral data of these compounds in the gaseous state. Therefore we shall first review the type of information obtained from such studies in the far infrared spectral region.

Symmetric Internal Rotors
It has long been established that while a double bond does not allow the rotation of a group of atoms at one end with respect to the group at the other, a single bond permits such rotation, at least to the extent that it is not generally possible for isomers to be isolated. Such rotation about the single bond was initially thought to be essentially free but, by the mid-1930's, it had been verified that barriers had to be surmounted in going from one conformation to another by rotation around a single bond. Although these potential barriers are only usually a few thousand calories, there are a number of thermodynamic properties which are markedly affected by them. Thus, the heat capacity, entropy, and equili-brium constants contain an appreciable contribution from hindered internal rotation.

Initially, and for several years, torsional barriers were determined from what can be described as indirect measurements, of which thermodynamics and microwave studies were the most common. The barriers were calculated from thermodynamic data by relating the difference in the observed and statistical entropies by tables involving the barrier height and the reciprocal of the partition function for free rotation. However, discrepancies often arose because many of the barriers were calculated from erroneously assigned or assumed normal vibrational frequencies. In the micro-wave spectrum, the observed perturbations on the pure rotational transitions are correlated to the torsional barrier height by either the splitting or intensity methods. The splitting method is by far the most exact and usually gives barriers to 1% but it is only applicable to rather low barriers and "light" rotors. The intensity method may give barrier heights to 15 to 30% accuracy but it is difficult to obtain quantitative intensities from the micro- wave spectra. Additionally, it is not always possible to identify the excited state transitions for the internal rotor from which to make the intensity measurements.

Far infrared spectroscopy is probably the most convenient method for studying barrier heights since one is actually obtaining the energy level separations for the rotor by vibrational spectroscopy. Experimentally one wishes to obtain the fundamental frequency for the torsional mode for the gaseous molecule so that the barrier in the isolated molecule can be ascertained. However, frequently the dipole moment change associated with the torsional mode may be quite small and consequently the resulting infrared band may often be quite weak. Therefore, assignments of the torsional modes from the infrared spectra of molecules in the gas phase were frequently in error in the initial investigations, and it is often necessary to use isotopic substitution to verify the torsional assignment for many molecules which have other low frequency vibrations. Also, it should be pointed out that torsional modes which give rise to B type band contours may have bands which are very broad and ill- defined because of unresolved excited state transitions. However, in favorable cases where the torsional mode gives rise to A or C type band contours with relatively strong Q branches, several excited state transitions may be resolved and not only the barrier height obtained but also the detailed shape of the potential well may be ascertained.

Raman spectroscopy can also be complementary to the far infrared studies on torsional modes since the overtones of these motions are symmetric and give rise to Q branches which are frequently several times more intense than the fundamentals. In the Raman spectrum, both mechanical and electrical anharmonicity allow the torsional overtones to be active and since they are totally symmetric, only isotropic polarizability terms can exist which lead to stronger Q branches than for anisotropic fundamentals. Finally, since $\Delta v = 2$ transitions are usually studied, the frequency spacing between peaks in a Raman torsional overtone series is approximately twice that in the infrared where $\Delta v = 1$ selection rules are obeyed, and therefore, resolution rarely presents a problem in the Raman spectrum. Therefore, when the torsional modes give rise to band contours in the infrared which contain no Q branches it may be possible to obtain reliable frequencies from the Raman data [8] (see Fig. 1.).

Finally, it should be pointed out that when a molecule has a too complicated far infrared spectrum for quantitative interpretation because of the difficulty in establishing the origin of the Q branches for the torsional mode, reliable barrier heights may still be obtained by studying the infrared and Raman spectra of the

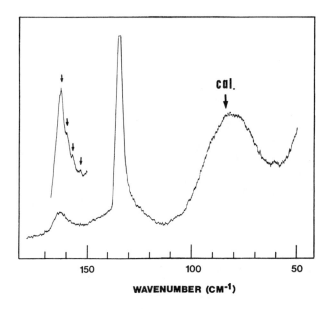

WAVENUMBER (CM⁻¹)

FIG. 1. Low frequency Raman spectrum of the gaseous phase of CF_3CH_2I. The main spectrum was recorded using 2 cm^{-1} and the insert with 0.65 cm^{-1} spectral bandwidth. 'Cal' designates the calculated 1 ← 0 torsional transition, other arrows mark the torsional overtones. Used by permission, Ref. [8].

sample in the crystalline state. At liquid nitrogen temperature, the upper vibrational states are effectively depopulated and only the 1 ← 0 transition is observed. Although it is possible to obtain the barrier height from this single piece of experimental datum, a detailed analysis of the shape of the potential well is not possible. Also it should be noted that the barriers obtained for the solid may be 15 to 25% higher than the barriers obtained from the gas phase data because of the frequency shifts due to the intermolecular forces in the crystal [9]. Additionally, it is sometimes possible to observe the torsional mode in the solid state when it is forbidden for the gaseous molecule [10].

Types of internal rotors fall into two categories -- symmetric tops and asymmetric tops. For symmetric tops, a rotation about the top-frame bond of $2\pi/n$ (where n is an integer) will bring the top to a position symmetrically equivalent to, or indistinguishable from, the original configuration. Therefore, it is usual to speak of the foldness of the rotor in terms of n. For example, a silyl group (SiH_3) is a threefold symmetric top (local C_{3v} symmetry) since a rotation about the silicon-to-frame bond of 120° will result in an orientation superimposable on the initial orientation. Twofold rotors include $-NO_2$, $-BF_2$, and phenyl groups (local C_{2v} symmetry). When a rotation of 360° (i.e., when n = 1) is the only

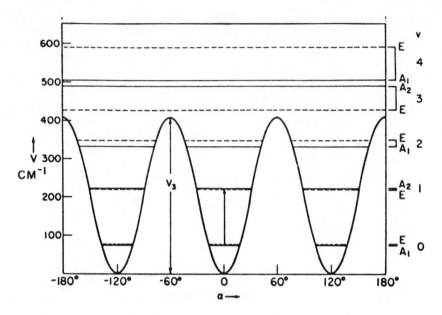

FIG. 2. Three rotor potential function. Used by permission,
Ref. [13].

operation that results in a symmetrically equivalent position for
the top, the rotor is known as an asymmetric top. Examples of
asymmetric tops include alcohol (-OH), thiol (-SH), amino (-NH$_2$),
and phosphino (-PH$_2$) groups when they are bonded to an asymmetric
frame, and these asymmetric tops will be covered later. For the
case of a symmetric frame, the top with the highest degree of
symmetry prevails, and when two tops of different foldness are
bonded directly to one another, the resultant foldness is the
product of the two individual tops' foldness. For instance, CH$_3$BH$_2$
would be classified as a sixfold internal rotor whereas ethane,
CH$_3$CH$_3$, would be threefold.

The energy minima and maxima for a symmetrical threefold group
(-CH$_3$) are 60° apart (Fig. 2). The simplest mathematical function
which will reproduce such a potential variance upon rotation is a
cosine function. By assuming the problem to be one-dimensional,
i.e., no coupling with any of the other normal modes, the quantum
mechanical energy solutions are readily obtainable, and they have
been discussed in detail by Kilpatrick and Pitzer [11], Herschbach
[12], and Fateley and Miller [13-15]. The eigenvalue problem will
be outlined and the interested reader is referred to these cited
references for additional details. The model employed is a rigid
symmetric top (CH$_3$ group) attached to a rigid frame which may be
completely asymmetric. There are four rotational degrees of
freedom, three for overall rotation and one for the hindered
rotation of the two groups. The axis of internal rotation coin-

cides with the unique axis of the symmetric top. Since the top has a threefold symmetry axis, the potential energy hindering rotation may be expressed by a Fourier expansion:

$$V(\phi) = (V_3/2)(1 - \cos 3\phi) + (V_6/2)(1 - \cos 6\phi) + \ldots$$

where V_3 is the height of the threefold barrier, and the sixfold term, V_6, merely changes the shape of the potential well. A positive V_6 makes the minima narrower and the maxima broader which results in the energy levels near the bottom of the well becoming somewhat more widely separated. A negative V_6 term has the opposite effect. Thus, the well is broader at the bottom and the levels are more closely spaced than when $V_6 = 0$. Experimentally, it has been found that $0 \leq V_6/V_3 < 0.05$ and V_9 and higher terms are not necessary. Thus to a good approximation

$$V(\phi) = (V_3/2)(1 - \cos 3\phi).$$

This potential function can be expressed in a more general manner so as to be applicable to an n-fold barrier, thus

$$V(\phi) = (V_n/2)(1 - \cos n\phi)$$

where V_n is the height of the n-fold barrier.

The Hamiltonian for this system may be expressed as

$$H = H_r + F(p - P)^2 + (V_n/2)(1 - \cos n\phi)$$

where H_r = the rigid rotor Hamiltonian; $F = h^2/8\pi^2 I_r$; I_a = moment of inertia of the internal top; $I_r = I_a[1 - \Sigma_g \lambda_g^2(I_a/I_g)]$ = the reduced moment of inertia for internal rotation; I_g = gth principal moment of inertia of the entire molecule; λ_g = cosine of the angle between the axis of the internal top and the gth principal axis of inertia of the entire molecule; p = the angular momentum of the internal top; and $(p - P)$ = the relative angular momentum of the top and the frame.

The eigenvalue problem associated with the torsional motion may be transformed into the well-established Mathieu equation, if the cross term $-2FP_p$ is neglected:

$$(d^2\phi_{v\sigma}/dx^2) + (b_{v\sigma} - s \cos^2 x)\phi_{v\sigma} = 0$$

where $b_{v\sigma} = (4/n^2)E_{v\sigma}/F$ = an eigenvalue of the Mathieu equation; $s = (4/n^2)V_n/F$; $2x = n\phi + \pi$; n = foldness of barrier; v = torsional quantum number; and σ = sublevel index. The energy levels for the hindered-rotational mode are

$$E_{v\sigma} = \frac{1}{4} n^2 F b_{v\sigma}.$$

These levels are n-fold degenerate and yield two sublevels; one is labeled +, and the other -, the signs corresponding to the symmetry of the wavefunction.

The observed vibrational frequency, $\bar{v}$, is the difference between two of the energy levels:

$$\bar{v} = E_{v'\sigma'} - E_{v\sigma} = \frac{1}{4} n^2 F (b_{v'\sigma'} - b_{v\sigma}) \ (\text{cm}^{-1}).$$

Clearly, if the $1 \leftarrow 0$ frequency is known and if there is enough structural information to determine F, $\Delta b_{v\sigma}$ can be calculated. From $\Delta b_{v\sigma}$ a dimensionless parameter s can be obtained from tables of solutions for the Mathieu equation. By employing the definition of s, the barrier height may be calculated:

$$V_n = \frac{1}{4} n^2 F s \ (\text{cm}^{-1}).$$

If the vibrational frequencies are known and have been assigned correctly, the principal source of error in the calculated barrier height will be in the value of F. This is quite obvious since V_n is a linear function of F and an error in the latter is reflected directly in V_n.

The barrier height may also be obtained by assuming that the torsional oscillation is <u>harmonic</u>. This assumption is valid only if the first excited state is well below the top of the barrier; consequently this method is applicable to problems where the barrier is ≥ 2.5 kcal/mol. The mathematical treatment of this case is quite straightforward once the cosine term is expressed in series form

$$1 - \cos x = \frac{1}{2!} x^2 - \frac{1}{4!} x^4 + \ldots$$

where $x = n\phi$. This x^4 and higher order terms are zero for small amplitudes, and the potential energy expression may now be expressed as

$$V(\phi) = \frac{1}{4} n^2 \phi^2 V_n.$$

The frequency of the one-dimensional oscillator is

$$v = (1/2\pi c)(k/I_r)^{\frac{1}{2}}$$

and upon rearrangement one observes

$$k = 4\pi^2 c^2 v^2 I_r.$$

The force constant, k, is by definition equal to the second derivative of the potential energy

$$k = d^2 V(\phi)/d\phi^2 = \frac{1}{2} n^2 V_n.$$

If these two expressions for k are equated, the resulting expression becomes

$$V_n = \frac{8\pi^2 I_r c^2 v^2}{n^2} \text{ (ergs).}$$

Division of this equation for V_n by hc will convert the units to wavenumbers

$$V_n = \frac{8\pi^2 c I_r \bar{v}^2}{n^2 h} = \frac{\bar{v}^2}{n^2 F} .$$

This expression is quite simple and an estimate of the barrier to internal rotation may be easily computed if the torsional fundamental is observed and the molecular geometry is reasonably well characterized so that F may be ascertained. One should be aware that two assumptions are inherent in these calculations: (1) the torsional mode is not mixed with other vibrations, and (2) the barrier is symmetrical. The experimental evidence accumulated thus far indicates that methyl torsions interact very little with the other fundamentals since they often shift by a factor of approximately 1.4 upon deuteration. Since the torsional fundamental is generally at a very low frequency, any perturbation would probably result in the absorption appearing at a higher wavenumber. This, of course, would result in an apparent higher barrier. If the barrier is not symmetric, in other words, if all the barriers of the particular torsion are not the same, the potential barrier can no longer be expressed by one parameter. The problem then becomes one of evaluating the contributions of the V_1, V_2, V_3, and possibly higher order terms. Obviously this requires at least three pieces of experimental data and this topic will be covered later.

We shall now demonstrate how far infrared spectroscopy is utilized to obtain barriers to internal rotation. The spectrum of ethyl-chloride is shown in Fig. 3 and the $1 \leftarrow 0$, $2 \leftarrow 1$, $3 \leftarrow 2$ transitions are clearly discernible. Additionally, the $4 \leftarrow 3$ transition shows the A-E splitting expected as the energy levels approach the top of the barrier. From these transitional frequencies, one calculates the V_3 to be 1291 cm^{-1} and the V_6 to be -12 cm^{-1}. Please note that the V_6 term is quite small and that it may either be positive or negative. The observed and calculated transitions are shown in Table 1. This example demonstrates the utility of far infrared data for the determination of barriers to internal rotation. Several compounds are listed in Table 2 for which the torsional barriers have been obtained and it should be noted that trends appear as additional halogens are added to one end of ethane. The addition of one chlorine atom raises the barrier 0.79 kcal/mol whereas the second raises the barrier 0.41 kcal/mol and the third one by 1.27 kcal/mol. Similar effects are observed in the bromine series, where the addition of one, two, and three bromine atoms to

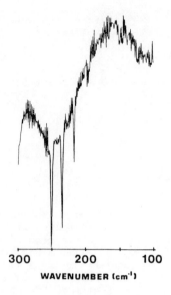

FIG. 3. Far infrared spectrum of gaseous ethylchloride recorded on a Digilab Model FTS-15B.

TABLE 1. Torsional Transitions for Ethylchloride

Observed Frequency (cm^{-1})	Assignment $v'' \leftarrow v'$	Calculated Frequency (cm^{-1})	Obs.-Calc.
251.0	$1 \leftarrow 0$	251.5	0.5
235.0	$2 \leftarrow 1$	235.5	0.5
217.3	$3 \leftarrow 2$	217.0	0.3
198.0	$4E \leftarrow 3$	197.0	1.0
196.0	$4A \leftarrow 3$	194.0	2.0

TABLE 2. Torsional Barriers (kcal/mol) for Some Haloethanes and
          Some Perfluoromethyl Haloethanes

Molecule	X = F	X = Cl	X = Br	X = CH_3
CH_3CH_3	2.93 [16][a]	2.93	2.93	2.93
CH_3CH_2X	3.33 [17]	3.72 [18]	3.71 [18]	3.28 [19]
CH_3CHX_2	3.18 [20]	4.13 [21]	4.33 [21]	3.94 [22]
CH_3CX_3	3.19 [23]	5.40 [21]	5.78 [21]	4.30 [10]
CF_3CH_2X	3.58 [8]	4.87 [8]	4.39 [8]	4.16 [8]

[a]Numbers in brackets indicate references.

ethane raises the barrier 0.78, 0.62 and 1.45 kcal/mol, respectively. Thus the addition of the third chlorine or bromine raises
the barrier considerably more than either of the first two halogens, but each halogen does raise the barrier. This observation
disproved the belief which apparently arose from the trends in the
fluorine series that, after the addition of one halogen, the
barrier would not be drastically affected by the substitution of
the second or third halogen. Therefore, with the amount of
reliable torsional data available, trends appear to be emerging
which are consistent with theoretical predictions for methyl
rotors.

For perfluoromethyl rotors the data are not nearly as extensive.
In general the far infrared data for these tops usually have
resulted in very broad bands without discernible Q branches. This
probably results from the fact that these molecules all have low
frequency CF_3 rocking modes which have appreciable excited states
populated at room temperature and the torsional modes for molecules
in these excited states have slightly different frequencies. Also
much of the data were taken in the late sixties and early seventies
when the instrumentation was much poorer so that it was not
possible to resolve the many closely spaced Q branches for the CF_3
torsions. Thus, at present there are not sufficient reliable data
for these tops to determine if trends are predictable (see Table
2). These tops are too heavy for barrier determinations from the
microwave splitting method and the barriers determined from
microwave intensity measurements have large uncertainties. As an
illustration of this point the torsional frequency in 3,3,3-tri-
fluoropropene has been predicted at 88 ± 25 cm^{-1} from microwave
intensity measurements [24] which gives a barrier of 4.69 ± 2.29
kcal/mol but from far infrared studies [25] the torsion was
reported at 60 cm^{-1} from which a barrier of 1.53 kcal/mol has been
calculated. We have also recorded the far infrared spectrum of
this molecule but no absorption was observed at 60 cm^{-1}. This
example is not unique but is representative of the state of the

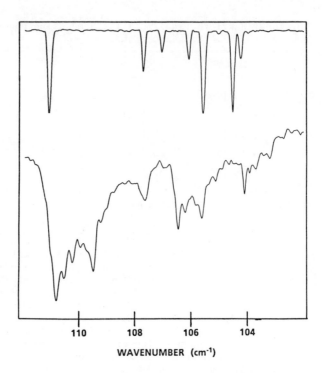

FIG. 4. Torsional transitions of benzaldehyde. Used by permission, Ref. [26].

determination of CF_3 barriers. Additionally, it should be noted that for some molecules with CF_3 tops the torsions are relatively strong in the far infrared spectrum whereas for other molecules it is not possible to observe them even with pathlengths up to 20 meters. Thus, there are very large variations in the relative infrared intensities of CF_3 torsional modes.

So far our illustrations have been for threefold rotors but there are some data available for twofold rotors. In Fig. 4 the torsional transitions for benzaldehyde are shown and it is clear that the observed transitions follow a nice pattern but there is some perturbation of the 2 ← 1 transition and additionally there are transitions which arise from molecules in the excited states of one or more of the low frequency bending modes. Thus the assignment of the torsional transitions may not be straightforward and, in fact, it frequently requires very careful analysis of a trained investigator to correctly assign the excited states.

Therefore most of the well determined symmetrical barriers have been for methyl rotors attached to the second row elements, i.e., carbon, nitrogen, oxygen. Very few barriers have been determined for the silyl rotors and even a smaller number for germyl tops. Also there are only limited data for methyl barriers for these tops attached to the third row elements such as phosphorus and sulfur. In general the data are not sufficiently extensive to be able to observe any trends or even predict barriers for compounds when they have not been determined. Thus, far infrared studies of molecules with symmetric tops have the promise of providing a considerable amount of torsional data which could help the chemists discern which factors are apt to effect the barriers the most and, additionally, how such barrier information may be transferred to larger molecules which may be biologically important or of industrial importance. However, it should be noted that one frequently has to deuterate the top in order to obtain confident barrier values from far infrared data!

For the case of two C_{3v} rotors the potential function becomes quite complex and is of the general form given below where it has been expanded into a Fourier series in two variables.

$$2V(\tau_0, \tau_1) = V_{30}(1 - \cos 3\tau_0) + V'_{30} \sin 3\tau_0 + V_{60}(1 - \cos 6\tau_0) +$$

$$V'_{60} \sin 6\tau_0 + V_{03}(1 - \cos 3\tau_1) + V'_{03} \sin 3\tau_1 +$$

$$V_{06}(1 - \cos 6\tau_1) + V'_{06} \sin 6\tau_1 + V_{33}(\cos 3\tau_0$$

$$\cos 3\tau_1 - 1) + V'_{33} \sin 3\tau_0 \sin 3\tau_1 + V''_{33} \sin 3\tau_0$$

$$\cos 3\tau_1 + V'''_{33} \cos 3\tau_0 \sin 3\tau_1.$$

This potential function is the most general, i.e., for a molecule with nonequivalent tops and of C_1 symmetry. A molecule with higher symmetry will impose restrictions on this potential function (see Refs. [27] and [28]).

It should now be useful to consider some representative examples. In many cases the spectra are very rich and complex and additional data were needed from the Raman spectra of the gases before confident assignments could be made. The relative intensities of the Q branches can also aid in their assignment. In this regard we have found that it is very important to record the spectra with a resolution of at least 0.25 cm^{-1} since the Q branches are very sharp and the relative intensities may change with a resolution of 0.5 cm^{-1}. Also, because of the limited number of points in the interferogram, we have found that a Q branch may be "cut off" if the point does not come at the maximum of the absorption peak. This may cause the Q branch to appear "weaker" than it really is and therefore the relative intensities will be incorrect. One further point should be made. Water vapor is a serious problem in the far infrared spectral region and at times we have attempted to subtract it from the sample spectrum. We have found that it is almost impossible to distinguish between the "spectral artifacts" from the improper canceling of the water bands and the torsional

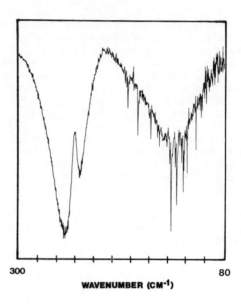

WAVENUMBER (CM⁻¹)

FIG. 5. Far infrared spectrum of gaseous dimethylphosphine-d_3 recorded at 0.25 cm⁻¹ resolution. Used by permission, Ref. [29].

transitions. Therefore, we should like to stress that it is very important to have both a dry sample and a dry cell simultaneously, which is no small task, if good torsional data are to be obtained. In fact, an interferometer with a vacuum bench is also desirable. Now we shall give an illustrative example.

From the far infrared spectrum [29] of dimethylphosphine-d_3, Fig. 5, it is clearly evident that two distinct series of torsional transitions exist. The higher frequency series begins at 183.2 cm⁻¹ with the lower frequency series beginning at 137.2 cm⁻¹. An initial assignment was made using the first three transitions in each series. The transitions (01) ← (00), (02) ← (01), and (03) ← (02) were assigned to bands at 183.2, 172.6 and 158.6 cm⁻¹, respectively, and the transitions (10) ← (00), (20) ← (10), and (30) ← (20) were assigned to bands at 137.2, 131.2 and 124.2 cm⁻¹, respectively. This assignment was used as the basis for the more complete assignment. It should be noted that splitting of several of the torsional transitions was observed. For example, the (50) ← (40) transition was predicted to be split by 6.4 cm⁻¹ whereas a splitting of 8.3 cm⁻¹ was observed. Similarly, the (40) ← (30) transition was split by 3.4 cm⁻¹ between the Γ^{10} and Γ^{11} levels whereas the calculated splitting was 2.4 cm⁻¹. A better fit of this splitting could not be obtained without drastically affecting the fit for the transitions lower in the well. The average barrier of 832 cm⁻¹ obtained agrees extremely well with the value of 811 cm⁻¹ obtained by the microwave splitting method [29]. However, it

TABLE 3. Torsional Potential Constants (cm^{-1}) for Some Molecules
with Two Internal C_{3v} Rotors

	$V_{30}=V_{03}$	$V_{60}=V_{06}$	V_{33}	V'_{33}	$V_{eff}{}^a$	Ref.
$(CH_3)_2NH$	1281.7	---	227.9	30.2	1053.8	[31]
$(CH_3)_2PH$	905.6	-29.4	95.5	-29.3	810.1	[29]
$(CH_3)_2O$	1215.28	-20.75	272.68	21.64	942.60	[32]
$(CH_3)_2S$	797.3	-11.8	51.4	12.4	745.9	[33]
$CH_3CH_2CH_3$	1323.4	---	176.6	-132.0	1146.8	[34]
$(CH_3)_2C=CH_2$	893.7	-4.0	165.7	-135.1	728.0	[35]

$^a V_{eff} = (V_{30} + V_{03})/2 - V_{33}$.

should be pointed out that the barrier of 711 cm^{-1} which was obtained in the initial far infrared study [30] of dimethylphosphine
was in error because of the misassignment of the "second" set of
transitions. Spectral "artifacts" resulting from the subtraction
of water led to the misassignment. In Table 3 are listed the
potential constants for several molecules with two internal C_{3v}
rotors.

For many of the molecules studied so far (i.e., dimethylphosphine
[29], propane [34], and isobutene [35]), rather large, negative
sine-sine coupling terms, V'_{33}, have been obtained (see Table 3).
Differences in frequency between that of the geared and antigeared
torsions (an A_2 and a B_2 in the case of C_{2v} molecular frame) are
determined by the g^{45} and V'_{33}. The effects of the kinetic term far
outweigh those of the sine-sine coupling term and the latter can
actually be thought of as a small perturbation of g^{45}. Potential
functions which have relatively large negative V'_{33} terms have been
obtained for molecules which have rather small negative g^{45} terms.
On the other hand, when large negative g^{45} terms were obtained from
the structural parameters as in dimethylether [32] and dimethylamine [31], the sine-sine coupling term, V'_{33}, was relatively small
in magnitude and in fact for these two molecules it was positive.

The terms V'_{60}, V'_{06}, V''_{33} and V'''_{33} in the potential function have been
found to have little effect on the calculated values of the tor-
sional transitions and are, therefore, ill-determined. These terms
have been excluded in all of the calculations for the molecules
cited above. Coefficients of this type have been considered in the
potential function of a semi-rigid $C_{3v}(T) - C_1(F) - C_{3v}(\bar{T})$ model by
Grant et al. [36] but they had not been considered for semi-rigid

molecules, $C_{3v}(T) - C_s(F) - C_{3v}(T)$, such as dimethylamine and dimethylphosphine, until Groner and Durig [32] introduced such terms with these coefficients into the potential function of certain types of semi-rigid two-top models. Using the fact that the potential function is invariant with respect to operations of the internal isometric symmetry group, they proved [32] that these coefficients may be non-zero by symmetry arguments. Without introducing these terms, the potential function would have the same symmetry for $(CH_3)_2NH$ as for $(CH_3)_2O$ and this seems unreasonable from a chemist's viewpoint.

It appears that vibrational spectroscopic data provide excellent promise for the determination of the potential coupling terms for molecules with two C_{3v} rotors. However, it should be emphasized that usually the Raman data for the two-quantum jumps are needed for a confident assignment of the far infrared transitions. There has been some difficulty of an energy level being perturbed by interaction with the bending mode in several of the molecules studied so far. However, there are usually eight to twelve transitions observed for these two-top molecules, so the perturbed level can be readily identified.

The cosine-cosine coupling terms, V_{33}, are usually taken to be zero in microwave studies but it has been found that these terms have values on the order of 15% of the V_3 term. This coupling term is obtained only from excited torsional states in the microwave studies, and frequently there are not sufficient splitting data to determine simultaneously both the sine-sine and cosine-cosine coupling parameters. The sine-sine coupling term is very much dependent on the g^{45} term, so a good structural determination is necessary for the evaluation of this coupling term. It is suspected that part of the differences between the microwave barrier values and those obtained by vibrational spectroscopy can be attributed to poor structural parameter values. Further work is needed on more two-top molecules before trends in the coupling parameters can be ascertained, if indeed they do exist. A combination of infrared and Raman spectroscopic studies appears to provide the best method for continuing such investigations.

Asymmetric Internal Rotors
The final type of torsional motion which we shall discuss is the asymmetric rotor. Such rotors usually lead to two or more stable conformers. One of the major goals of conformational analysis is the calculation of the energy difference between the two conformers, ΔE, as well as the energy necessary for interconversion. Once these data are available, calculation of the thermodynamic functions, G, H, S and C_p, is possible. The calculation of these energy quantities is facilitated by using a potential function which describes the vibrational motion, or internal rotation (torsion), as a function of the dihedral angle, α. The potential function is called asymmetric because both the frame and the top portions of the molecule have no symmetry element higher than a plane. This can be illustrated by methyl vinyl ketone where the

frame portion of the molecule is the methyl ketone moiety, CH_3CO, and the lighter top portion is the vinyl group.

Historically, conformational analyses of molecules have been carried out by a large number of techniques--both experimental and theoretical. While theoretical methods currently being used, such as CNDO or molecular mechanics, have great utility in accounting for observed molecular structures or conformations, they are not at present completely accurate in their a priori predictions. Therefore, experimental studies must be conducted, both to establish the stable structures and to verify or contradict the theoretically predicted results.

Practically all of the conformational barriers reported in the literature have been obtained from NMR data on the liquids or in solutions, or from microwave data where the asymmetric torsional frequencies are estimated from the relative intensities of the microwave lines in the excited vibrational states and the relative intensities of the conformer lines. However, the measurement of the relative intensities of microwave lines is notoriously difficult and, in the few cases which we have checked, the observed frequencies have been found to differ by as much as 20 to 25 cm^{-1} so that the potential functions determined from these microwave data alone are rather poor. Also, the results from solutions or the liquid state differ so significantly from the gas phase value that they can not be used to check quantum mechanical calculations or be used as estimates for larger biological or polymeric molecules. Therefore, vibrational spectroscopy may well be the most generally applicable method to be used in the study of the conformations of certain types of small molecules with few substituents. Infrared and Raman spectra can be recorded in all phases, and variable temperature experiments can also be conducted in all phases. Additionally, gas phase band contours in both the infrared and Raman spectra, along with Raman depolarization data, provide considerable structural information and lend confidence to conclusions arrived at from vibrational studies. Of course, the limitation of the vibrational spectroscopy technique when used for gas phase studies is that it is best applied to relatively simple molecules which often contain an element of symmetry (or more), and one or perhaps no more than two portions of the molecule capable of producing different conformations upon internal rotation. Thus, a significant portion of our research efforts over the past few years have involved the determination of the asymmetric torsional frequencies and associated potential functions, by far infrared and gas phase Raman spectroscopy, for several important classes of compounds. Before discussing a few recent results, we will present the theory used in determining potential functions from vibrational data.

In order to characterize an asymmetric potential function, four types of information are required: (1) the approximate dihedral (torsional) angle of each conformer, because the number of torsional energy levels is directly related to the number of potential minima; (2) the approximate relative enthalpy difference between the high and low energy conformers, since this is one of the constraints defining the potential function; (3) the change in molecular kinetic energy as a function of torsional angle; and (4) the accurate observation of torsional transition frequencies with their correct assignment of the energy levels involved with the correct conformer. This latter requirement, in many cases, is not a simple one.

Several important consequences of molecular symmetry properties should be mentioned since they have a bearing on the correct assignment of the observed data necessary to calculate the potential function. The most common type of molecule being studied has at least one planar conformation, which has at least C_s symmetry. For the conformer with C_s symmetry the normal vibrations are either in-plane or out-of-plane. The out-of-plane vibrations, of which the asymmetric torsion is one, have the dipole change along the principal axis out-of-plane. When this corresponds to a B-type contour, problems arise, since usually only the torsional fundamental is observed. However, no such problems arise for A- or C-type band contours. In the case of non-planar (gauche) conformers, C_1 symmetry is usually present, giving rise to A/B/C hybrid band contours for the asymmetric torsion. In summary, bands which give rise to A- or C-type contours are most desirable as they exhibit strong Q-branches, whereas B-type bands may have no Q-branches and thus only allow the location of the fundamental torsional frequency at the band center. In fact, it is sometimes difficult to find the band center because of the overlapping contours of the hot bands. In this case, it is usually difficult to obtain a potential function with confidence that it is the correct one.

A molecule undergoing hindered internal rotation about a single bond changes its overall energy as the torsional angle, α, changes. This energy change is manifested in the torsional vibrational energy as well as the overall molecular energy. The change in the molecular rotational energy arises from the alteration in geometry of the molecule as α changes, which in turn alters the principal moments of inertia which are dependent upon α. In the analysis of the torsional motion, calculation of the asymmetric potential function is a necessary requirement for the understanding of the energy restrictions and molecular dynamics involved in hindered internal rotation. The theoretical treatment used most often is a rigorous numerical method involving the calculation of torsional eigenfunctions and corresponding eigenvalues which are then iterated, in a least-squares manner, to give the best possible fit to the experimentally observed torsional transition frequencies [37].

If the molecule is totally asymmetric, with no plane or rotational axis present in any conformer, then the potential function must be expressed as a Fourier series in both sine and cosine terms. A potential of this type results in all the minima and all the barriers being nonequivalent. Simpler potential functions result when both the asymmetric top and the molecular frame have planes of symmetry. In these types of systems at least one value of α results in the entire molecule having a plane of symmetry. A much more insignificant consequence of symmetry in these systems is that a rotation of $+\alpha$ or $-\alpha$ away from the planar configuration results in energetically indistinguishable conformations since the potential function itself now has a plane of symmetry. A Fourier series expansion in one term is now adequate to describe the potential function. The most common type of potential function, where at least one of the minima coincides with a plane of symmetry, is of the type

$$V(\alpha) = \tfrac{1}{2}\Sigma_i V_i (1 - \cos\ i\alpha).$$

In general it has been found that a six term expansion in V_i is sufficient to handle almost any molecular system [37]. This is because V_1, V_2, and V_3 are in general significantly larger than V_4, V_5, and V_6, and thus, terms greater than V_6 should be negligible.

The kinetic energy term, $T(\alpha)$, is rather complicated. Several different approaches have been made to solve this expression using either ab initio or semi-empirical quantum mechanical methods. In general, if the only data being used to calculate the potential function are torsional transitions, and if one continues within the boundary conditions of a one-dimensional problem, then the following treatment is adequate. Calculation of the kinetic energy term requires evaluation of the internal rotation constant, F, commonly called the "F number," which is related to the reduced moment of inertia of the top, I_r, by $F = h/8\pi^2 c I_r$ but unlike the similar quantity for the symmetric rotor this F will vary with conformation. In order to determine the angular dependence of $F(\alpha)$, F is calculated at various values of α, the results of which are then curve fitted to a Fourier expansion in α via a least-squares method:

$$F(\alpha) = F_0 + \Sigma_i F_i \cos\ i\alpha.$$

The kinetic term, $T(\alpha)$, can now be written as

$$T(\alpha) = P_\alpha F(\alpha) P_\alpha$$

where P_α is the conjugate momentum operator given by

$$P_\alpha = -(1/i)(\partial/\partial\alpha).$$

The two Fourier expansions, one in $F(\alpha)$ and one in $V(\alpha)$, are now substituted into the Schrödinger equation for the general torsional Hamiltonian,

$$H_\tau \Psi = E_\tau \Psi$$

where

$$H_\tau = P_\alpha F(\alpha) P_\alpha + V(\alpha).$$

The solution to the Schrödinger equation is obtained by solving the secular equation for the Hamiltonian matrix, to give both the torsional eigenvalues and eigenfunctions. In order to symmetrize the Hamiltonian matrix, it is rewritten using the definition of P_α:

$$H_\tau = \tfrac{1}{2}[P_\alpha^2 F(\alpha) + F(\alpha) P_\alpha^2] + f(\alpha) + V(\alpha),$$

where the new variable, $f(\alpha)$, is assumed to be negligible since the angular dependence of F is very small.

Once these substitutions have been completed, every element in the Hamiltonian matrix is then set up usually using a free-rotor basis set,

$$\Psi_m = (2\pi)^{-\frac{1}{2}} e^{im\phi}, \text{ where } m = 0, \pm 1, \pm 2...,$$

with about seventy-five functions. In the case of the cosine-based potential function, each eigenfunction is either symmetric or asymmetric with respect to rotation about $\alpha = 0$. These eigenfunctions are more commonly denoted as even (+) or odd (-) for symmetric and asymmetric reflection about $\alpha = 0$, respectively. To preserve these symmetry properties, a Wang transformation is applied to the free-rotor basis functions to express them as cosine or sine functions of even or odd symmetry, respectively:

$$\Psi_n^{even} = \pi^{-\frac{1}{2}} \cos n\phi \qquad n = 1, 2...$$

$$\Psi_0^{even} = (2\pi)^{-\frac{1}{2}}$$

$$\Psi_n^{odd} = \pi^{-\frac{1}{2}} \sin n\phi \qquad n = 1, 2...$$

The Hamiltonian matrix elements derived using the symmetry basis set are divided into two blocks, even and odd. These matrix elements can be used to calculate the infrared intensity of a band due to a transition from one eigenvalue, v'', to a higher level, v'. The relative intensity of this band is

$$I_{v'',v'} = (E' - E'') < v'|M|v''>^2 [g''e^{-E''/kT} - g'e^{-E'/kT}],$$

where E' and E'' are the energies of the v' and v'' levels, respectively, $<v'|M|v''>$ is the dipole transition matrix element, and g' and g'' are the degeneracies of the v' and v'' levels, respectively. It should be pointed out that those intensities are relative, and hold only for transitions of the same conformer. The selection rules for infrared and Raman activity are

Infrared $+ \leftrightarrow -$

Raman $+ \leftrightarrow +, - \leftrightarrow -$

where the notations are + and - for the cosine and sine symmetry
blocks, respectively. Hence, when performing calculations using
infrared active transitions, the sine operator must be used to
calculate the relative intensities.

The symmetry properties of each wavefunction are very important,
and a few points about the manner in which the eigenvalues are
arranged in the symmetry blocks should be clarified. In the more
common situation of a cosine based potential function, the
potential wells are either singly present, denoted cis or trans and
occurring at α = 0° or 180°, or appear as pairs which denote the
gauche conformer. The gauche wells may occur at any equal angles
less than or greater than 180°. Although the two gauche wells are
equivalent the energy levels within them are not degenerate except
for the trivial case of an infinite barrier. The gauche energy
levels are arranged in pairs with eigenfunctions of every level
present in both wells. The spacing between the levels in a pair is
small at the bottom of the well, increases progressively with each
pair, and becomes larger near the top of the barrier. The symmetry
of each successive eigenfunction alternates with the lowest being
symmetric. In the cis or trans wells, each energy level is singly
degenerate and the lowest eigenfunction is symmetric with the
symmetry of each subsequent level alternating minus and plus, etc.
Above the barrier to internal rotation the levels are distributed
in the manner of the free rotor system.

At this point several additional comments seem in order. First,
the uncertainty in the values of the potential constants diminishes
significantly as the observed number of transitions in each well
increases. If only one transition is observed for a given well
then the potential constants are usually poorly determined unless
other data are available. Second, since the V_1 term is multiplied
by only one whereas the V_2 is multiplied by four and the V_3 term is
multiplied by nine, the V_1 term is usually more poorly determined
than the V_2 and V_3 terms. Also one frequently finds that the V_1
and V_2 terms are highly correlated. Third, the value of ΔH can
significantly effect the values of the potential terms since it is
usually possible to fit the observed torsional transitions with
different potential terms and with ΔH values that differ by a
factor of two. Finally, it should be mentioned that the less
stable conformer is usually closer to the top of the "barrier" so
that the asymmetric torsional modes of this conformer fall at lower
frequencies than those for the more stable conformer. However, we
would like to reiterate that it is essential that the asymmetric
torsional transitions be appropriately assigned to the correct
conformer.

It is evident from the previous discussion that one of the most
important pieces of data necessary to calculate an accurate
torsional potential function is the enthalpy difference between the
conformers in the gaseous phase, ΔH. The various experimental
methods of calculating ΔH are based on the thermodynamic relationship between the temperature and the Gibbs free energy change
of the conformational equilibrium

$\Delta G = -RT \ln K$ where

$K = [\text{high energy conformer}]/[\text{low energy conformer}].$

Substituting for ΔG in the above equation and rearrangement of terms results in a form of the Van't Hoff isochore,

$\ln K = -(\Delta H/R)(1/T) + \Delta S/R.$

The usefulness of this equation is that if some experimental observation can be related to $\ln K$, then a variable temperature study of $\ln K$ leads to a graphical relationship to $1/T$. The slope of the graph is then $-\Delta H/R$ and the intercept is $\Delta S/R$.

Several different techniques have been used to determine ΔH by this relationship. This equation has been used directly by electron diffraction workers [38]. In order to fit the diffraction pattern of a gas which exists in a conformational equilibrium, the theoretical patterns for both conformers are mixed together in varying concentrations to achieve the best fit. If the pattern is recorded at a number of different temperatures, the change in $\ln K$ with $1/T$ can be measured. However, the error limits on K are normally high due to the manner in which K is obtained, but nevertheless this method usually gives an acceptable value of ΔH but a poor value of ΔS.

This equation is also used by microwave spectroscopists to obtain the ΔH between conformers. In this case a more complicated situation arises, because the calculation of K depends on a number of factors which govern the absolute intensity of the rotational lines. The calculation is simplified if the following conditions are followed: a) the same rotational transition is examined for both conformers, thereby avoiding the need for degeneracy and lineshape corrections and the calculation of different wavefunctions, and b) the lines to be examined should be similar in frequency to avoid frequency corrections to the line intensity. Of prime importance in the calculations are the dipole components for the different conformers; it is necessary to know the component for each conformer along the same principal axis as the angular momentum changes during the transition. If the above restrictions have been observed then a nonrigorous expression for K may be written,

$$K = \frac{I_i(x)}{I_i(y)} \cdot \frac{\mu_i^2(y)}{\mu_i^2(x)}$$

where $I_i(x)$ is the observed intensity of the specified i dipolar line of conformer x, and $\mu_i(x)$ is the dipole component of conformer x along the i axis. K is measured at several different temperatures, using as many different pairs of lines as possible, in order to overcome the problems of interference from other lines as well as Stark lobes, to give the best accuracy. As in the case of the electron diffraction determinations of ΔH, the accuracy achieved is quite acceptable and can provide considerable restrictions on the determined potential function from the observed asymmetric torsional transitions.

The calculation of ΔH and ΔS from vibrational spectroscopy is also quite complicated because of the manner in which $\ln K$ is calculated. For infrared absorption the Beer-Lambert Law holds for both conformers in the mixture so that

$$A_i = e_i c_i l$$

$$K = c_B/c_A = (A_B e_A)/(A_A e_B)$$

where B is the high energy conformer. Utilizing the earlier equation for $\ln K$ one obtains

$$\ln (A_B/A_A) = -\Delta H/R(1/T) + (\Delta S/R) - \ln (e_A/e_B).$$

This expression results in the direct calculation of ΔH from variable temperature measurements of absorbance. However, the intercept now contains two terms so ΔS cannot be directly evaluated.

Hartmann et al. [39] proposed a simple method for calculation of $\ln (e_A/e_B)$, which allows for a value of ΔS to be obtained. An expression similar to the one for infrared absorption may be written for variable temperature Raman results [40] as well. However, due to the more complex nature of the Raman effect, the last expression of the equation, equivalent to the term $\ln (e_A/e_B)$, cannot be easily evaluated and must be assumed negligible in order to obtain ΔS. The linearity of this equation rests on the assumption that three terms, i.e., ΔH, ΔS, and e_A/e_B, will not vary with temperature, but ΔH and ΔS will be temperature independent only if ΔC_p is zero. Since ΔC_p arises from the difference in the vibrational frequencies of the two conformers, the effects of this on both ΔH and ΔS can be significant, particularly if the temperature range is large.

It is the opinion of the authors that in many cases incorrect values of ΔH have led to significant errors in the determination of the internal rotation potential constants, particularly if the ΔH is used as one of the constraints during the calculation of the potential function. It is usually easier to obtain the ΔH of the liquid by infrared and Raman spectroscopy because of the greater ease in measuring the band areas or peak heights in the condensed phase, but values in the liquid may not even be closely related to ΔH values in the gas. This is particularly true for polar molecules. In fact, the sign of ΔH may change in going from the gas to liquid, i.e., the conformer stability may be reversed with change of phase. Therefore, it is imperative that the ΔH in the gas phase be used for potential function determinations. This is one area where significant improvement can be expected in the future. With the utilization of computers for signal averaging in

Fourier transform infrared spectroscopy or Raman spectroscopy, it should be much easier to obtain reproducible results for band areas. Nevertheless, one frequently finds that the measurement of ΔH utilizing two or more bands for the same molecule will lead to different results for each set, which points to the problem of a possible other band underlying the band or bands being measured. Thus, one should use as many of the doublets observed in the spectra as possible when determining the ΔH from the vibrational spectrum of a molecule. However, there is usually only one or two pairs of bands which are suitable for the determination of ΔH for most molecules.

Another possible source of error in the determination of the potential constants arises from a lack of knowledge of the appropriate structural parameters for the respective conformers. Frequently one finds that there is structural "relaxation" when the dihedral angle changes so that the same structural parameters utilized for one conformer may, in fact, not be appropriate for the second conformer. In most calculations, the F number is simply determined by assuming that the structural parameters of the two conformers are the same, except for the change in the dihedral angle. This certainly is not a good approximation, but it frequently leads to a simpler computational problem. Additionally, there are a large number of small molecules where the structural parameters have not been accurately determined. If only the microwave spectra are available, it is usually found that a limited number of moments of inertia have been obtained and certain assumptions were made concerning some of the structural parameters, with the remaining structural parameters obtained from the moments of inertia. Therefore, such r_0 structural parameters are no better than the parameters which have been assumed to obtain the remaining ones. If, however, a sufficient number of different isotopes have been studied then the r_0 or r_s structural parameters are usually of sufficient accuracy that these will lead to no errors in the determination of the potential constants. Structural parameters obtained from electron diffraction work are somewhat more suspect, particularly when there is more than one conformer present in the gas phase. Nevertheless, by a combination of electron diffraction data with microwave rotational constants, one can frequently obtain excellent structural parameters which, of course, are needed for the appropriate calculation of the F number. Finally, it should be mentioned that there have been a sufficient number of bond distances and bond angles determined so that it is sometimes possible to transfer such angles and distances to the conformers under study. It is probable that the errors in the structural parameters are the least likely ones to lead to errors in the determination of the potential constants.

In principle, one can obtain the dihedral angle from the determined potential function, but a considerable uncertainty usually exists. The confidence in the determined potential constants is significantly raised if the symmetry of the two conformers can be determined by other than the potential function, i.e., gauche versus trans or cis. It should also be mentioned that the ΔH, in principle, can be determined by the potential function, but in all cases it should be checked by an experimental technique. It is probable that ab initio calculations are at the point where the predicted ΔH's may provide useful constraints in the initial value of ΔH used in the potential function calculation. Thus, a combination of theoretical and experimental determinations will probably be the most accurate way of determining the potential functions. It should be stressed that experimental verification of the theoretical values is still quite desirable.

Our research efforts in the area of conformational analyses have involved the determination of the asymmetric potential function for a number of three-membered ring carbonyl compounds, substituted allylic compounds, propenoyl halides and alkyl halides. Although asymmetric potential functions had been previously proposed for some of these molecules, relatively few of these can be given a significant level of confidence due to a number of experimental problems which have been previously encountered. For example, far infrared spectroscopy has been used in the past for the determination of asymmetric torsional frequencies but we have found that a resolution of at least 0.1 cm^{-1} needs to be used to record the low frequency data because many torsional bands are extremely sharp and therefore were not even observed with the relatively low resolution utilized earlier. Additionally, insufficient resolution may result in misleading relative intensities of the Q branches so that incorrect assignments could have resulted. In many cases, due to symmetry, both infrared and Raman spectroscopic results are needed to identify the high and low energy conformers. In order to demonstrate how asymmetric torsional data can be used to obtain potential constants we shall provide two examples.

All of the available asymmetric torsional data for propionyl fluoride were obtained from the far infrared spectrum which is shown in Fig. 6 between 86 and 45 cm^{-1} at 0.12 cm^{-1} resolution [41]. There are clearly two series of Q branches with the higher frequency series beginning at 82.17 cm^{-1} with five additional pronounced hot bands falling to lower frequency, and a second series beginning at either 56.42 or 54.13 cm^{-1} with at least two additional hot bands at lower frequencies. These two series of bands are too low in frequency to be due to the methyl torsional modes. Thus, the higher frequency series beginning at 82.17 cm^{-1} is assigned to the asymmetric torsion of the s-cis conformer which is consistent with the predicted frequency of 85 ± 3 cm^{-1} for this mode from the microwave study [42].

In addition to the main series of six bands for the asymmetric torsion of the s-cis conformer which, of course, is the fundamental (1 ← 0 at 82.17 cm^{-1}) and five successively higher transitions falling at lower frequencies, i.e., 2 ← 1 at 78.79, 3 ← 2 at 75.34 cm^{-1}, etc., there are at least two additional series of weaker Q branches which fall on the high frequency side of these main bands. For example, there are clearly three Q branches on the high frequency side of the fundamental and we believe these bands are due to the s-cis asymmetric torsion in successive excited states of

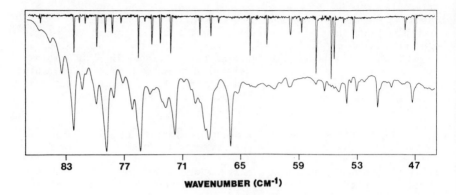

FIG. 6. Far infrared spectrum of gaseous propionyl fluoride at a resolution of 0.10 cm⁻¹. The top trace is the spectrum of water at the same resolution recorded with 40 scans. Used by permission, Ref. [41].

TABLE 4.   Observed Frequencies (cm^{-1}) and Proposed Assignments[a] for the Asymmetric Torsional Transitions of Gaseous Propionyl Fluoride

Conformer	Obs.	Transition	Calc.	Δ
s-cis	82.17	1 ← 0	82.60	-0.43
	78.79	2 ← 1	78.34	-0.55
	75.34	3 ← 2	75.90	-0.56
	71.78	4 ← 3	72.31	-0.53
	68.29	5 ← 4	68.45	-0.16
	66.04	6 ← 5	64.30	1.74
	61.50	7 ← 6?	61.96	-0.46
gauche	54.13	$1\bar{+} \leftarrow 0\pm$	54.64	-0.51
	50.91	$2\bar{+} \leftarrow 1\pm$	51.14	-0.23
	47.26	3+ ← 2±	46.89 46.82	0.40
	42.80	$4^{+} \leftarrow 3^{-}$	41.85	0.95
	38.53	$4^{-} \leftarrow 3^{+}$	40.69	--

[a]Data taken from Ref. [41].

the CFO rocking mode (ν_{15}) which is at 251.3 cm-^{1}. There also appears to be an additional series of Q branches on the low frequency side of the main bands and these Q branches probably result from the asymmetric torsion in the first excited state of the methyl torsion (ν_{23}) which is at 203.9 cm-^{1}. It is not possible to follow this latter series of excited states beyond the 3 ← 2 transition because it appears to overlap with bands from the series which falls on the high frequency side of the main transitions.

The asymmetric torsional series for the gauche conformer was assigned as either beginning at 56.42 or 54.13 cm-^{1} which again is consistent with the predicted value of 54 ± 5 cm-^{1} from the microwave study [42]. By choosing the lower frequency band, a reasonable progression was obtained for the successive excited states where the 2± ← 1± and 3± ← 2± transitions were assigned at 50.91 and 47.26 cm-^{1}. The reasonable regularity of both the high and low frequency series indicates that the higher energy levels of the asymmetric torsions are not perturbed significantly by their coincidence with the low frequency bending modes. The potential function for the asymmetric torsion was calculated with the torsional transitions listed in Table 4 and the potential function is shown in Fig. 7.

The main difference between the potential function obtained from the far infrared data and that previously reported from the microwave data [42] is the value of the V_4 term which is about one-half the value previously obtained. It should be noted that the ΔH values and the gauche to gauche barrier values are the same within experimental error. The value of the s-cis to gauche barrier is somewhat lower than the value previously reported [42] but this barrier will be the most effected by the V_4 term. It should be noted that the predicted splitting of 0.4 cm-^{1} for the 3± ← 2± transition of the gauche conformer was not observed which suggests that the barrier is slightly higher than that obtained from the potential function. The fit for the gauche transitions could possibly be improved if better structural parameters were available for this conformer. In both this study and the previous microwave study [42], it has been assumed that the structural parameters are not affected by the rotation of the CFO group with respect to the CCC plane. Overall the agreement between the microwave and infrared results is remarkably good when one considers the difficulty in measuring the relative intensities in the microwave spectrum.

As another example, no low frequency spectral data for gaseous 1-fluoropropane had been reported prior to our study [43], although a previous microwave study [44] indicated that the frequency of the asymmetric torsion for this molecule is below 200 cm-^{1}. In the infrared region below 200 cm-^{1} of gaseous 1-fluoropropane (Fig. 8), we observed two sets of transitions, one beginning at 140.76 cm-^{1} and the other at 128.88 cm-^{1}. The presence of a Q-branch at 140 cm-^{1} in the Raman spectrum of the gas established that the higher frequency series is associated with the gauche asymmetric torsion.

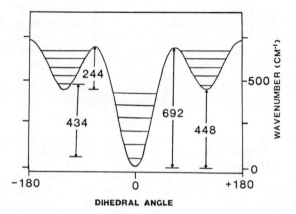

FIG. 7. Torsional potential function, including observed
energy levels, for the asymmetric torsion of propionyl
fluoride. The dihedral angle of zero corresponds to the s-cis
conformer. The gauche energy levels shown are doubly
degenerate, where the splittings between these levels are too
small to be apparent on the scale used. Used by permission,
Ref. [41].

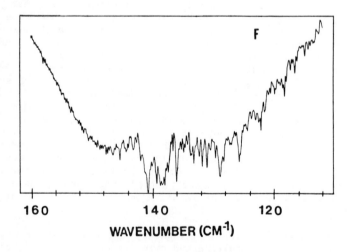

FIG. 8. Far infrared spectrum of the asymmetric torsional
region of gaseous 1-fluoropropane. Used by permission, Ref.
[43].

TABLE 5. Asymmetric Torsional Transitions for 1-Fluoropropane[a]

Conformer	Assignment	Infrared	Raman	Calc.	Δ
trans	1 ← 0	128.88		129.69	-0.81
	2 ← 1	125.74		125.60	0.14
	3 ← 2	122.10		121.22	0.88
	4 ← 3	118.23		116.55	1.68
gauche	1 ← 0	140.76	140	140.19	0.57
	2 ← 1	136.04		136.54	0.50

[a]Data taken from Ref. [43].

The lower frequency series can then be assigned to the trans rotamer of 1-fluoropropane (Table 5).

The plausible structural parameters given in the microwave study of 1-fluoropropane [44] were used to determine F_o and F_i (i = 1 to 6) in order to approximate the internal rotation constant as a function of the dihedral angle for this molecule. In the initial calculations of the potential function, both the 1 ← 0 and 2 ← 1 torsional transitions were included for each conformer, and the value of ΔH was taken from the microwave study [44] (164 ± 108 cm^{-1}). In subsequent calculations, the ΔH value was allowed to vary to give the best frequency fit. It should be mentioned that the observed transitions could be calculated equally well by using a ΔH of 122 or 250 cm^{-1}, either of which would be consistent with that reported in the microwave study [44]. Therefore we attempted to obtain a value for ΔH by using a variable temperature study of conformer peaks assigned to the CH_3 rock at 901 and 896 cm^{-1} in the Raman spectrum of the gas. We found no measurable change in the relative intensities of these peaks over a range of temperatures from 23-70°C, which indicates that the ΔH is reasonably small. Since it is likely that a ΔH of 250 cm^{-1} would be measurable over the temperature range studied, we felt that the value of 124 cm^{-1} is more reasonable and our data was interpreted as such (Fig. 9).

The present V_3 value of 1280 ± 4 cm^{-1} for the 1-fluoropropane molecule appears to more accurately reflect the potential function as compared to the previously reported [44] experimental V_3 value of 2266 ± 752 cm^{-1}, which was suspiciously large. Additionally, it should be noted that no V_4 or V_5 term was necessary to fit the data and only a very small V_6 (-12 cm^{-1}) was needed, which is in contrast to the very large V_4, V_5, V_6, and V_7 terms previously proposed [44]. The potential constants for 1-fluoropropane have also been calculated [45] using standard LCAO-SCF molecular orbital theory with an extended 4-31G basis set. The calculated V_1 and V_3 potential constants are somewhat larger than the present experimentally determined quantities but the V_2 coefficient, as well as

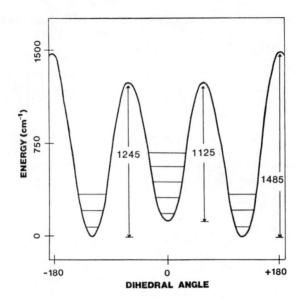

FIG. 9. The asymmetric potential function of 1-fluoropropane showing the energy levels of the observed transitions. The dihedral angle of zero corresponds to the high energy trans conformer. Used by permission, Ref. [43].

the energy difference, are smaller. However, the relative magnitudes of the potential constants and the resulting barriers are in reasonable agreement between the two studies.

These two examples can be used to illustrate several important points in the use of low frequency vibrational data for conformational analysis. First, in the case of propionyl fluoride, the spectrum clearly indicates that water may no longer be a problem for many cases in the far infrared, particularly if a resolution on the order of 0.1 cm^{-1} is used. Also, in the spectrum of this molecule, the asymmetric torsional transitions for the two conformers are well separated and there is no confusion where the transitions from one conformer end and the other begin. However, in the case of 1-fluoropropane, the torsional transitions are interdispersed and, without the Raman data, it would not have been possible to determine which Q branch is the first transition for the gauche conformer. Frequently, one finds that the "side" bands on the actual asymmetric torsional transitions have intensities comparable to those of the main transitions, or difference bands may appear interdispersed with the asymmetric torsional transitions. Therefore, the Raman spectra with its $\Delta v = 2$ transitions can be used to check the assignment of the asymmetric torsional modes. Also, it should be mentioned that the higher torsional transitions have a better chance for interactions with the higher

frequency normal modes so that perturbations can be expected higher
in the well. Also, in the case of propionyl fluoride, transitions
were observed near the top of the barrier which provides additional
confidence that the experimental potential function will nearly
approach that of the actual potential function experienced by the
molecule. However, in the case of 1-fluoropropane, only a limited
number of transitions were observed, and most of these are quite
deep in the potential well, particularly those for the gauche
conformer. Thus, the potential constants determined for propionyl
fluoride are certainly more accurate than those in 1-fluoropropane.
In summary, the utilization of the observed asymmetric torsional
transitions can provide sufficient data so that well determined
potential constants can be obtained.

For an extensive compilation of potential functions determined for
asymmetric rotors from low frequency vibrational data the reader is
referred to Ref. [46]. As an example of such compilation the
potential constants for propionyl fluoride and three related
molecules are listed in Table 6. Several interesting comparisons
may be made between the asymmetric potential function for propionyl
fluoride [41] and the functions obtained for propanal [47],
1-butene [48], and 2-methyl-1-butene [49]. The much larger V_2 term
in propanal, as compared to propionyl fluoride, moves the gauche
dihedral angle away from the normal 120° to a larger angle of 135°
for propanal. Examination of the values of the V_3 terms and the
various barriers listed in this table indicates an increase for the
four compounds, all of which have the rotation about =C-C-
linkages. The observed trend may be attributed to the electron
density at the sp^2 carbon. A comparison of the V_3 terms in
1-butene and 2-methyl-1-butene indicates that the methyl group
increases the electron density, whereas the electron density is
decreased by the polarity of the carbonyl bond in propanal and even
further by the fluorine atom in propionyl fluoride. This trend is
reflected in the calculated barriers. Thus, the V_3 term should be
a reasonably good measure of the electron density at the sp^2
carbon. It would be interesting to determine the asymmetric
potential function for propionyl chloride and compare it to the one
determined for the corresponding fluoride since the chlorine atom
is expected to give rise to a greater steric interaction.
Additionally, far infrared studies on isopropyl-carbonyl fluoride
should be carried out to obtain the conformer stabilities and to
check the barriers which have been proposed [42] for this molecule
based on the barrier differences between acetyl and propionyl
fluoride which result from the replacement of a hydrogen by a
methyl group. Thus, it is clear that additional research in this
area has considerable potential for providing the chemist with
information on intramolecular forces which cannot be easily
obtained by other methods.

Ring-Bending Modes
Chemists have been interested in the equilibrium conformation of
four-, five-, six-, and seven-membered rings for some time. Many
of the ring-bending vibrations in such molecules have normal modes

TABLE 6. Comparison of the Potential Coefficients and Barriers to
Internal Rotation (cm^{-1}) of the Asymmetric Torsional Mode
of Propionyl Fluoride with those of Related Molecules[a]

	Propionyl fluoride	Propanal	1-Butene	2-Methyl-1-butene
V_1	341	0[b]	-243	0[b]
V_2	236	559	185	79
V_3	390	651	921	1484
V_4	21	-105	165	108
V_5	0[b]	0[b]	-36	-95
V_6	0[b]	-46	-76	-29
ΔH	434	241	26	61
s-cis/gauche	692	1038	1114	1600
gauche/gauche	283	425	591	1328
gauche/s-cis	244	803	1088	1549
Ref.	[41]	[47]	[48]	[49]

[a]All molecules exist as a mixture of low energy planar and high
energy gauche conformers. The planar forms are all structurally
equivalent and denoted as s-cis except for 2-methyl-1-butene
where the planar form is denoted as s-trans.

[b]Calculated to be insignificant and so held at zero.

in the far infrared region of the spectrum and, thus, there are
several vibrational energy levels populated at room temperature.
From the analysis of the frequencies for these ring-bending motions
it is possible to obtain the potential surfaces for the interconversion of the different conformers. Fourier transform spectroscopy has been extensively used along with conventional spectrometers to obtain the frequencies for these ring-bending modes.

The equilibrium conformation of four- and unsaturated five-membered
rings is determined by two major opposing forces. The first is the
ring strain which tends to keep the ring skeleton planar. Puckering the ring decreases the already strained angles, thereby increasing the angle strain. The second is the torsional forces.
The torsional repulsions of adjacent groups are at a maximum for a

TABLE 7. Ring-Puckering Potential Energy Functions: $V = aX^4 - bX^2$

Molecule	a(cm^{-1}/A^4)	b(cm^{-1}/A^2)	Barrier	Dihedral angle	Ref.
Cyclobutane	7.6×10^5	39.6×10^3	515	33°	[50]
Silacyclobutane	3.5×10^5	24.9×10^3	440	37°	[51]
1,3-Disilacyclo- butane	2.3×10^5	9.0×10^3	87	24°	[52]
Cyclobutanone	4.3×10^5	2.9×10^3	5	~0°	[53]
Methylene cyclobutane	4.1×10^5	15.2×10^3	141		[54]
Oxetane	7.2×10^5	6.6×10^3	15	~0°	[55]
Thietane	5.1×10^5	23.5×10^3	274	33°	[53]
Selenetane	4.3×10^5	25.4×10^3	378	32.5°	[56]
Cyclopentene	12.34×10^5	33.8×10^3	232	23°	[57]

planar ring since the groups are eclipsed; therefore, bending the ring out of the plane will reduce these repulsive forces. It is the delicate balance between these two large forces which determine the ground state structure of the molecule.

The inversion barriers and dihedral angles for several four-membered rings and cyclopentene are listed in Table 7. The parameters a and b in the potential functions for these molecules have been shown to arise primarily from angle strain and torsional forces, respectively, and for an in-depth discussion the interested reader should consult Refs. [58] and [59].

A single substituent on the cyclobutane ring leads to conformational isomers and introduces asymmetry into the potential function for ring inversion. As the ring inverts, the substituent goes from the axial conformation to the equatorial, or vice versa. These two conformations will of course have different energies. The observed spectra for this type of conformational inversion can be interpreted using a potential function of the form: $V(x) = ax^4 - bx^2 - cx^3$, where x is the coordinate of the ring inversion. For the four-membered rings, 2x corresponds to the distance between the two ring bisectors. The cubic term is added to the usual quartic-harmonic potential function because of the asymmetry. For a review of the interpretation of the spectra of such molecules, the reader should see Ref. [60]. Also it should be noted that Jonvik and Boggs [61,62] have predicted the conformational stabilities.

In five-membered ring molecules, there are two low frequency out-of-plane ring motions. These are usually qualitatively described as the ring-twisting (radial mode) and the ring-puckering (pseudo-rotational mode) motions. Initially, in order to handle the interpretation of the low frequency far infrared data, assumptions were usually made as to the form of these normal vibrations. If the five-membered ring contains an endocyclic double bond, it has usually been assumed that there is no interaction between the "high" frequency ring-twisting mode which falls around 400 cm^{-1} and the "low" frequency ring-puckering mode which is near 100 cm^{-1}. Therefore, the two out-of-plane ring modes are handled as two one-dimensional problems with the anharmonic low frequency ring-puckering transitions interpreted in terms of a one-dimensional potential function of the form $V(X) = AX^4 \pm BX^2$. A review of the spectral studies of these molecules has been given by Wurrey et al. [63].

If the five-membered ring contains no endocyclic double bonds, then the two out-of-plane ring motions can usually couple and the Hamiltonian for this motion is generally written as a two-dimensional problem in Cartesian coordinates. The two-dimensional Cartesian Hamiltonian is then transformed to polar coordinates, q and ϕ. If one assumes that the barrier to planarity is very high, this two-dimensional Hamiltonian can be simplified by separating it into two one-dimensional Hamiltonians as suggested by Kilpatrick, Pitzer and Spitzer [64]. One of these coordinates is essentially a ϕ-dependent pseudorotational motion and the other coordinate is a q-dependent radial mode which is a pseudorotational amplitude coordinate and initially assumed to be harmonic. The anharmonic low-frequency transitions observed in the far infrared spectral region for this type of ring have been explained in terms of a ϕ-dependent Hamiltonian of the form:

$$H = \beta P_\phi^2 + \sum_n (V_n/2)(1 - \cos n\phi)$$

and these transitions are commonly referred to as pseudorotational transitions.

Carreira et al. [65] have shown that the assumption of separability of the twisting and puckering modes is not entirely valid even when the ring has an endocyclic double bond. They have shown that the fourth power cross term connecting the two vibrations is significantly large and should probably be used for all cyclopentene-like rings. Ikeda et al. [66] have investigated the effects of a finite barrier to planarity on unhindered pseudorotation in saturated five-membered rings. They concluded that a finite barrier would show negative curvature from a straight line in a plot of the pseudorotational energy level spacings versus the pseudorotational quantum number. Also the q-dependent radial mode should show negative anharmonicity.

TABLE 8.   Comparison of Barriers Between the One-Dimensional and
Two-Dimensional Treatments

Molecule	One-dimensional barrier (cm^{-1}) to interconversion	Two-dimensional barrier (cm^{-1}) to planarity	Barrier (cm^{-1}) to interconversion	Ref.
Cyclopentane	---	1824	---	[69]
Silacyclopentane[a]	1390	1558.6	1414	[70]
Germacyclopentane[a]	2043	1474.4	1454	[71]
Selenacyclopentane[a]	1882	1693	1693	[72]
Cyclopentanone[a]	1303	750		[73]

[a]The interconversion for these molecules is through the planar
conformation whereas the pathway for interconversion for the other
molecules takes place through a slightly puckered configuration.

The ring-puckering vibrations are usually observed in the far
infrared spectra whereas the ring-twisting modes are observed in
the Raman effect.   The first saturated five-membered rings to be
studied were 1,3-dioxolane [67] and tetrahydrofuran [68].   In these
early studies it was assumed that the barrier to pseudorotation was
zero or near zero and as previously pointed out, the radial angular
equations were separable.   Several recent studies have shown that
even though the twisting and bending motions are fairly
independent, the one-dimensional approximation leads to a much too
high energy for interconversion and also exaggerates the actual
molecular path for interconversion.

The two-dimensional analysis has been given for cyclopentane [69],
silacyclopentane [70], germacyclopentane [71], selenacyclopentane
[72], and cyclopentanone [73] and a comparison of the two different
treatments is given in Table 8.   One can then see that, by a
combination of far infrared data ($\Delta v = 1$) for the ring-puckering
vibration (pseudorotational mode) and the $\Delta v = 1$ transitions for
the ring-twisting mode (radial mode) obtained from the Raman effect
for saturated five-membered rings, it is possible to observe the
effects of a .finite barrier to planarity and to calculate the
two-dimensional potential surface for the two low frequency
out-of-plane modes for these molecules.   On the basis of a very
limited number of cases it would seem that the one-dimensional
model breaks down as the path to interconversion approaches one
which involves a planar intermediate.   Unfortunately, it is usually
not possible to determine a priori if a twisted molecule
interconverts through a planar or puckered intermediate.   For a
more detailed discussion on this topic, the reader is referred to
Ref. [74].   It should be noted that sometimes difficulties arise
when employing the two-dimensional model as was found recently for
thiacyclopentane [75].

Six-membered rings which contain one double bond have two large-amplitude, low frequency ring deformations: a ring bend and a ring twist, along with a third out-of-plane motion which is a torsion about the double bond. This torsion has a much higher frequency and therefore does not couple with the other two out-of-plane ring motions. However, the two out-of-plane ring modes are strongly coupled and the two-dimensional treatment must be used. Six-membered rings with two double bonds or one double bond and a fused epoxide ring have one low frequency puckering mode which does not appreciably couple with the other ring-bending modes so the one-dimensional Hamiltonian is appropriate for the ring inversions. There have been many studies of such molecules utilizing FT-IR spectroscopy and usually the ring-puckering motion is relatively strong in the far infrared. However, the higher frequency ring modes may not be observed in the far infrared but difference bands with the low frequency ring-puckering motions may give rise to far infrared absorptions. From such studies it has been possible to determine the most stable conformation as well as the barriers to interconversion and the lower energy path for inversion.

For seven- and eight-membered rings, most of the far infrared studies have been done by Strauss and coworkers [76] at Berkeley utilizing an FT-IR spectrometer. For example, cycloheptanone was found to be nonrigid with a pseudorotational energy surface whereas cycloheptane, oxepane, and 1,3-dioxepane showed no noticeable effects of nonrigidity. For the eight-membered rings, the far infrared spectra have helped to confirm the boat-chair conformations determined by other methods but the far infrared data provide a means for obtaining the parameters for a model of the conformations. For further details on the stable conformation of seven- and eight-membered rings, the interested reader should refer to the review article on this subject by Rounds and Strauss [76].

CONDENSED PHASE STUDIES

In the first part of this chapter, the applications of low frequency spectral data for chemical information reflect the authors' personal interests but there are several other areas where low frequency data yield unique information. These include such things as nonspecific absorption in liquids, intermolecular vibrations of molecular crystals, lattice modes of ionic crystals, and weak intermolecular interactions. For this latter topic, most of the studies have been on hydrogen bonded molecules or charge transfer complexes and a review of this topic has been given by Griffiths [6]. There have been several far infrared studies of polar liquids which show strong, broad absorptions below 100 cm^{-1}. This absorption cannot be attributed to intramolecular vibrations or pure rotations of the molecules. However, other explanations have been proposed such as: (1) stretching modes of dipole-dipole complexes; (2) vibrations of a "liquid lattice"; and (3) hindered rotation of the molecule in a cage of its neighbors. The first explanation seems unlikely but the latter two have considerable merit and studies are continuing in this area.

Most of the advances in the understanding of the vibrational spectra of solids with its ramifications in related fields have been
made for simple inorganic ionic crystals. Because of the weaker
binding in molecular crystals, the optical lattice modes occur at
considerably lower frequencies and have thus remained largely un-
explored to date. Moreover, because of the relative complexity of
many of the molecular crystals, the understanding of their spectra
has been quite superficial. Thus studies of the frequencies for
the optical lattice modes of molecular crystals are particularly
interesting, since to date only limited information exists concerning these crystals. Such studies may provide information on
the optical constants (refractive index and extinction coefficients) in the low frequency region of the spectrum, on the
molecular motion giving rise to a particular intermolecular mode,
on the occurrence of multiphonon transitions, and finally on the
elucidation of intermolecular forces for the crystals. Sometimes
it may be possible to distinguish between crystal space groups
proposed from X-ray investigations by a study of the optical
lattice modes.

A molecular crystal composed of N primitive cells with m atoms per
cell has 3Nm normal modes of vibration. These 3Nm modes are distributed on 3m branches; three are acoustical branches and the
other 3m - 3 are optical branches. The three acoustical modes have
essentially zero frequency at k = 0, the Brillouin zone center, and
are not observed in either the infrared or in the Raman spectrum.
These three acoustical modes represent the translational motions of
all particles in the same sense in the primitive cell in the three
spatial directions. The 3m - 3 optical branch frequencies corresponding to k = 0 (where k is the wave vector) constitute the
lattice fundamental modes of vibration. In each branch the normal
mode frequency depends upon the wave vector, $k = 2\pi/\lambda$, where k can
assume all values in the Brillouin zone. However, all modes of
vibration except the ones in which equivalent atoms are moving
identically in-phase are forbidden as fundamentals both in the
infrared absorption and in the Raman scattering process. This
results from the fact that the wave vector of the phonon is near
zero compared with that of most of the phonons of the same energy,
except for the ones near the center of the Brillouin zone [77].
Thus, the excited phonons have small wave vectors. The distribution of the normal modes among the various symmetry classifications
and their optical activity can be enumerated by a consideration of
the primitive cell instead of the entire crystal.

For definite molecular crystals which have strong binding within
the molecules and relatively weak intermolecular forces holding the
molecules together, the vibrational spectrum breaks up into two
different parts. This results from the fact that the individual
atoms of the molecule are held together by chemical bonds whereas
the individual molecules in the crystal are held together mainly by
van der Waals-type forces. For example, in a single crystal of a
hydrocarbon, the forces holding together the carbon and hydrogen
atoms are nearly two orders of magnitude larger than the forces
holding together the molecules. Thus the normal vibrations arising

from the internal motions of the individual atoms of the respective chemical group give rise to bands in the spectrum which correspond closely to the vibrational frequencies of the isolated molecule. There will be a slight distortion of the spectrum on account of the molecular actions of the crystal. These normal vibrations are usually referred to as internal modes of the group or molecule.

The motions of the molecules relative to one another give rise to the lattice modes which are frequently called external modes. Since the forces among the molecules are relatively weak, these normal modes fall in the region of the spectrum from 200 to 10 cm^{-1} or at considerably lower frequencies than for the internal modes. These external modes can be further subdivided into motions of translatory and rotatory origin which, in the limit of vanishing intermolecular forces, correspond to pure translations and rotations. The rotatory lattice modes are usually referred to as librational modes.

The determination of the selection rules and the symmetry classifications of the primitive cell fundamentals of a crystal can be done by two different methods. The first method, by Bhagavantam and Venkatarayudu [78], considers the atoms in the primitive cell as a "large" molecule, and the usual method for deriving the optical activity and symmetry classification of the normal modes is applied. The second method, by Halford [79] and Hornig [80], considers the site symmetry or the "local symmetry" of the molecule in the primitive cell, and then the factor-group symmetry is considered. For many molecular crystals only the first approximation of the site symmetry needs to be considered for the internal modes; however, the factor-group analysis can be readily made if the site and crystal symmetry are known.

If the crystal is disordered such as that found for phase II of tert-butyl chloride then the site symmetry will not be a subgroup of the molecular symmetry. For such cases one can readily determine the symmetry of the external modes by the consideration of the symmetry for the translations and rotations in the character table for the site symmetry. However, in principle, one obtains a density-of-states spectrum, and the selection rules for disordered crystals have been discussed by Bertie and Whalley [81]. For a thorough discussion of the symmetry considerations of crystals one should consult the original papers by Halford [79] and Hornig [80] in which the results are clearly presented. Additionally, Fateley and coworkers [82] have published a book on the infrared and Raman selection rules for molecular and lattice vibrations by the correlation method and the interested reader is referred to this text for a more in-depth discussion of this topic.

For a number of molecular crystals, the internal modes of the molecules may be at a sufficiently low frequency to be in the same region of the spectrum as the lattice modes. Sometimes it is possible to distinguish the lattice modes from the internal modes by the stronger temperature dependence of their width, intensity, and position. Also lattice bands will not appear in solution. The dichroic ratio and depolarization factors are the most important data for assigning the lattice modes in the infrared and Raman spectra, respectively. However, for some cases the isotopic shift factor can be used to distinguish between translatory and rotatory motions and, under favorable circumstances, the librational frequency can be assigned to a specific rotational axis. An empirical rule has been proposed by Bhagavantam [83] that the rotatory lattice modes give rise to strong Raman bands whereas the translatory types may give weak ones, but the rule should be employed with considerable caution.

As an example of low frequency studies of molecular crystals, we shall consider the chlorocyclopentane molecule. We [84] found a very broad band centered at 50 cm^{-1} in the spectrum of this molecule in crystal phase I; the spectrum of this crystalline phase is very similar to the spectrum of the liquid in this spectral region. With the transition to crystal phase II, this band sharpens and pronounced bands were observed at 51, 69 and 75 cm^{-1} (see Fig. 10). This behavior is consistent with a transition from a disordered to an ordered crystalline state. Changes in the 622 and 584 cm^{-1} doublet in the mid-infrared spectrum of chlorocyclopentane on the transition from solid phase I to solid phase II clearly showed the transition involves the change from a mixture of axial and equatorial conformations to an axial conformation only. It is interesting to note that there is still controversy as to the conformational stability of this molecule in the gaseous state where the electron diffraction data [85] have been interpreted as resulting from an equilibrium mixture of the axial and equatorial conformers whereas from the microwave study [86] it was concluded that the stable molecular conformation is the bent axial form and no evidence was found for a second stable molecular conformer.

For molecules which contain hydrogen, it is frequently possible to distinguish between the translational and librational motions with deuteration. For example, the translational modes will shift by the factor $(M_H/M_D)^{\frac{1}{2}}$ where M_H and M_D are the total masses of the "light" and heavy molecules, respectively, whereas the librational modes will shift by the factor $(I_H/I_D)^{\frac{1}{2}}$ where the I_H and I_D are the different moments of inertia of the "light" and heavy molecules, respectively. This method is only useful if the translatory and libratory modes are not appreciably mixed.

There have been a fair number of far infrared spectral studies of molecular crystals utilizing both conventional grating spectrometers and FT-IR interferometers. Additionally, there have been numerous Raman studies on the lattice modes of molecular crystals but, because of the lack of far infrared data, this information has not been as useful as if all of the lattice frequencies were known. With the availability of FT spectrometers, we believe this is one area of study which will receive greater attention in the future. In fact, the first far infrared spectral data for metal atoms and clusters entrapped in a real catalyst support for silver zeolites A and Y has been recently reported [87]. These data were collected on a Fourier transform interferometer and the results show that

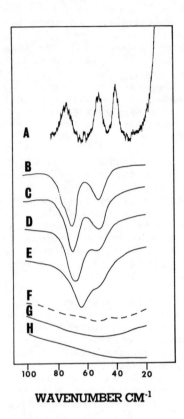

FIG. 10. Low frequency vibrations of chlorocyclopentane: (A) Raman spectrum of crystal II; far infrared spectra of crystal II at (B) 25K, (C) 110K, (D) 128K, (E) 155K, (F) crystal I, (G) cooled liquid (190K), and (H) liquid at room temperature. Used by permission, Ref. [84].

far infrared spectroscopy offers a direct and sensitive in situ probe of metal atoms and metal clusters entrapped in the cavities of zeolite supports.

SUMMARY AND CONCLUSIONS

The utilization of FT-IR spectroscopy in the far infrared spectral region has made it possible to study many problems of chemical interest. Initial applications involved the study of pure rotations of small molecules, the skeletal bending modes and heavy atom vibrations, as well as the study of both molecular and ionic crystals. Studies in the latter two areas have continued, and this spectral region can now be investigated with about the same ease as one investigates the mid-infrared spectral region with modern commercial interferometers. With the improvements in instrument design, there has been increased application to conformational analysis in small ring compounds and torsional oscillations of asymmetric rotors attached to asymmetric frames. It is expected that as more commercial instruments become available to the chemist there will be increased interest in the far infrared spectral region. Experimental difficulties have been largely overcome so that the chemist, in addition to the spectroscopist, can utilize this spectral region to solve their problems; thus more activity in far infrared spectral studies is anticipated in the future.

REFERENCES

1. E. D. Palik, J. Opt. Soc. Amer., 50, 1329 (1960).
2. H. Rubens and R. W. Wood, Phil. Mag., 21, 249 (1911).
3. P. Fellgett, J. Phys. Radium, 19, 187 (1958).
4. P. Jacquinot and C. J. Dufour, J. Rech. C.N.R.S., 6, 91 (1948).
5. H. A. Gebbie, Appl. Opt., 8, 501 (1969).
6. P. R. Griffiths, Chemical Infrared Fourier Transform Spectroscopy, Wiley (Interscience), New York, 1975.
7. J. R. Durig and A. W. Cox, Jr., in Fourier Transform Infrared Spectroscopy, Vol. 1, (J. R. Ferraro and L. J. Basile, eds.), Academic Press, New York (1978).
8. A. D. Lopata and J. R. Durig, J. Raman Spectrosc., 6, 61 (1977).
9. J. R. Durig, S. M. Craven and W. C. Harris, in Vibrational Spectra and Structure, Vol. 1, (J. R. Durig, ed.), Marcel Dekker Inc., New York (1972).
10. J. R. Durig, S. M. Craven and J. Bragin, J. Chem. Phys., 52, 2046 (1970).
11. J. E. Kilpatrick and K. S. Pitzer, J. Chem. Phys., 17, 1064 (1949).
12. D. R. Herschbach, J. Chem. Phys., 31, 91 (1959).
13. W. G. Fateley and F. A. Miller, Spectrochim. Acta, 17, 857 (1961).
14. W. G. Fateley and F. A. Miller, Spectrochim. Acta, 18, 977 (1962).
15. W. G. Fateley and F. A. Miller, Spectrochim. Acta, 19, 611 (1963).
16. S. Weiss and G. E. Leroi, J. Chem. Phys., 48, 962 (1968).
17. G. Sage and W. Klemperer, J. Chem. Phys., 39, 371 (1963).
18. J. R. Durig, W. E. Bucy, L. A. Carreira, and C. J. Wurrey, J. Chem. Phys., 60, 1754 (1974).
19. J. R. Durig, P. Groner, and M. G. Griffin, J. Chem. Phys., 66, 3061 (1977).
20. D. R. Herschbach, J. Chem. Phys., 25, 359 (1956).
21. J. R. Durig, W. E. Bucy, and C. J. Wurrey, J. Chem. Phys., 60, 3293 (1974).
22. J. R. Durig, S. M. Craven, and J. Bragin, J. Chem. Phys., 53, 38 (1970).
23. J. R. Durig, W. E. Bucy, and C. J. Wurrey, J. Chem. Phys., 63, 5498 (1975).

24. S. Saito and F. Makino, Bull. Chem. Soc. Jpn., $\underline{47}$, 1863 (1974).
25. D. Christen, V. Hoffmann, and P. Klaeboe, Z. Naturforsch., $\underline{34a}$, 1320 (1979).
26. J. R. Durig, H. D. Bist, K. Furic, J. Qiu, and T. S. Little, J. Mol. Struct., in press (1985).
27. J. R. Durig and S. D. Hudson, in Analytical Applications of FT-IR to Molecular and Biological Systems, (J. R. Durig, ed.), D. Reidel Publishing Co. (1980).
28. P. Groner, J. F. Sullivan, and J. R. Durig, in Vibrational Spectra and Structure, Vol. 9, (J. R. Durig, ed.), Elsevier, Amsterdam (1981).
29. J. R. Durig, S. D. Hudson, M. R. Jalilian, and Y. S. Li, J. Chem. Phys., $\underline{74}$, 772 (1981).
30. J. R. Durig, M. G. Griffin, and W. J. Natter, J. Phys. Chem., $\underline{81}$, 1588 (1977).
31. J. R. Durig, M. G. Griffin, and P. Groner, J. Phys. Chem., $\underline{81}$, 554 (1977).
32. P. Groner and J. R. Durig, J. Chem. Phys., $\underline{66}$, 1856 (1977).
33. J. R. Durig, M. R. Jalilian, J. F. Sullivan, and D. A. C. Compton, J. Chem. Phys., $\underline{75}$, 4833 (1981).
34. J. R. Durig, P. Groner, and M. G. Griffin, J. Chem. Phys., $\underline{66}$, 3061 (1977).
35. J. R. Durig, W. J. Natter, and P. Groner, J. Chem. Phys., $\underline{67}$, 4948 (1977).
36. D. M. Grant, R. J. Pugmire, R. C. Livingstone, K. A. Strong, H. L. McMurry, and R. M. Brugger, J. Chem. Phys., $\underline{52}$, 4424 (1976).
37. J. D. Lewis, T. B. Malloy, Jr., T. H. Chao and J. Laane, J. Mol. Struct., $\underline{12}$, 427 (1972).
38. K. Hagen and K. Hedberg, J. Amer. Chem. Soc., $\underline{95}$, 1003 (1972).
39. K. O. Hartmann, G. L. Carlson, R. E. Witkowski and W. G. Fateley, Spectrochim. Acta, $\underline{24a}$, 157 (1968).
40. D. A. C. Compton, W. O. George, and W. F. Maddams, J. Chem. Soc., Perkin Trans. II, 1666 (1976).
41. G. A. Guirgis, B. A. Barton, and J. R. Durig, J. Chem. Phys., $\underline{79}$, 5818 (1983).
42. O. L. Stiefvater and E. B. Wilson, J. Chem. Phys., $\underline{50}$, 5385 (1969).
43. J. R. Durig, S. E. Godbey, and J. F. Sullivan, J. Chem. Phys., $\underline{80}$, 5983 (1984).
44. E. Hirota, J. Chem. Phys., $\underline{37}$, 283 (1962).
45. L. Radom, W. A. Lathan, W. J. Hehre, and J. A. Pople, J. Amer. Chem. Soc., $\underline{95}$, 693 (1973).
46. D. A. C. Compton, in Vibrational Spectra and Structure, Vol. 9, (J. R. Durig, ed.), Elsevier, Amsterdam (1981).
47. J. R. Durig, D. A. C. Compton, and A. Q. McArver, J. Chem. Phys., $\underline{73}$, 719 (1980).
48. J. R. Durig and D. A. C. Compton, J. Phys. Chem., $\underline{84}$, 773 (1980).
49. J. R. Durig, D. J. Gerson, and D. A. C. Compton, J. Phys. Chem., $\underline{84}$, 3554 (1980).
50. J. M. R. Stone and I. M. Mills, Mol. Phys., $\underline{18}$, 631 (1970).
51. J. Laane and R. C. Lord, J. Chem. Phys., $\underline{48}$, 1508 (1968).

52. R. M. Irwin, J. M. Cooke, and J. Laane, J. Amer. Chem. Soc., 99, 3273 (1977).
53. T. R. Borgers and H. L. Strauss, J. Chem. Phys., 45, 947 (1966).
54. J. R. Durig, A. C. Shing, L. A. Carreira, and Y. S. Li, J. Chem. Phys., 57, 4398 (1972).
55. S. I. Chan, T. R. Borgers, J. W. Russell, H. L. Strauss, and W. D. Gwinn, J. Chem. Phys., 44, 1103 (1966).
56. A. B. Harvey, J. R. Durig, and A. C. Morrissey, J. Chem. Phys., 50, 4949 (1969).
57. J. Laane and R. C. Lord, J. Chem. Phys., 47, 4941 (1967).
58. J. Laane, J. Chem. Phys., 50, 1946 (1969).
59. J. Laane, Quant. Rev., 25, 533 (1971).
60. C. S. Blackwell and R. C. Lord, in Vibrational Spectra and Structure, Vol. 1, (J. R. Durig, ed.), Marcel Dekker, Inc., New York (1972).
61. T. Jonvik and J. E. Boggs, J. Mol. Struct., 85, 293 (1981).
62. T. Jonvik and J. E. Boggs, J. Mol. Struct., 105, 201 (1983).
63. C. J. Wurrey, J. R. Durig, and L. A. Carreira, in Vibrational Spectra and Structure, Vol. 5, (J. R. Durig, ed.), Elsevier, Amsterdam (1976).
64. J. E. Kilpatrick, K. S. Pitzer, and R. Spitzer, J. Amer. Chem. Soc., 69, 2483 (1947).
65. L. A. Carreira, I. M. Mills, and W. B. Person, J. Chem. Phys., 56, 1444 (1972).
66. T. Ikeda, R. C. Lord, T. B. Malloy Jr., and T. Ueda, J. Chem. Phys., 56, 1434 (1972).
67. G. G. Engerholm, A. C. Luntz, W. D. Gwinn, and D. O. Harris, J. Chem. Phys., 50, 2446 (1969).
68. J. A. Greenhouse and H. L. Strauss, J. Chem. Phys., 50, 124 (1969).
69. L. A. Carreira, G. J. Jiang, W. B. Person, and J. N. Willis Jr., J. Chem. Phys., 56, 1440 (1972).
70. J. R. Durig, W. J. Natter, and V. F. Kalasinsky, J. Chem. Phys., 67, 4756 (1977).
71. J. R. Durig, Y. S. Li, and L. A. Carreira, J. Chem. Phys., 58, 2393 (1973).
72. J. R. Durig and W. J. Natter, J. Chem. Phys., 69, 3714 (1978).
73. T. Ikeda and R. C. Lord, J. Chem. Phys., 56, 4450 (1972).
74. L. A. Carreira, R. C. Lord, and T. M. Malloy Jr., in Topics in Current Chemistry, Vol. 82, Springer-Verlag, Berlin (1979).
75. T. L. Smithson and H. Wieser, J. Mol. Spectrosc., 99, 159 (1983).
76. T. C. Rounds and H. L. Strauss, in Vibrational Spectra and Structure, Vol. 7, (J. R. Durig, ed.), Elsevier, Amsterdam (1978).
77. C. Kittell, Introduction to Solid State Physics, Wiley, New York (1966).
78. S. Bhagavantam and T. Venkatarayudu, "Theory of Groups and Its Application to Physical Problems," Andhra Univ., Waltabi, India, 1951.
79. R. S. Halford, J. Chem. Phys., 14, 8 (1946).
80. D. F. Hornig, J. Chem. Phys., 16, 1063 (1948).
81. J. E. Bertie and E. Whalley, J. Chem. Phys., 46, 1264 (1967).

82. W. G. Fateley, F. R. Dollish, N. T. McDevitt, and F. F. Bentley, Infrared and Raman Selection Rules for Molecular and Lattice Vibrations: The Correlation Method, Wiley (Inter-science), New York, (1972).

83. S. Bhagavantam, Proc. Indian Acad. Sci., 13a, 543 (1941).

84. J. R. Durig, B. A. Hudgens, G. N. Zhizhin, V. N. Rogovoi, and J. M. Casper, Mol. Cryst. Liq. Cryst., 31, 185 (1975).

85. R. L. Hilderbrandt and Q. Shen, J. Phys. Chem., 86, 587 (1982).

86. R. C. Loyd, S. N. Mathur, and M. D. Harmony, J. Mol. Spectrosc., 72, 359 (1978).

87. M. D. Baker, G. A. Ozin, and J. Godber, J. Phys. Chem., 89, 305 (1985).

AUTHOR INDEX

Numbers in brackets are reference numbers and indicate that an author's work is referred to although his name is not cited in the text. Underlined numbers give the page on which the complete reference is listed.

A

Abney, W. de W., 11, 12, 45

Acquista, N., 8, 44

Adams, R. G., 182 [14], 187 [18], 197

Adar, F., 102 [34], 109

Albrecht, M. G., 129 [14], 137

Alexander, G., 80 [16], 86

Allara, D. L., 129 [8], 137

Alley, E. G., 278 [4,5,6,7,8, 9,10,13,14], 294 [9,10], 299

Ambrose, E. J., 27, 48

Ampère, A., 3, 43

Anderson, M. E., 102 [35], 110

Andrews, D. H., 19, 46

Angell, C. L., 255, 275

Angelotti, N. C., 73, 85

Ångström, K., 12, 45

Antoon, M. K., 115, 128

Aschkinass, E., 12, 45

Astbury, W. T., 27 [161], 48

B

Baghadi, A., 88 [14], 109

Bahl, S. K., 163 [18], 172

Baker, M. D., 375 [87], 380

Barber, P. G., 141 [6], 171

Barbillat, J., 102 [38], 110

Barnes, R. B., 25, 26, 29, 30, 31 [175], 48

Bartges, B., 231 [37], 251

Barton, B. A., 361 [41], 362 [41], 364 [41], 367 [41], 368 [41], 378

Bell, A. T., 255 [3], 275

Bell, R. J., 301 [1], 306, 334

Bellamy, L. J., 17 [80,81], 35, 57, 217 [30], 226 [30], 250 [30], 45, 49, 72, 251

Benham, V., 42 [211], 50

Bentley, F. F., 88 [9], 374 [82], 109, 380

Beny, J. M., 148 [9], 171

Bertie, J. E., 40, 42 [211], 374, 50, 379

Berube, G. N., 41 [208], 49

Beutelspacher, H., 112 [1], 128

Beveridge, D., 141 [5], 171

Beyer, W., 74 [8], 86

Bhagavantam, S., 374, 375, 379, 380

Bishop, E. R., 292 [23], 299

Bist, H. D., 348 [26], 378

Bjerrum, N., 11, 18, 19, 45, 46

Blackwell, C. S., 369 [60], 379

Blout, E. R., 39, 40, 49

Boerio, F. J., 163 [18], 172

Boggs, J. E., 76 [11], 369, 86, 379

Borgers, T. R., 369 [53,55], 379

Boys, C. V., 8, 44

381

SUBJECT INDEX

A

Abney, 11
adsorbed CO, 258
adsorbed molecules, 253
albumin, 200, 202
alumina, 265
amide I, 199
amide II, 199
amorphous silicon films, 88
Ångström, 12
asphalts, 68
asymmetric tops, 341
attenuated total reflectance
  (ATR), 203, 296

B

band shapes, 154
Barnes, 25, 29, 30
Bellamy, 35, 57
benzaldehyde, 348
beta-sheet, 205
bismuth molybdates, 94, 95
Bjerrum, 11, 18, 19
bolometer, 8
Brandmüller, 22
1-butene, 367, 368

C

carbon monoxide, 262
carbonyls, 249
catalysts, rhodium, 264
CF_3CH_2X, 347

CH stretch, 75, 175, 177
CH_3CH_2X, 347
CH_3CH_3, 347
CH_3CHX_2, 347
CH_3CX_3, 347
clathrin, 192
coal, 225, 231
Coblentz, 4, 13
Cole, 34
Colthup, 35, 57
complex mixture identification, 286
Connes, 40
contaminant detection, 285
Cooley, 41
Crawford, 19, 34
Cross, 19
cyclic peptide, 189
cyclobutane, 369
cyclobutanone, 369
cyclopentanone, 371
cyclopentene, 369, 371

D

data collection, 315
data sorting, 311
Decius, 19
degradation analysis, 277
Delhaye, 148
Dennison, 19
dichroic ratio, 305
diffuse reflectance spec-
  trometry (DRIS), 88, 215,
  244, 261, 277, 294